Brandt, Fuß (eds.)
Form, Structure, and Grammar

studia grammatica

Herausgegeben von Manfred Bierwisch

unter Mitwirkung von

studia grammatica 63

Patrick Brandt and
Eric Fuß (eds.)

Form, Structure, and Grammar

A Festschrift Presented to Günther Grewendorf on Occasion of His 60th Birthday

Akademie Verlag

ISBN-10: 3-05-004224-9
ISBN-13: 978-3-05-004224-4

ISSN 0081-6469

Das eingesetzte Papier ist alterungsbeständig nach DIN/ISO 9706.

Druck und Bindung: MB Medienhaus Berlin

Printed in the Federal Republic of Germany

Contents

Contributors

Werner Abraham
University of Vienna
Institut für Sprachwissenschaft
1090 Wien
Austria

Josef Bayer
University of Konstanz
Lehrstuhl für Allgemeine und
Germanistische Sprachwissenschaft
78457 Konstanz
Germany

Adriana Belletti
University of Siena
Facoltà di Lettere e Filosofia
53100 Siena
Italy

Manfred Bierwisch
Humboldt University
Projektgruppe Strukturelle Grammatik
10117 Berlin
Germany

Rainer Dietrich
Humboldt University
Institut für deutsche Sprache und
Linguistik
10117 Berlin
Germany

Gisbert Fanselow
University of Potsdam
Institut für Linguistik
14415 Potsdam
Germany

Sascha W. Felix
University of Passau
Lehrstuhl für Allgemeine Linguistik
94030 Passau
Germany

Werner Frey
Zentrum für Allgemeine
Sprachwissenschaft
10117 Berlin
Germany

Hans-Martin Gärtner
Zentrum für Allgemeine
Sprachwissenschaft
10117 Berlin
Germany

Katharina Hartmann
Humboldt University
Institut für deutsche Sprache und
Linguistik
10117 Berlin
Germany

Tilman N. Höhle
University of Tübingen
Deutsches Seminar
72074 Tübingen
Germany

Manfred Krifka
Humboldt University
Institut für deutsche Sprache und
Linguistik
10117 Berlin
Germany

Georg Meggle
University of Leipzig
Institut für Philosophie
04107 Leipzig
Germany

Gereon Müller
University of Leipzig
Institut für Linguistik
04107 Leipzig
Germany

Cecilia Poletto
Istituto di Scienze e Tecnologie della Cognizione
35121 Padova
Italy

Monika Rathert
Saarland University
FR 4.1. – Germanistik
66041 Saarbrücken
Germany

Luigi Rizzi
University of Siena
Centro Interdipartimentale di Studi Cognitivi sul Linguaggio
53100 Siena
Italy

Tom Roeper
University of Massachusetts
Department of Linguistics
Amherst, MA 01003
USA

Joachim Sabel
Catholic University of Louvain
Département d'études germaniques
1348 Louvain-la-Neuve
Belgium

Mamoru Saito
Nanzan University
Department of Anthropology and Philosophy
Nagoya 466-8673
Japan

Markus Steinbach
University of Mainz
Deutsches Institut
55128 Mainz
Germany

Dietmar Zaefferer
University of Munich
Institut für Deutsche Philologie
80799 Munich
Germany

Jochen Zeller
University of Kwazulu-Natal
Department of Linguistics
Durban 4041
South Africa

Malte Zimmermann
Humboldt University
SFB 632 – Informationsstruktur
10099 Berlin
Germany

Thomas Ede Zimmermann
University of Frankfurt
Institut für Kognitive Linguistik
60629 Frankfurt
Germany

Introduction

Eric Fuß and Patrick Brandt

FestSchrift Grewendorf

It is a long-standing academic tradition to honor outstanding personalities of the field with a Festschrift, which is usually presented in connection with some milestone birthday.[1] Taking up this noble custom, this collection of papers is dedicated to Günther Grewendorf and offered to him on the occasion of his sixtieth birthday.

During his entire career, Günther has been among the foremost scholars in the field of Generative Grammar who established the German branch of what is now the leading paradigm in the study of natural language. Over the last thirty years, Günther has contributed substantively to the enterprise of generative linguistics (his achievements in this area are too numerous to be mentioned here in detail, but cf. the list of publications at the end of this volume). As a teacher and organizer, he has built a research environment that nurtured and shaped many younger academics, particularly in the field of theoretical syntax. It seems fair to say that Günther has a talent to attract people who do good work as linguists and get things off the ground and often make a career in linguistics. This is reflected by the fact that among the contributors, there are many former students of his, a number of them now leading figures in the field as well.

Günther's work encompasses the disciplines of philosophy, pragmatics, and theoretical linguistics. After studying philosophy and linguistics with Wolfgang Stegmüller, Eike von Savigny, Max Käsbauer, Johannes Bechert and Richard Hare, Günther's early publications deal with issues of ordinary language philosophy and pragmatics, in particular the analysis of speech acts. During the 1970s, he then turned to problems of formal pragmatics, focusing on a formal analysis of question-answer pairs. At the same time, he developed a keen interest in theoretical syntax when he became acquainted with the contemporary work of Chomsky and other generative linguists of that period. With the wake of the Principles & Parameters theory, Günther advanced to one of the most prominent German linguists with his work on reflexivization in A.c.I constructions (Grewendorf 1983b). Based at the University of Frankfurt since 1984, Günther has been and still is among the most productive, influential and internationally renowned German linguists, constantly shaping and reshaping the understanding of the syntax of German. Among his many influential publications, *Aspekte der deutschen Syntax* (1988), *Ergativity in German* (1989), *Minimalistische Syntax* (2002) as well as a number of papers in *Linguistic Inquiry* and *Natural Language and Linguistic Theory* figure prominently. In addition to his research work, and apart from his work as editor of the principal German

---

1 With some notable exceptions, cf. e.g. *Studies Out in Left Field: Defamatory Essays Presented to James D. McCawley on the Occasion of his 33rd or 34th Birthday*, ed. by A. Zwicky, P. Salsus, R. Binnick, and A. Vanek (Edmonton, Alta.: Champaign, Ill.: Linguistic research, 1971)

linguistics journal, *Linguistische Berichte*, Günther almost single-handedly turned the previously rather low-profile linguistics chair of the German department in Frankfurt into a centre for generative linguistics, at times producing more promising young linguists than any other linguistics department in Germany. Recently, his efforts culminated in the founding of a separate *Institut für Kognitive Linguistik* ('Institute of Cognitive Linguistics') at the University of Frankfurt, which after all provides an appropriate setting for his work as a linguist and teacher.

The articles in this volume aim at doing some justice to his wide interests. They reflect concerns that dominate Günther's work and thinking, providing an up to date view on topics ranging from key questions of theoretical syntax to pragmatic and philosophical issues. These include the relation between word order and the structuring of information, pro-drop phenomena, verbal syntax, the interface between syntax and the interpretative components of grammar, clausal typing and the encoding of mood as well as questions of language acquisition and evolution, semantic interpretation and the relation of language and the law.

We should acknowledge, though, that the present selection of papers is not representative in at least two respects. First, given the number of Günther's students, friends, and collaborators and the amount of influence he exerted on the field, we easily could have collected more than the 23 papers you'll find below. However, to keep this book from exceeding the usual page limits of a single-volume publication, at a certain point we simply had to stop asking people for contributions. Second, it is quite obvious that this collection cannot cover the whole breadth of Günther's work. Thus we feel that many aspects of his work on pragmatics (cf. e.g. Grewendorf 1979b, 1981a, 1983a, 1984a, 2002a) as well as Günther's work on the intricacies of the German tense system (cf. e.g. Grewendorf 1982a, 1984c, 1995a), or topics like the analysis of (multiple) *wh*-constructions (cf. e.g. Grewendorf 2001a and his 2002 book) are underrepresented here. Still, we think that there is considerable overlap between the topics dealt with in the present volume and Günther's research interests.

The introduction to a Festschrift is the place as well, of course, for some more personal words, including anecdotes (and possibly even some juicy gossip) that reveal some characteristic traits of the one to be honored.

As already briefly mentioned above, Günther has a striking ability to attract people to linguistics. Moreover, due to his talent to organize and get things off the ground, he is often in the position to provide jobs for people he believes in, which enabled many to turn an interest into a profession. In fact, there is a vast number of present-day linguists in Germany which started their career as assistants of Günther's. As a boss, he always tries to establish an amicable relationship with his co-workers (he really dislikes it if someone calls him *Chef* 'boss'), which includes regular visits to his favorite place in Frankfurt, the *Pielok*. At times, he can be quite demanding towards his associates. He is used to high standards and seeks them in his co-workers as well, both personally and linguistically. He appreciates loyalty and diligence among his staff. If he believes in someone and/or the work done by that person, he is usually very supportive. On the other hand, he takes it very personal if someone disappoints him (personally or linguistically). Due to his many occupations, his time is quite limited and he is often not willing to spend it on things or people that once didn't meet his expectations. Still, he will always agree to discuss any weird theoretical idea that comes to your mind if you approach him directly.

A characteristic which many of his co-workers might have noticed over the years is Günther's highly ambivalent attitude towards technology. On the one hand, he's always keen on the latest (IT-) devices (such as notebooks, cell phones, etc.). Often, this enthusiasm is also to the benefit of his co-workers – in fact Günther is quite proud of the fact that his chair is at the fore of technology (Günther was among the first within humanities at the University of Frankfurt who had computers, e-mail and all that, due to a research project funded by IBM). On the other hand, Günther is deeply suspicious and skeptical of technological developments and changes affecting his gear, in particular if these "innovations" frustrate his efforts to tame the technological beast. And the facts often seem to prove him right – the installation of new devices or software regularly doesn't lead to the desired results or, even worse, shuts down his computer, which usually sends him into a frenzy about the shortcomings and inadequacies of modern technology (the AutoFormat function of MS Word is another regular cause of similar fits of rage). On such occasions, Günther wholeheartedly condemns the wonders of modern information technology and proclaims that he will go back to DOS (when we still knew why things didn't work), or, if disaster strikes really badly, to the typewriter. His particular experience in this area might also be the reason why he still clings to his beloved and trusty pocket diary, resisting all temptations to switch to a much more fashionable palmtop (like Ede Zimmermann has).

Coupled with this fraught relationship is his equally frustrating relationship to other conveniences of modern life such as cars and bank accounts (e.g., his new Volvo with its leaking roof, and the intricacies of managing the cash flow between various research and university related accounts).

At times, his skepticism about progress seems to translate into a gloominess about the world in general and his day-to-day business at the university in particular. When Günther is in this particular mood, he often seems to entertain the suspicion that things aren't running as smoothly as they may seem to be and that disaster may be lurking behind every corner. And often enough, his seemingly groundless concerns appear to be justified by the developments in the real world. Then again, Günther is always prepared to roll up his sleeves and fight off the threatening troubles, which regularly ends those periods of gloom. This kind of *Hemdsärmeligkeit* ('shirt-sleevedness', see below for some explication) is not only typical of Günther, it is also a quality which he appreciates greatly in his co-workers (who sometimes have to struggle to live up to these high standards).

This particular trait of Günther's can also be traced in his approach to science sometimes. To him, it seems, things start to get interesting when there is real disagreement that sparks fervent discussion. Again, it is when he "rolls up his sleeves" that he appears to be really in his element. Among students and fellow linguists, Günther is famous (and sometimes feared) for his critical, plain-spoken, no-nonsense way of dealing with theoretical ideas and linguistic reasoning. This characteristic of his can be best experienced during his colloquium (the *Oberseminar*), every Wednesday from 6 to 8 pm, where the latest developments in generative syntax are discussed (and often shredded). The Frankfurt linguistics *Oberseminar* is also widely known for its critical audience, and it has seen many pilot runs of talks later delivered to much larger crowds (who can make it there will make it anywhere – God bless those who try sell something that goes against Günther's most deeply felt theoretical convictions.).

Instead of rambling on about purported character traits, we'll rather try to give an impression of a particular evening at the *Pielok*, Günther's favorite place in Frankfurt where he can refuel on Wednesday evenings after the *Oberseminar*. This occasion was particular because it was the evening of Günther's first appearance at the senate as newly elected dean. The meeting of the senate was busted up by students protesting against the newly introduced study fees; a glass window had been broken with a fire-extinguisher and the president of the university had locked himself up in his room, possibly because he was afraid of suffering physical damage. Instead of fleeing the scene, Günther once again rolled up his sleeves and engaged in a discussion with the students, eventually managing to cool down the situation and pave the way for negotiations with the president. On the evening of this Wednesday, then, Günther sat in the Pielok with us, reporting the turbulent events of the day, which he would finally sum up with a long sigh: "It's like the good old days", and a sentimental smile would cross his face.

Of course, now that Günther is dean of the faculty, his schedule is even tighter than it used to be. In view of his broad research interests and his newly acquired duties as dean, one is not surprised that such industriousness has repercussions in other areas of his life as well. Thus, there is often a clash between his affection for Italy (including his fondness of the Italian way of life) and his cramped schedule. At times when he is frustrated and tired by the negative aspects of academic life such as endless meetings, grim competition, lazy students, cuts in the budget, and the seemingly endless chase for the best theory, he announces that it is his utter wish to retire to his house in Italy, to become a vintner and olive farmer. On the other hand, things don't always go so smoothly in Italy, either. Returning to Frankfurt after his last sabbitical, Günther appeared a bit worn out from fighting a population of small flies that infested his Italian residence and tormented him at night (or was it bugs? it took him quite some time and energy to only figure out what was actually bugging him, in any event). Moreover, olive cultivation is hard and occasionally even dangerous work – recently, when Günther was cutting down some branches and trees he sustained a serious injury which required a trip to the hospital and nine stitches. Be this as it may, having to meet all the different demands that he sees himself confronted with naturally requires some external assistance. None of his co-workers is surprised that his business is always very urgent – as Günther would put it: *schwerpunktmäßig wird hier immer noch gearbeitet.*

Form, Structure, and Grammar

The papers collected in this volume are grouped into three parts which we take to reflect three major strands of interest in Günther's work and thinking.

The first part, *Form*, addresses issues pertaining to morphology, lexical semantics, and their interface to syntax, a topic area which is touched upon in quite a number of Günther's publications (to name only a few, cf. Grewendorf's 1989 book on the licensing of *pro* and the analysis of ergative constructions in German, Grewendorf 1995a on the syntax and semantics of present and perfect tense forms in German, Grewendorf and Sabel 1999 on the relation between properties of inflectional systems and the availability of scrambling, Grewendorf 2001a on morphological properties licensing *wh*-cluster formation, or Grewendorf 2003a on the question of how the shape of reflexive pronouns influences their binding properties).

The part *Form* includes contributions by Werner Abraham (deriving the interpretations and the syntactic behavior of different participial forms), Manfred Bierwisch (accounting for properties of German reflexive forms in terms of the distinction between improper, pseudo- and regular arguments), Sascha W. Felix (presenting an experimental study on the acquisition of Japanese word structure), Tilman N. Höhle (analyzing morphological and syntactic properties of verb clusters in a variety of German dialects), Manfred Krifka (providing a novel analysis of the pronoun system of Tok Pisin), and Gereon Müller (offering a new explanation for the "pro-drop parameter" based on an in-depth analysis of inflectional features and their relation to syntax).

The second part, *Structure*, features a set of new and original syntactic analyses, mainly concerning the syntactic encoding of information structure, and in particular the structure of the middle field and the left periphery. This topic has been at the heart of Günther's research interests for many years (cf. e.g. Grewendorf 2002, 2002c, 2003b, 2005a, 2005b, forthcoming; Grewendorf and Poletto 1989, Grewendorf and Sternefeld 1990; Grewendorf and Sabel 1994, 1999).

This part includes contributions by Josef Bayer (presenting new evidence for the hypothesis that different types of A'-movement target different left-peripheral positions of the German clause), Adriana Belletti (proposing a uniform analysis of clitic left dislocation and relative clauses involving resumption in Italian), Gisbert Fanselow (providing an account of word order variation in the German middle field in terms of rules of accent attraction), Werner Frey (offering a new account of the distribution of object-*es* in German), Katharina Hartmann and Malte Zimmermann (investigating the make-up of the left periphery in the Chadic language Dghweɗe), Cecilia Poletto (analyzing scrambling in Old Italian as movement to an IP-internal Focus projection), Luigi Rizzi (offering a new perspective on inversion phenomena in Italian and Romance), Joachim Sabel (linking the availability of non-finite relatives and *wh*-questions to morphological properties of the C-system), Mamoru Saito (explaining expletive verb constructions in Japanese in terms of the minimalist principles of Full Interpretation and Last Resort), and Jochen Zeller (who develops a minimalist analysis of applicative constructions in Kinyarwanda).

The third part, *Grammar*, collects papers that deal with the interfaces of core grammar with the systems of interpretation and use as well as papers addressing the role and nature of the language faculty within a broader cognitive and evolutionary perspective. In Günther's oeuvre, this subject matter is addressed from two different angles: first, in a theoretical manner, dealing with issues in the theory of (universal) grammar and the way the phenomenon of language is rooted in properties of the human mind (cf. e.g. Grewendorf 1985c, 1993a, 1995, 1996, 2006; Grewendorf and Zaefferer 1991, Grewendorf and Wilder 1994). Second, in a way that emphasizes practical applications of theoretical linguistics, focusing on the relation between language and the law in particular (cf. e.g. Grewendorf 1985a, 1990c, 1992a, 2000a, 2002b).

The third part contains contributions by Rainer Dietrich (providing evidence that the language faculty is subject to a circadian rhythm of its own), Hans-Martin Gärtner and Markus Steinbach (arguing against the feasibility of the analysis of speech acts and point of view in terms of functional projections), Georg Meggle (arguing that apparent paradoxes concerning the evolution of language dissolve if a proper distinction is made between the development of non-verbal communicative abilities and genuine linguistic abilities), Monika Rathert (proposing that the notion of comprehensibility in forensic linguistics is better approached from the perspective of Frame Semantics), Tom Roeper (developing an account of focus binding in *Only I* and related constructions in terms of point of view features and Relativized Minimality), Dietmar Zaefferer (reanalyzing sentence mood in terms of "lean specification" and "reconceived content" and deemphasizing the role of truth for the definition of speech acts), and Thomas Ede Zimmermann (arguing for a distinction of the "internal" and "external" semantic roles of linguistic expressions and thereby solving certain substitution fallacies).

For help in the production of this book we wish to thank Ortrud Bruchelt, Anne Fuß, Katharina Hartmann, Shin-Sook Kim, Cécile Meier, Christian Plunze, Andreas Runkel, Tanja Schmid, Matthias Schulze-Bünte, Rachel Szekely, Martin Urban, Jenny Wilcke, and Jochen Zeller. Special thanks are due to Manfred Bierwisch and Peter Heyl, who offered us the opportunity to have this volume published in the series *Studia Grammatica* of the Akademie Verlag.

Summaries of the Individual Contributions

To help the reader navigate through this volume according to her or his interests, we provide in the following an alphabetically ordered list of abstracts which briefly summarize the content of the individual contributions.

Werner Abraham: Distinct Diathesis and the Mechanics of Underspecification

This article sets out to derive the different interpretation and syntactic behavior of superficially identical participial forms from differing underlying structures these participles find themselves in. More specifically, it is argued (a) that the relevant verbal/adjectival forms are underspecified with respect to formal features including phi features and event features in particular, and (b) that the particular interpretation of the forms in question is determined by the functional structure embedding these forms. Special attention is paid to the encoding of the inner temporal structure of passive forms as determining their distribution. The analysis is eventually extended to impersonal passives.

Josef Bayer: A Note on Targets of A'-Movement in the Left Periphery of German Sentences

In his contribution, Josef Bayer pursues two principal goals. First, he presents further empirical support for the hypothesis put forward in Grewendorf (2002) that in German, different types of A'-movement target different positions in the left periphery of the clause. Second, he challenges the hypothesis (cf. Grewendorf 2002) that this difference also accounts for the fact that in German, only long *wh*-movement – in contrast to short *wh*-movement – leads to WCO effects. Based on the observation that successive-cyclic *wh*-movement does not give rise to WCO effects in the embedded clause, it is argued that the absence of WCO effects with short *wh*-movement should better be attributed to the availability of (A-) scrambling which circumvents WCO effects in German. In addition, it is suggested that the operator status of a phrase is not determined by the kind of position the phrase moves to. Rather, it is taken to be an inherent property of the phrase itself (cf. e.g. *wh*-in-situ approaches like Reinhart 1998). Finally, it is shown that so-called emphatic topicalization in Bavarian (in which fronting of an element in an embedded clause triggers pied-piping of the whole embedded clause to the left periphery of the matrix clause) provides evidence for a further position (encoding emphasis) in the left periphery of the clause.

Adriana Belletti: Extending Doubling to Non-Local Domains: Complete vs. Partial Copying + Deletion and Related Reconstruction Issues

This paper proposes a uniform analysis of clitic left dislocation (CLLD) constructions and relative clauses involving resumption in Italian. It is shown that in both constructions, extraction out of (strong) islands is marginally possible. This is argued to be due to a configuration where the left-dislocated element originates in a big DP, stranding the clitic head of the big DP within the island; this type of partial copying and deletion contrasts with cases without doubling, where the copy in the base position is fully deleted. It is noted, in addition, that there is still a contrast in acceptability between the two structures. Belletti proposes that island violations in relative clauses involving resumption are felt to be less severe because the contrast with the 'concurring' relative construction involving an island violation but no resumption is very sharp. Island violations in clitic left dislocation structures, on the other hand, are felt to be relatively strong since the 'concurring' Hanging Topic structures are in fact fully grammatical. In contrast to widely held beliefs, the latter are argued to involve movement as well, but this movement is assumed to occur only within the island, affecting an empty DP double of the clitic pronoun. The non-local relation between the overt Hanging Topic, which is directly merged in the left periphery, and the empty DP is established not in the syntax but by discourse mechanisms which are not sensitive to the same kind of locality conditions.

Manfred Bierwisch: German Reflexives as Proper and Improper Arguments

Reflexive pronominal forms in German have special and relatively extensive properties. They are utilized, among other, to derive unergative forms (e.g., *sich setzen* 'seat oneself') from ergative verbs (*setzen* 'seat'), as well as in the formation of middle constructions (*die Tür öffnet sich leicht* 'the door opens easily'). This paper motivates an analysis treating reflexives uniformly as regular arguments that identify the pertinent argument position with the one of the antecedent.

Rainer Dietrich: Modularity and the Tic-Toc of Language

This paper deals with the question of whether the human language faculty is subject to a circadian rhythm of its own or whether it is clocked along with other cognitive abilities. After presenting the issue of the modularity of language and giving a survey of related studies, Dietrich reports on an experimental study that was designed to test the time dependency of general attention as compared to language comprehension abilities. To test time-dependent performance in the respective domains, the "d2 paper and pencil letter cancellation test" and a self-paced reading task of sentence comprehension were used. The results support the hypothesis that language obeys its own rhythm: while general attention appears to reach its peak around 5.a.m., language comprehension has its peak around 7.p.m.

Gisbert Fanselow: On Pure Syntax (Uncontaminated by Information Structure)

In his contribution, Gisbert Fanselow argues that the encoding of information structure does not involve functional projections such as TopicP or FocusP. This claim is supported by both conceptual considerations and empirical evidence. On the conceptual side, it is reasoned that the presence of features such as [topic] and [focus] would violate the inclusiveness condition (Chomsky 1995), since lexical entries that enter the syntactic derivation cannot be specified for these information-structural distinctions. In addition, it is noted that the optionality of topic and focus movement sets it apart from typical syntactic movement processes such as *wh*-or Case-driven movement, which are usually obligatory. It is then shown that the placement of focused phrases is better analyzed in terms of syntactic operations that attract elements that bear an accent (feature) that is added to lexical items prior to their insertion. In a similar vein, it is argued that the distribution of topics in the German "middle field" receives a more satisfying analysis if their position is attributed to prosodic factors. The empirical findings are corroborated by the results of an experiment in which the acceptability of pre-sentence adverb placement of topics was tested.

Sascha W. Felix: The Acquisition of Japanese Word Structure

This paper deals with potential learnability problems in the acquisition of Japanese, presenting an experimental study of one particular type of problem. In Japanese, different meanings are often encoded by homophonous forms, where disambiguation is achieved through the writing system that preschool children have not learnt yet, however. A related second problem that is more to the point of the experimental study that is presented consists in the fact that particular meanings may be encoded by different (Sino-Japanese vs. Japanese) forms, depending on whether the meanings in question form part of simplex vs. compound words. The forms are not systematically related to each other but are the result of historical accident; it is the writing system, again, that helps to identify the intended meaning for a particular form.

Werner Frey: How to Get an Object-*es* into the German Prefield

This paper offers a new perspective on the availability of the object pronoun 3sg.neut. *es* in clause-initial position. It is shown that, in contrast to widely held beliefs, object-*es* may occupy the so-called "prefield" and that its availability in this position is determined by the structural position of the subject in the IP domain (the "middle field"). More precisely, an object-*es* may occupy the prefield only if the subject does not assume topic character and stays in a position to the right of sentence adverbials such as *leider* 'unfortunately'. The author develops an analysis of this finding which is based on the idea that the prefield can be targeted by two different types of movement: (a) "genuine" A'-movement which may affect any constituent and leads to stress and a contrastive interpretation of the moved element, and (b) so-called "Formal Movement" (FM) which may attract only the highest element in the IP domain (and which has no effects on the interpretation of the clause). By assumption, object-*es* cannot undergo A'-movement due to its inherent prosodical weakness. Thus, it may occupy the prefield only as a result of FM (when it is the highest element in the IP domain). Accordingly, the relevant restrictions on the placement of object-*es* in the prefield are reconstructed as restrictions on the structure of the middle field, based on the assumption that in the IP domain, topics occupy a designated structural position to the left of sentence adverbials (SpecTopP).

Hans-Martin Gärtner and Markus Steinbach: A Skeptical Note on the Syntax of Speech Acts and Point of View

The paper by Hans-Martin Gärtner and Markus Steinbach deals with the relation between syntax and pragmatics. They critically examine a recent proposal by Speas and Tenny (2003) that aims at modelling illocutionary force (i.e., sentence mood) and point of view related phenomena via purely syntactic means and makes use of a richly structured clausal left periphery. Gärtner and Steinbach argue that this particular approach to the syntax/pragmatics interface is not adequate, since it faces a number of shortcomings, both conceptually and empirically. They show that under Speas and Tenny's analysis, the inventory of theoretically possible sentence moods is a product of stipulation rather than formal syntactic deduction. In addition, it is demonstrated that the interpretation of indexicals and logophoric elements cannot be successfully captured by this model. The authors conclude that an adequate theory of these facts cannot be based on purely syntactic means but must also incorporate insights from other theoretical modules (e.g., (formal) pragmatics).

Tilman N. Höhle: Observing Non-Finite Verbs: Some 3V Phenomena in German-Dutch

In his contribution, Tilman N. Höhle presents a vast variety of data from mostly Middle German dialects which shed new light on the inventory and ordering possibilities of non-finite verb forms. It is shown that there is much more variation in three-verb clusters than usually assumed. For example, Höhle demonstrates that in a larger Middle German dialect area, it is not the infinitive that replaces the participle in IPP (*infinitivus pro participio*) contexts, but rather a special non-finite form which Höhle identifies as a supine. Upon closer inspection, it becomes clear that in many dialects, the inventory of non-finite forms is much larger than in the standard language, including up to eight different non-finite verb forms as e.g. in the Middle German dialect of Barchfeld. The paper discusses the distribution of these additional forms, showing that there is quite some amount of variation, even between neighboring dialects. In addition, Höhle investigates a set of intricate "displacement" phenomena, which involve shifts in the satisfaction of selectional requirements inside the verb cluster (as in German *ohne das Buch haben lesen zu wollen*, where the infinitival marker *zu* does not appear on the highest auxiliary *haben* but rather on the selected modal *wollen*).

Katharina Hartmann and Malte Zimmermann: Information Structure and the Left Periphery in Dghweɗe

Hartmann and Zimmermann discuss a number of syntactic phenomena that result from the interaction with information structure in Dghweɗe, a Central Chadic VSO-language, spoken in Northeastern Nigeria. These phenomena include (a) *wh*-question formation; (b) the fronting of information-structurally prominent constituents; and (c) the effects of fronting on the functional systems of the left periphery, in particular on the T- and C-systems, which encode a clause's temporal and illocutionary properties. It is shown that in *wh*-questions, *wh*-arguments and *wh*-adjuncts behave differently. *Wh*-arguments move to a left-peripheral position, whereas *wh*-adjuncts occupy a position immediately to the right of the verb. A similar asymmetry between arguments and adjuncts can be observed in the case of (fronted) focus and topic constituents: while arguments (focused subjects, focused objects (optionally), subject topics) are fronted to clause-initial position, focused adjuncts must occur in immediately postverbal position. In addition, Dghweɗe has a clause-external left-peripheral position, in which frame-setting adverbials and topicalised nominal constituents can be base-generated. Completing their overview of the make-up of the left periphery in Dghweɗe, the authors note that there is a structural position immediately to the left of the verb that can host various head-like elements, and that encodes temporal and illocutionary information, arguing for a matching of CP- and TP-domain in Dghweɗe. Interestingly, fronting to Spec,CP/TP sometimes leads to the suppression of the functional marker in T/C. Finally, it is shown that Dghweɗe shows some surprising similarities with Celtic VSO-languages, which may open new perspectives on the study of the syntax of VSO-languages in general.

Manfred Krifka: A Note on the Pronoun System and the Predicate Marker in Tok Pisin

This paper offers a new analysis of the rich systems of pronouns found in Tok Pisin. The proposed account is based on four central assumptions: first, the category [person] is decomposed into two binary features [±Speaker], [±Hearer]. Second, Krifka argues that the pronoun system of Tok Pisin should be analyzed in terms of a minimal/augmented system (similar to e.g. Ilocano). Third, it is claimed that there are "transnumeral" pronominal forms which lack an explicit number specification. Fourth, it is assumed that the use of a less specific form is blocked if a more specific form exists in the paradigm (i.e., some form of "elsewhere" or "blocking" principle is invoked). It is then shown that these assumptions not only make available an adequate account of the pronoun system of Tok Pisin, but also offer new insights into the distribution of the predicate marker *i*. More specifically, it is shown that *i* is not used with subjects that occupy a high position in a hierarchy of person/number distinctions constructed from the feature distinctions proposed previously.

Georg Meggle: The First Traces of Language. On Darwin's Package Hypothesis

Georg Meggle offers a philosophical perspective on certain issues relating to the evolution of human language, and in particular to Darwin's Package Hypothesis (DPH, the idea that characteristics unique to humans evolved in a single evolutionary step). The problem at hand has to do with the fact that according to archaeological findings, the development of an advanced brain organization is rather ancient (having taken place about 2.5 million years ago), while the evolution of the speech apparatus and the development of advanced tools and arts took place much later. This gap not only seems to contradict the DPH, but also raises questions concerning the origin of language. Meggle argues that the divergent evidence can be reconciled with one another if a proper distinction is made between (the development of) non-verbal communicative abilities and genuine (verbal) linguistic abilities. Based on this distinction, it is possible to link the early evolutionary step leading to an advanced brain organization to the development of basic (non-verbal) communicative skills and the production and use of first tools. In a subsequent gradual development, the evolution of brain size and skull form then could have led to a certain point at which a cognitive threshold was crossed, giving rise to genuine verbal linguistic abilities, advanced tools, and the evolution of artistic facilities.

Gereon Müller: Pro-Drop and Impoverishment

This paper offers a new account of the often-noted correlation between properties of verbal inflection and the availability of pro-drop, focusing on German. In contrast to previous proposals, the explanation does not refer to the make-up of inflectional paradigms, but, making use of recent advances in theoretical morphology (i.e., Distributed Morphology, in particular the notions of fission and impoverishment), to abstract properties of the inflectional system of a given grammar. More specifically, Müller argues that referential subject *pro* is licensed only if the relevant inflectional morpheme with which *pro* enters into a syntactic Agree relation – i.e., T – is not affected by impoverishment rules which delete a subset of T's φ-set and give rise to systematic syncretisms across inflectional paradigms/conjugation classes (e.g., homophony of 1pl and 3pl in the preterit paradigm of many Germanic languages). This analysis requires that the effects of impoverishment must be detectable at the point when *pro* is inserted/merged. Accordingly, it is assumed that the morphological component operates presyntactically. It is then shown that in addition to German, the account successfully captures the behavior of (referential) pro-drop in Icelandic, Irish, and Russian.

Cecilia Poletto: Old Italian Scrambling: The Low Left Periphery of the Clause

In her contribution, Cecilia Poletto argues that OV (with V = participle) patterns in Old Italian are to be analyzed as the result of XP-movement to the specifier of a lower, IP-internal Focus projection (accompanied by movement of the participle to $Focus^0$). The availability of movement to a lower SpecFocP is linked to the phenomenon of participle agreement (in the sense that overt object movement requires agreement on the participle). Furthermore, it is claimed that the featural content of one and the same type of functional head is uniform across phases. That is, by assumption, a low IP-internal Focus head and a high, left-peripheral Focus head must have identical feature values. If the relevant features are strong, they give rise to V2 effects in the left periphery and OV patterns within IP (and similar word order effects inside DP, if DP is analyzed as a phase that may contain a separate Focus projection). Under these assumptions, it is expected that there is not only a synchronic, but also a diachronic link between V2 effects, OV patterns (and participle agreement), and DP-internal element reordering. Poletto shows that this expectation is borne out by the historical facts, that is, V2 effects and IP- and DP-internal scrambling are lost at the time (and rate) in the history of Italian.

Monika Rathert: Comprehensibility in Forensic Linguistics – New Perspectives for Frame Semantics

The principal objectives of this paper are as follows. First, it is demonstrated that most previous attempts to evaluate the comprehensibility of (legal) texts suffer from serious shortcomings (lack of explicitness, lack of objective criteria for comprehensibility etc.). Second, it is argued that a "shallow" semantic approach based on Frame Semantics (Fillmore 1976) offers a promising alternative that combines formal criteria for the evaluation of text comprehensibility (e.g. in terms of Frames that cannot be integrated) with the possibility of implementation in an automated software tool.

Luigi Rizzi: Selective Residual V-2 in Italian Interrogatives

This paper offers a new perspective on inversion phenomena in Italian (and Romance in general). Following the idea that in *wh*-questions, a checking relation must be established between the *wh*-phrase and a (criterial) Q feature (Rizzi 1991), the parametrical difference between languages that require inversion in interrogatives and others that do not is attributed to the position where the Q feature is inserted: languages in which Q is located in T require inversion (T-to-C movement) to establish a checking relation between Q and the *wh*-phrase; alternatively, Q may be inserted directly in the C-system (e.g. Fin), giving rise to a grammar lacking inversion in *wh*-questions. A complication is raised by the fact that in Italian interrogatives, both the auxiliary and the participle must appear to the left of the subject. This kind of inversion is analyzed in terms of fronting the whole TP (containing the Q feature) to the left clausal periphery, leaving the subject – which has moved out of TP (to SpecSubj) previously – behind in final position. In addition, Rizzi addresses the question of why Italian lacks inversion in *wh*-questions involving *perché* 'why' or (marginally) in *wh*-questions with D-linked *wh*-phrases. Following Rizzi (2001a), he assumes that *perché* is base-generated in the left periphery as the specifier of $Int_{[+Q]}$. As a result, there is no verb movement into the C-system and accordingly no inversion in questions featuring *perché*. The Q feature of Int^0 is also exploited to account for the (marginally acceptable) lack of inversion in the context of D-linked *wh*-phrases, by assuming that $Int^0_{[+Q]}$-to-Top^0 movement creates a possible landing/checking site for D-linked *wh*-phrases (which are specified for +Top(ic), +Q).

Tom Roeper: Not *Only I*: Notes on the Syntax of Focus Binding

In his contribution, Tom Roeper argues that the indexical binding of *only I got a question I could answer* generalizes to all categories (cf. e.g. *only Jane looks like Jane in these old pictures*). It is proposed that this binding involves focal movement which is subject to Relativized Minimality at a much subtler level than usually assumed. By assumption, any perspectival shift causes a conflict in Point of View features creating a barrier to movement. In this way, the paper creates a model for an interface relation between the semantics of set-induction and the effects of feature-attraction under phase theory.

Joachim Sabel: Impossible Infinitival Interrogatives and Relatives

This paper discusses the observation that languages differ in whether they allow non-finite interrogatives and relatives. It is shown that the (non-) availability of these non-finite constructions is linked to certain morphological properties of the C-system. More precisely, the author demonstrates that the absence of *wh*-infinitives (and non-finite relatives) in German and other Indo-European languages correlates with the lack of base-generated material (such as non-finite complementizers, infinitival markers etc.) in the left periphery of non-finite clauses. In contrast, languages that license *wh*-infinitives (such as English or Dutch) do exhibit non-finite complementizers. The relevant generalization (dubbed the "The *Wh*-Infinitive-/Infinitival-Relative-Correlation") is then explained by assuming that languages without *wh*-infinitives have a "defective" C-system in non-finite clauses, which renders *wh*-movement impossible.

Mamoru Saito: Expletive Replacement Reconsidered: Evidence from Expletive Verbs in Japanese

This article offers empirical evidence from Japanese for an analysis of *there*-sentences in terms of associate raising/expletive replacement (Chomsky 1986). The core idea is that the element *su* when functioning as an 'expletive verb' must be replaced at LF so that the principle of Full Interpretation is obeyed. *Su* will be replaced only if there is independent motivation for this replacement (last resort). The assumption that theta assigning heads move to the position of *su* in order to discharge theta roles (but not otherwise) is used to explain Grimshaw and Mester's generalization that one internal argument of theta assigning nouns must be realized outside the projection of the N head that licenses it thematically. Saito further argues that the analysis can account for the fact that VP scrambling disambiguates structures between a 'main verb' and 'expletive verb' interpretation of *su*, as well as for the fact that VP scrambling is disallowed with unaccusative structures containing 'expletive *su*'.

Dietmar Zaefferer: Conceptualizing Sentence Mood – Two Decades Later

In this paper, Dietmar Zaefferer argues for a view that differs from the one advocated by him and Günther Grewendorf in a joint paper written in the mid-eighties mainly in three respects: (a) lean specification: basic speech act concepts, the ones that are coded by sentence mood, are much less specific than canonical speech act concepts; (b) reconceived content: speech act contents are best characterized in terms of Cognitivized Austinian Propositions and their values in terms of congruence; (c) deemphasizing truth: the truth bearers fallacy shows that the relation between truth and linguistic action must be conceived as much less close than previously assumed. This clears the way to a fresh view of explicit performatives analyzing them as direct transparent epistemics whose performance makes congruent their propositional content.

Jochen Zeller: Agreement and the EPP in Kinyarwanda Applicatives

This contribution investigates the structure of (locative) applicatives in the Bantu language Kinyarwanda. Based on an in-depth study of the syntactic properties of this construction, it is argued that in contrast to recent proposals, applicatives in Kinyarwanda are best analyzed in terms of preposition incorporation, as originally proposed by Baker (1988). Furthermore, Zeller suggests that the word order in this type of applicative is derived by moving the applied object to an outer specifier of VP. This operation is required to establish agreement (and Case checking) between the moved DP and v^0, which agrees with the closest DP in its c-command domain. Zeller argues convincingly that this movement step, which is necessary to circumvent a locality violation, is not triggered by a feature associated with v, but by an EPP-feature associated with the verb itself, possibly inherited from the phase head v, along lines proposed recently in Chomsky (2005).

Thomas Ede Zimmermann: The Values of Semantics

This paper is about the status and significance of semantic values that ('realistic') semantic theories assign to linguistic expressions. Starting with some puzzles concerning substitutivity, it is argued that sameness of semantic values does not coincide with synonymy, which should instead be construed as sameness of the roles these values play: their internal role, determined by their interaction within semantic theory; and their external role(s), determined by their interpretation (if any) in terms of neighboring (sub-) disciplines.

Part I: Form

Distinct Diatheses and the Mechanics of Underspecification

Werner Abraham

0. Introduction

The present article* sets out to derive the different interpretation and syntactic behavior of superficially identical participial forms from differing underlying structures these participles find themselves in. More specifically, it is argued that

(a) the relevant verbal/adjectival forms are underspecified with respect to formal features including in particular phi features and event features, and that

(b) the particular interpretation of the forms in question is determined by the functional structure embedding these forms. Special attention is paid to the encoding of the inner temporal structure of passive forms as determining their distribution. The analysis is eventually extended to impersonal passives.

The general idea of category underspecification as employed here has been taken up in recent times by Distributed Morphology, but dates in fact back, at least for longitudinal European methodological thinking, to the mediaeval Modistae (Thomas of Erfurt and others; see Jakobson 1985, Leiss 1998).

1. Source categories and derivative categories

The term "diathesis" co-classifies all verbal genders/diatheses and their pre-derivative roots, including the non-verbal (viz. de-adjectival verbs), presupposing that all such categories be represented on a common basis along distinct and non-ad hoc derivational mechanisms. (1) lists which data such a representational strategy should minimally unpack into categorial components as well as jointly employed derivational steps in the context of passive diathesis. The sample of underspecified categories in (1a,bi-iii) is not exhaustive.

(1) a. Adjectives: root sources for adjectival verbs such as German *offen* – (*sich*) *öffnen* 'open$_A$ – (REFL) open$_V$', *0* – (*sich*) *schließen* '0$_A$ – (REFL) close$_V$' etc.
b. Past Participles (PP) with the following selectional distinctions:
(i) STATE PASSIVE PARTICIPLE – eV participle:
PP+*sein* 'be': *geöffnet* 'opened', *geschlossen* 'closed'

---

* Thanks to Eric Fuß and Patrick Brandt who have helped make things more legible. With Elisabeth Leiss I was able to discuss details of linguistic pre-category concepts in the schools of mediaeval thinking.

(ii) ONGOING EVENT PASSIVE PARTICIPLE:
PP+*werden* 'become': *geöffnet* 'opened', *geschlossen* 'closed'

(iii) ONGOING EVENT ACTIVE PARTICIPLE:
PP+*haben* 'have': *geöffnet* 'opened', *geschlossen* 'closed'

"eV participle" presupposes that ergative verbs, eV, are perfective intransitives. Stative, or resultative, passives (unambiguously identifiable in German), not to be missed from the list of ergative/unaccusative diagnostics, can be formed only from perfective intransitives (Abraham 2004).

c. Resultative participles: only perfective
(iv) intransitives: *(müde) gelaufen sein*
(tired) run be
(v) transitives: *(seine Füße platt) gelaufen haben*
(one's feet flat) run have

d. Impersonal passive participle: only imperfective
(vi) de-intransitive: *Es wird gelaufen/*Es wird hineingelaufen.*
it becomes run *it becomes into-run
(vii) unpromoted direct object: *Es wird (??den) Marathon gelaufen.*
it becomes (the.ACC) marathon run

e. Ongoing imperfective passive participle:
Der Hund wird (von Konrad) gepeitscht.
the dog becomes (by Conrad) whipped

Passivization of perfectives will receive extra attention here. Notice that not all derivation chains are supported by lexical fillers for each chain member; see *0 – (sich) schließen* '0_A – (REFL) $close_V$' in (1a) which has no derivation-chain inceptive adjectival (anti-inchoative) lexical entry. Nevertheless, given the numerous derivative realizations, adjectives are considered in general to be sources for lexical inchoative and transitive derivations. They will consequently be considered in the ensuing derivative formats. Notice further that intransitive resultative predicates as in (1civ) are taken to be equivalent to what are called unaccusatives, or ergatives, in other languages (Abraham 2000, 2004).

Naturally, an explanatory account for the commonalities and the differences in (1a-c/i-v) will seek to assume syntactic representations that are independently motivated. The following partial tree structure in (2) will be assumed to accommodate the syntactic categories required for the phenomena to be accounted for.

(2)

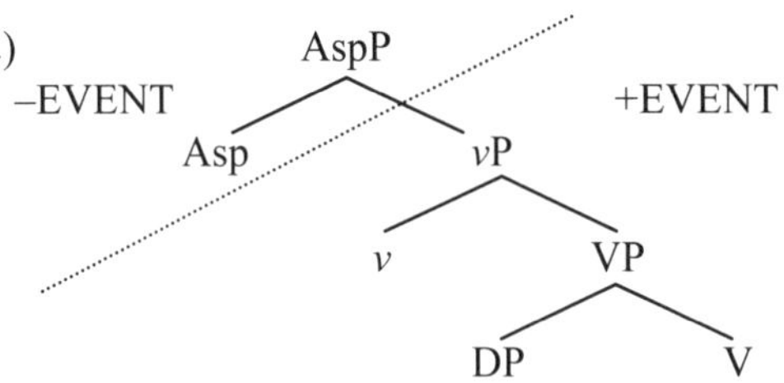

The category Asp(ect) hosts perfective features of the verb, e.g., separable particles and adjectives and affixoids that lead to a bounded interpretation of an event and that are typical of the perfective transition of the imperfective verbal stem in the Germanic languages and particularly in Dutch and German. See (3a,b) – SMALL CAPS signal distinct word accent.

(3) a. separable particles:
essen – AUFessen 'eat – up-eat', *trinken – AUStrinken* 'drink – out-drink'
b. bounding adjectives:
laufen – sich MÜDE laufen 'run – REFL tired run'
c. bounding affixoids:
gehen – HINÜBERgehen 'walk – DEIXIS-over-walk'

The assumption that perfective, resultative Asp is located outside of *v*P accounts for the fact that both adjectives and the adjectival, stative passive participle – see (1a,b-i) – do not carry event properties diagnosable, e.g., by an instrumental Agent phrase. Cf. the oblique dividing line in (2) above.

In order to represent both relations and differences between the categories in (1a-c), a neutral "root" category is assumed to underlie each of the categories in (1a-c), the specifics being that it is stripped of all distinguishing features determining the realizable category lexemes in (1a-c). The following different types of underspecification appear to account for the differences of the homonyms in (1a-c) (aside from *offen* 'open' vs. *geöffnet* 'opened', of course; see (5) below).

(4) a. Adjectives: void of the event feature characterizing actional verbs; no person, number and agreement features.
b. Past Participles (PP):
(i) STATE PASSIVE PARTICIPLE – eV participles:
void of the event feature characterizing actional verbs (no prepositional Agent phrase projectable); no person, number and agreement features.
(ii) ONGOING EVENT PASSIVE PARTICIPLE:
event feature present (prepositional Agent phrase projectable); void of person, number and agreement features; designated valence reduced.
(iii) ONGOING EVENT ACTIVE PARTICIPLE:
event feature present; void of person, number and agreement features; designated valence maintained.
c. Resultative participle (RPP) constructions:
(iv) intransitives (iRPP): no event feature present; void of person, number and agreement features; designated valence reduced.
(v) transitives (tRPP): no event feature present; void of person, number and agreement features; designated valence reduced.

With the feature distribution in (4) in mind, in order to achieve syntactic distinctions among the categories in (1a-c), the following syntactic distributions will be assumed to lie at the basis. [The arc notation, A⌒Asp, indicates sisterhood in the syntactic tree.]

(5)	a.	*offen* (*sein*) – 'be open'	Adjective:	$A^{\frown}Asp$
	b.i	*geöffnet* (*sein*) – 'be opened'	Stative participle:	$V^{\frown}Asp$
	b.ii	*geöffnet* (*werden*) – 'be opened'	Ongoing passive participle:	$V/A^{\frown}V$
	b.iii	*geöffnet* (*haben*) – 'have opened'	Ongoing active participle:	$V/A^{\frown}v$

(5a) and (5bi) can be jointly represented in the stative category by A/V. The two resultative participle complexes, (4c-iv-v), cannot be categorized so simply. Their structurally distinct representation will be presented later.

Let us take the neutral unprojected root category (in the sense of Distributed Morphology; see Halle and Marantz 1993 and, more recently, Embick 2004) to be signaled by A/V. It will be assumed that the content of underspecified A/V is whatever is left after the specific categorial features of verbs (eventivity; tense and mood; agreement for person and number) and of adjectivals (stativity) have been left unassigned (i.e., undecided between + and –).

2. Distributional diagnostics

This section illustrates the differing morphological, syntactic, and semantic properties of the categories under discussion. Different combinations of participle and auxiliary verb(s) are related to different temporal schemata representing their semantics. A partial discussion follows of how the results of established tests for the different categorial realizations (e.g., statives vs. resultatives) can be understood against the background of the proposed analysis.

It is particularly telling to see which ambiguities exist for the English past participle in sentences like (6) below when compared with highly form-content disambiguating languages like German.

(6) *The door was opened.*

(7)	a.	*Das*	*Tor*	*wurde*	*geöffnet.*		...	ongoing event passive in the past
		the	door	got	opened			
	b.	*Das*	*Tor*	*war*	*geöffnet*	*worden.*	...	ongoing event passive pluperfect
		the	door	had	opened	gotten		
	c.	*Das*	*Tor*	*war*	*geöffnet.*		...	stative passive in the past
		the	door	was	open(ed)			

(6) is usually called "eventive passive". As such it would translate more specifically as 'the door was in a state of becoming, or getting, open' corresponding to German (7a), whereas the equivalent of German (7c) would be 'the door was in a state of having become open' (result of a preceding incremental event). See the graphs in (8a) for the ongoing event in the past and in (8c) for that of the stative passive. Keep in mind that the verbs in participial form are perfective. (7b) appears to be more or less equivalent to (7a) only for perfective verbs, not for imperfective ones.

Notice the following graphic distinctions between (a-c). The t_E-induced downward arrow signals which phase of the biphasic Aktionsart property is directly denoted (while the remaining phase, emerging (|>>>|) or resultative (|---|), is always implied). No such bi-implication holds for imperfective predicates as in (9) below. The combined aspect/tense reference graphs in (8)-(9) make use of Reichenbachian time variables, t_E, t_R, t_S, distributing them (by vertical double arrows, ⇓) in accordance with the tensing predication expressed in the illustration.

(8) a. *Das Tor wurde geöffnet.* ... ongoing event passive in the past perfective
the door was opened
t_E t_S ....t_S
|>>>>>⇓>>>>|-----------------|

b. *Das Tor war geöffnet worden.* ... ongoing event passive pluperfect
the door was opened/had been opened
t_E t_R ....t_S
|>>>>>⇓>>>>|-----------------|

c. *Das Tor war geöffnet.* ... stative passive in the past
the door was open(ed)
t_E t_S ....t_S
|>>>>>>>>>>|--------⇓--------|

(9) *Das Tor wurde gestrichen.* ... ongoing event passive in the past imperfective
the door was painted
t_E t_S ....t_S
|~~~~~~~~~~~~⇓~~~~~~~~~~~|

The following diagnostic tests to distinguish statives (adjectives, A) from resultative passives (RP, German *Zustandspassiv*) and ongoing event passives (OEP) have been suggested in the literature over and over again (Kratzer 1994, Wilhelm 2001, Embick 2004, Rathert 2006 (in press), and many others).

(10) adverbial modification of the past participle (PP) is possible only for RP:
a. *Das Fenster blieb (/ist) sorgfältig gesäubert.* ... RP
the window remained carefully cleaned
b. **Das Fenster blieb sorgfältig gesäubert worden.* ... OEP
the window remained carefully cleaned become
c. **Das Fenster blieb durch den Fensterwascher sorgfältig*
the window remained by the window cleaner carefully
gesäubert (worden).
cleaned (become)
d. ?? *Das Fenster blieb sorgfältig sauber.* ... A
*the window remained carefully clean

blieb gesäubert 'remained cleaned' can only be an RP, since *bleiben* 'remain' is a stative verb, which is incompatible with the ongoing event passive of a perfective predicate. See the ungrammatical (10b). The reason for the distributional restriction (10a) vs. (10d) is that the adverb implies agentivity, which is out for adjectives. Notice that the very fact that RPs turn out to be compatible with agentive adverbials, but not with demoted agents as in (10c), indicates interpretive compatibility, but not syntactic compatibility with agents.

(11a-d) lead us through the derivative variants of N-attribution with respect to German *offen/geöffnet* 'open/opened'. Asterisks mean 'derivation not possible under specific source category'.

(11) attributive modification of NPs by the bare past participle (PP) is possible only for RP:

a.	*das geöffnete Tor*	<	*das Tor ist/ war geöffnet*	... RP
	the opened door		the door is was opened	
		<	**das Tor wird geöffnet*	... OEP
			the door is being opened	
b.	*das offene Tor*	<	**das Tor wird/ war geöffnet*	... RP
	the open door		the door is was being opened	... OEP
c.	*das geöffnet (worden) seiende Tor*			... RP
	the opened (become) being door			
d.	*das geöffnet werdende Tor*			... OEP
	the opened becoming door			

(11c,d) show that the attributive OEP readings are possible (albeit stylistically stigmatised) if the Aux/Cop is overt in present participial form derived either from *sein* 'be' or from *werden* 'become/get'.

If a temporal adverb modifies the attributive RP and A, the RP-readings are different.

(12) a.	*die kürzlich eröffnete Taverne*	E	S......S
		|>>>>>>>>>|--------------|	
	the recently opened tavern		
b.	*die kürzlich offene Taverne*	E	S......S
		|----------------------------|	
	the recently open tavern		

The reading in (12b) dates back the event state described to the past, while the time reference of being opened refers to the verging point between the two phases in (12a). Verbs of building and creation, such as *(er)bauen* 'build', *erschaffen* 'create', *machen* 'make', *gründen* 'found', can be expected to create a perfective context (Kratzer 2003, Embick 2004: 357), which is not expected to be compatible with another perfective predicate. This is borne out. See (13)-(14).

(13) a.	*die Türe war/wurde offen erbaut ??(worden)*	... A
	the door was/got built open	
b.	**die Türe war geöffnet erbaut *(worden)*	... RP
	the door was built *opened	

(14) a. *Der Verein war* <u>*groß*</u> *gegründet* *(*worden*) ... <u>A</u>
the club was founded big
b. **Der Verein war* <u>*vergrößert*</u> *gegründet* *(*worden*) ... <u>RP</u>
the club was founded enlarged

Notice that, in line with Embick's appreciation of the English equivalents (Embick 2004: 357) – see the English interlinear glosses in (13a,b) –, the (b)-versions in (13)-(14) do not work with formal resultatives. They can only be OEPs – something that cannot be shown easily in English for lack of Aux distinctions (but see German *sein* vs. *werden*). This is clearly indicated by the German (a)-versions as the Aux *werden*/*worden* cannot be gapped. In other words, the (a)-adjectives *offen*/*groß* 'open/big' in (13)-(14) can only function as stative depictives.

(15) The privative prefix *un-* is usually adjoinable only to A, not to verbs (at least in German – though not without exceptions). In the present context, however, this turns out not to be true since RPs are amenable to *un*-prefigation by contrast to the infinitive and all finite verbal forms.

a. *ungeöffnet*, *ungesäubert*, *ungehindert* ... RP of perfective verbs
unopened, uncleaned, unimpeded
b. **ungelaufen*, **ungeschoben*, **ungezogen* ... PP of imperfective verbs
unrun, unpushed, undrawn
c. **unöffnen*, **unsäubern*, **unhindern* ... present infinitives
unopen, unclean, unimpede
d. ?? *unoffen*, **ungroß*, **unklein* ... **un*-A
unopen, unlarge, unlittle
e. *unschön*, *unsauber*, *unehrlich* ... *un*-A
unbeautiful, unclean, dishonest

The fact that adjectives and RPs can be prefixed in such a way, though not without lexical restrictions on the A-category, is in line with the attributive property and the SEIN/BE-Aux selection as a stative (both resultative and category-underdetermined).

(16) The ability to occur as secondary resultative constructions (object predicative) is open for A, not for RP:
a. *Der Verein gründete sich* <u>*groß*/**vergrößert*</u>.
the club founded itself big/enlarged
b. *Jupiter stieß die Wolken* <u>*auf*/**geöffnet*</u>.
Jupiter pushed the clouds open/opened
c. *Der Schmied schlug das Eisen* <u>*flach*/**abgeflacht* –</u>
the smith beat the iron flat/flattened –
<u>*hauchdünn*/**ausgedünnt*</u>.
air-thin/thinned out

What we discussed above was a number of derived verbal and adjectival forms and their mutual relations in distributions bringing to the fore more clearly schemata representing their semantics. The main categorial realizations, i.e., that of statives vs. resultatives and the respective event types, ongoingness and states, was seen to be dependent upon the use of the German Aux *werden* vs. the Copula *sein*. English turned out to be less disambiguable in this respect. The next section will fathom out more deeply how we fare in the description of these verb-related categories with the notion of "categorially underspecified".

3. Syntax and underspecified components

This section assigns concrete syntactic structures to the various forms under discussion in line with (2) above. This means for the eventive (ongoing) passive (here of a perfective predicate, *öffnen*) that the agentive reading is associated with the feature [Agent] on *v* (cf. Embick 2004: 364 referring further to Marantz 1984 and Kratzer 1996, 2003). The bold structural component in (17) marks the difference to the stative passive participle in (18) below. The perfectivizing particle is zero (0) for the deadjectival verb *öffnen* in (17) as well as in (18). A/V is the underspecified category of the past participle neutral between A and V.

The ongoing passive predicate in (17) is derived from the adjective *offen* 'open'.

(17) *Die Tür wird* (*vom Zimmermann*) *geöffnet.*
'The door becomes/gets/is being opened by the carpenter.'
$[_{\text{PERF Asp}}$ 0- $[_{\boldsymbol{v}\mathbf{P}}$ **Agent** $[_{\mathbf{A/V}}$ *offen* $[_{\text{DP}}$ *die Tür*]]]]

In the stative (resultative, non-ongoing) passive, no agentive PP is adjoinable to *v*; see (18).

(18) *Die Tür ist* (**vom Zimmermann*) *geöffnet.*
'The door is (*being) opened (by the carpenter).'
$[_{\text{PERF Asp}}$ 0- $[_{\mathbf{A/V}}$ *offen* $[_{\text{DP}}$ *die Tür*]]]

As for the inchoative component, both the eventive (ongoing) passive of perfectives in (17) and the stative (non-ongoing) passive (restricted to perfective predicates) in (18) involve an inchoative component (incremental phase in (17), resultative phase in (18)), which has a structure in its own right. Let us spell these inchoative structures out in more detail in (19)-(22).

(19) Resultative intransitives: in German and Dutch ergative/unaccusative, since perfective (Abraham 2000, 2004):

Die Wolken lockern auf. 'The clouds dilute.'
$[_{\text{PERF Asp}}$ *auf-* $[_{\text{SpecvP}}$ *die Wolken* $[_{v'}$ BECOME $[_{\mathbf{A/V}}$ *locker*]]]

The structure of any adjective-derived inchoative/resultative intransitive without a perfectivizing prefix or particle is then the following:

(20) *Der Abend dunkelt/dämmert.* 'The evening darkens.'
$[_{v\text{P}}$ *der Abend* $[_{v'}$ BECOME $[_{\mathbf{A/V}}$ *dunkel/dämmrig*]]]]

The adjective or the neutralized, underspecified category, *dunkel/dämmrig* 'dark', is the complement of the atomic category BECOME.

The difference between the intransitive perfective (= unaccusative) in (19) and resultative transitives is represented by the structure in (21). The DP *der Sturm* 'the storm' functions as an agent that is licensed by an abstract CAUSE predicate. See also Embick (2004: 379).

(21) *Der Sturm lockert die Wolken auf.* 'The storm dilutes the clouds.'
[$_{\text{PERF ASP}}$ *auf-* [$_{vP}$ *der Sturm* [$_{v}$ CAUS [$_{\text{SpecvP}}$ *die Wolken* [$_{v}$ BECOME [$_{\text{A/V}}$ *locker*]]]]]]

The structure of any adjective-derived resultative transitive without a perfectivizing prefix or particle, but with a perfectivizing adjective, is the following:

(22) *Der Hufschmied hämmert das Eisen flach.* 'The farrier hammers the iron flat.'
[$_{\text{PERF ASP}}$ [$_{\text{vP}}$ *der Schmied* [$_{\text{v}}$ CAUS [$_{\text{SpecvP}}$ *das Eisen* [$_{\text{v}}$ BECOME [$_{\text{A/V}}$ *flach*]
[$_{\text{V}}$ *hämmer*]]]]]

The adjective *flach* 'flat' as well as the main verb, *hammer-*, will raise to the perfAsp-node to materialize the perfective property. In the prior case of a perfectivizing verbal particle, *auf-* 'up', the BECOME-*locker* derived the main verb, *lockern* 'dilute', will raise and adjoin to the particle in perfAspP.

This ends this section on inchoative structures contained in various transitive and intransitive constructions. The main insight was that restrictions are placed on the category status and perfective meaning of the lexical items to mark the goal denotation of the inchoative result.

4. Conclusion

We have seen that the description and analysis of transitive as well as intransitive perfectives stands and falls with the introduction of the intransitive 'grow'-predicate BECOME. In the same vein, the category status of the result of the BECOME-predicate remains undecided until the atomic semantic predicates are spelled out lexically. Furthermore, the result predicate, whether in adjectival (participial), adverbial, or nominal category status, must not overdetermine resultativity already implied by the BECOME-predicate.

In the remainder I extend the ideas developed so far to impersonal passive forms focusing on their semantics.

(23) The impersonal passive has imperfective structure, |~~~~~~~~~|, since it can be formed only for the verb-aspectual feature [–perf]. This is in itself an indication of the aspect/Aktionsart contingency of the impersonal passive.

(24) The impersonal passive is thus contrary to the stative passive; the latter can be formed only from perfectives/terminatives: |>>>>>>>>>>>>|------------------|.

(25) What is the semantics of the imperfective impersonal vs. the resultative passive participle? Cf. Kratzer 1996, Embick (2004: 361); Mohr (2005: 169-171):

a. (*Das Fenster wird*) *geöffnet*. 'The window is being opened.':
$\lambda x \lambda s[$ *geöffnet* $(x)(s)]$

b. (*Das Fenster ist*) *geöffnet.* 'The window is opened.':
$\lambda x \lambda s \exists e[$ *geöffnet* $(x)(s) \wedge s = f_{target}(e)]$

c. (*Es wird*) *geöffnet*. 'There is opening.':
$\exists e[$ *geöffnet* $(e)]$

(26) Quantification and impersonal passive: indefinite durative event.

(27) Demotion of the eA in the impersonal passive serves the indefiniteness of the argument only. It indicates full event predication without overt argument assignment apart from
$\exists x[AG(x) \wedge (e \rightarrow x)]$
or from
$\lambda e \exists x[e(t_1\text{-}t_n) \wedge [e(t_1) - e(t_2)] \wedge e(x)]$

(28a-c) relates argument projection to the types of passives discussed earlier. [eA, iA = external vs. internal argument.]

(28) The semantics of the participle, theta role projection, and homonymic forms: *ongoing active perfect = Aux selection only for verbs of movement and a few other selected verb lexemes. The orthogonal arrow signals which structural arguments are pivotal in the formal grid of the predicate.

a. stative passive = eA, iA |>>>>>>>>>>>>|--------⇓---------| iA
(reduced argument assignment; i.e., reduced valence)

b. ongoing active perfect(ive) event = eA, iA |>>>>>⇓>>>>>>|--------------------| iA
(full argument assignment; i.e., unreduced, designated valence)

c. ongoing active perfect, but imperfective event = eA, iA |~~~~~~⇓~~~~~~~~~~~~|
(full argument assignment; i.e., unreduced, designated valence)

d. ongoing passive perfect, but imperfective event = iA |~~~~~~~~~⇓~~~~~~~~~|
(full argument assignment; i.e., unreduced, designated valence)

e. ongoing passive perfect, but imperfective event = 0 |~~~~~~~~~⇓~~~~~~~~~~|
(no argument assignment; i.e., zero valence)

Finally, (29) to (33) summarize and extend the evidence in favor of the categorial underspecification of past participle forms. What speaks in favor of categorial underspecification of past participles?

(29) V-status indicated by by-PP for Cop+PP

(30) Adjectival status because of *un-* prefixation etc.; see (6)-(16) above.

(31) The option between Anterior-P and passive P is contingent upon the choice between types of Aux and COP.

(32) The underspecified A/V-status is to be linked to Aktionsart/Aspect status which controls the empirical distribution in German (as well as in Russian and Italian).

(33) A/V status is codetermined by tV ≠ iV ≠ impersonal passive: the impersonal passive is incompatible with the resultative/adjectival passive (durativity ≠ perfectivity).

Underspecification of category status, it appears, is not only a useful tool in the analysis and description of A and V derived forms. In the assumption of the mediaeval Modistae, it helped make lighter the burden of the grammatical device in the human brain and keep void the lexicon of grammatical information not belonging there. In this sense, the discussion above may well be a methodological exercise in the attempt to keep apart more purely the tasks of the different linguistic modules of analysis and description. And we may perhaps spare us the criticism extended by Roman Jakobson (1985), namely that modern linguistics still has not reached the status achieved by the speculative grammarians in the late Middle Ages.

References

Abraham, W. 2000. The aspect-case typology correlation: perfectivity triggering (split) ergativity. Burzio's generalization explained. In: E. Reuland (ed.), *Arguments and Case: Explaining Burzio's Generalization*, 129–190. Amsterdam: John Benjamins [Linguistics Today – Linguistik Aktuell 34].

Abraham, W. 2004. VP-internal subjects as 'unaccusatives': Burzio's 'Object Account' vs. the 'Perfectivity Account'. In: A. ter Meulen and W. Abraham (eds.), *The Composition of Meaning: from Lexeme to Discourse,* 1–16. Amsterdam: John Benjamins.

Alexiadou, A., M. Rathert, and A. von Stechow (eds.). 2003. *Perfect Explorations.* Berlin: Mouton de Gruyter.

Embick, D. 2004. On the structure of resultative participles in English. *Linguistic Inquiry* 15/3, 355–392.

Halle, M. and A. Marantz. 1993. Distributed morphology and the pieces of inflection. In: K. Hale and S.J. Keyser (eds.), *The View from Building 20: Essays in Linguistics in Honor of Sylvain Bromberger*, 111–176. Cambridge, Mass.: MIT Press.

Jakobson, R. 1985. Glosses on the Mediaeval insight into the science of language. In: R. Jakobson, *Selected Writings* 7, 185–198. Berlin: Mouton de Gruyter.

Kratzer, A. 1994. The event argument and the semantics of voice. Ms., University of Massachusetts, Amherst.

Kratzer, A. 1996. Severing the external argument from ist verb. In: J. Rooryck and L. Zaring (eds.), *Phrase Structure and the Lexicon*, 109–138. Dordrecht: Kluwer.
Kratzer, A. 2003. Building statives. In: L.J. Conathan, J. Good, D. Kavistkaya, A. Wulf, and A.C.L. Yu (eds.), *Proceedings of the 26th Annual Meeting of the Berkeley Linguistic Society*, 385–399. Berkeley: University of California, Berkeley Linguistic Society.
Leiss, E. 1998. Aristotelische Linguistik. Der Neubeginn einer philosophischen Grammatik durch Jean-Marie Zemb. *Sprachwissenschaft* 23/2, 141–165.
Mohr, S. 2005. *Clausal Architecture and Subject Positions. Impersonal Constructions in the Germanic Languages*. Amsterdam: John Benjamins [Linguistik Aktuell/Linguistics Today 88].
Marantz, A. 1984. *On the Nature of Grammatical Relations*. Cambridge: MIT Press.
Rathert, M. 2006 (in press). Simple preterit and composite perfect tense: The role of the adjectival passive. In: W. Abraham and L. Leisiö (eds.), *Passivization and Typology: Form and Function*. Amsterdam: John Benjamins.
Wilhelm, A. 2001. Categories of lexical aspect in German. Paper presented at *GLAC* 7, April, 21, 2001 at Banff, Alberta, University of Calgary.

German Reflexives as Proper and Improper Arguments

Manfred Bierwisch

0. The issue

Reflexive pronouns as central anaphoric elements are subject to general principles determined by Universal Grammar and shared by all languages that use reflexives as part of their grammatical structure. In addition to these general conditions, there are language-particular properties, which different languages can exhibit on the basis of different regulations. One variation of this sort is the particular role of German reflexives as improper arguments, which are subject to standard syntactic and morphological conditions, but do not represent a real argument of the head they belong to. The phenomena I want to explore in this contribution are related to the following observation.

In addition to so-called ergative verbs like (A) with the standard transitive/intransitive alternation, German has a (much larger) group (B) of verbs that realize the unergative variant by reflexivization:

(A)	*Paul zerbrach die Rute.*	'Paul broke the rod.'
	Paul hat die Rute zerbrochen.	
	Die Rute zerbrach.	'The rod broke.'
	Die Rute ist zerbrochen.	
(B)	*Paul bog die Rute.*	'Paul bent the rod.'
	Paul hat die Rute gebogen.	
	Die Rute bog sich.	'The rod bent.'
	Die Rute hat sich gebogen.	

The reflexive version of group (B) verbs is similar in many respects to so-called absolute reflexive verbs as in (C):

(C)	**Jemand verbeugte Paul.*	'*Someone bowed Paul.'
	Paul verbeugte sich.	'Paul bowed.'
	**Paul verbeugte.*	

A general account of these facts emerges, if we assume that reflexives determine a specific operation on semantic variables in accordance with standard requirements of c-selection, case-assignment and syntactic conditions on argument positions and binding.

1. Some general observations

To begin with, we notice that German reflexives occur in three types of positions, which might be called proper, improper, and pseudo-argument positions, shown in (1), (2) and (3), respectively:

(1) a. *Ich*$_i$ *rasiere mich*$_i$ *nicht regelmäßig.*
I shave myself not regularly
'I don't shave regularly.'

b. *Ihr*$_i$ *konntet euch*$_i$ *im Spiegel sehen.*
you.PL could yourselves in.the mirror see
'You could see yourselves in the mirror.'

(2) a. *Wir*$_i$ *bedanken uns*$_i$ *beim Veranstalter.*
we thank ourselves at.the organizer
'We thank the organizer.'

b. *Seine Annahmen*$_i$ *ändern sich*$_i$ *oft.*
his assumptions change themselves often
'His assumptions frequently change.'

(3) a. *Es*$_i$ *handelt sich*$_i$ *hier um Prinzipien.*
it concerns itself here with principles
'Principles are at stake here.'

b. *Es*$_i$ *fragt sich*$_i$, *ob das geht.*
it asks itself whether that works
'One wonders whether that will work.'

Reflexives in proper argument positions (proper reflexives, for short) can be replaced by or conjoined with non-anaphoric DPs under appropriate conditions, as shown in (4):

(4) a. *Ihr könnt Eva im Spiegel sehen.*
you.PL can Eva in.the mirror see
'You can see Eva in the mirror.'

b. *Er*$_i$ *sah sich*$_i$ *und Eva im Spiegel.*
he saw himself and Eva in.the mirror
'He saw Eva and himself in the mirror.'

Improper reflexives are obligatory reflexives, they cannot be replaced by non-anaphoric elements or conjoined with other DPs, cf. (5).

(5) a. *Er*$_i$ *bedankt sich*$_i$ / **uns*$_k$ *beim Veranstalter.*
he thanks himself / *us at.the organizer
b. **Wir*$_i$ *bedanken sich*$_k$ *und uns*$_i$ *beim Veranstalter.*[1]
we thank himself and us at.the organizer

Pseudo-reflexives are a special case of improper reflexives. They are governed by a small number of verbs with improper subjects, i.e. subjects without referential capacity, restricted to the pronoun *es*. Pseudo-reflexives are therefore restricted to *sich*. Some general observations about proper and improper reflexives can be stated as follows:

First, proper reflexives can be arguments of all major lexical categories – verbs in (1) and (6a), prepositions in (6b), (6c) and (6d),[2] and adjectives in (6e) –, while improper (and pseudo-) reflexives are licensed only by verbs, as indicated in (7).

(6) a. *Du*$_i$ *glaubst dir*$_i$ *vermutlich nicht.*
you believe yourself presumably not
'Presumably, you$_i$ don't believe yourself$_i$.'
b. *Du*$_i$ *glaubst bestimmt an dich*$_i$.
you believe certainly in yourself
'You$_i$ certainly believe in yourself$_i$.'
c. *Dein*$_i$ *Glaube an dich*$_i$
your$_i$ belief in yourself$_i$
d. *Neben sich zu stehen, ist schwierig.*
next.to onself to stand is difficult
'It is difficult to stand beside oneself.'
e. *Vera*$_i$ *ist sich*$_i$ *nicht unangenehm.*
Vera$_i$ is to.herself not unpleasant
'Vera isn't unpleasant to herself$_i$.'

(7) a. *Er*$_i$ *hat sich*$_i$ *über sich*$_i$ / *ihn*$_k$ *geärgert.*
he$_i$ has himself about himself / him upset
'He was upset about himself.'
b. *Sein*$_i$ (**sich*$_i$) *Ärger über sich*$_i$ / *ihn*$_k$
his$_i$ (*himself) vexation at himself$_i$ / him$_k$

Second, for all reflexives alike, the antecedent is the subject of the head. If the reflexive depends on a proposition or adjective, which cannot provide a subject by its own, the antecedent is subject of the governing category, as illustrated in (6b), (6c), and (6e). If

---

1 Notice that due to fact that for a large group of verbs, improper reflexives indicate the inchoative variant corresponding to regular causatives (cf. (B) above), the obligatory character of the reflexives is obscured by the ambiguity of the verb, as illustrated in (i). It still does not allow to conjoin the reflexive with other DPs. Thus (ii) is deviant, unless one interprets the subject DP as an agent causing an event that involves it as the theme or patient.

(i) *Eva*$_i$ *legte sich*$_i$ / *ihre*$_i$ *Tochter hin.*
Eva laid herself/ her daughter down
(ii) * *Eva*$_i$ *legte sich*$_i$ *und ihre*$_i$ *Tochter hin.*
* Eva laid herself and her daughter down

2 That nouns cannot govern reflexives directly, but only as part of a PP, as in (6c) or (7b), follows from a gap in the morphology of reflexives. I will return to this problem below.

there is no overt subject meeting this condition, the reflexive remains unbound, as in infinite VPs like (6d), nominalized infinitives like (8a)[3] or impersonal passives like (8b) and (8c):

(8) a. *Das wiederholte sich im Handumdrehen.*
that repeated itself in.the twinkling.of.an.eye
'It repeated itself in an instant.'
b. *Hier wird sich nicht unterhalten*!
here is each.other not talked.to
'One doesn't talk to each other here.'
c. *Nun wurde sich ernsthaft angestrengt.*
now was oneself seriously exerted
'One seriously exerted oneself now.'

Third, proper and improper reflexives are subject to similar conditions syntactically, except that only proper reflexives can be in Topic- or Focus-Position, as can be seen from the contrast in (9):

(9) a. *Sich*$_i$ *haben sie*$_i$ *nicht registriert.*
themselves$_i$ have they not registered
'They didn't register'
b. **Sich*$_i$ *haben sie*$_i$ *nicht geirrt.*
themselves have they not erred
'They didn't make a mistake.'

Otherwise, positions of reflexives are remarkably free, as will be seen below.

Fourth, morphologically, proper and improper reflexives alike have features of case, number, and person, which are subject to two conditions:

(i) Case is determined by the head governing the anaphor.
(ii) Person and number must agree with features of the antecedent.

In constructions without an overt antecedent, the default reflexive *sich* shows up, as illustrated in (6d), (8) and (10):[4]

---

3 Derived nouns do not allow for reflexives, as this would require a genitive case, which is not available. Thus for cases like (i), only the non-reflexive version can be nominalized, as indicated in (ii):

(i) *Sich / ihn zu retten war schwierig.*
oneself / him to rescue was difficult
(ii) *Seine Rettung / *die Rettung seiner (selbst) war schwierig.*
his rescue / *the rescue his (self) was difficult

4 It might be noted that pseudo-reflexives as in (3) cannot occur without an antecedent. Hence constructions like (i) and (ii) are out:

(i) * *Die Idee, sich um Prinzipien zu handeln*
the idea itself with principles to concern
(ii) * *Der Gedanke, sich nicht zu gehören*
the thought itself not to be.appropriate

(10) a. *Der Versuch, sich hinzusetzen*
the attempt oneself down.to.sit
'the attempt to sit down'

b. *Die Möglichkeit, sich zu freuen*
the possibility oneself to rejoice
'the possibility to rejoice'

Most of the properties of proper and improper reflexives can be derived from the assumption that proper reflexives saturate a syntactic argument position which corresponds to a regular semantic variable, while improper reflexives saturate a syntactic argument position that does not correspond to a semantic variable.

A further peculiarity of German proper reflexives must be noted. Under appropriate semantic conditions, reflexives can be interpreted as reciprocals, as indicated in (11) and (12):

(11) *Die Studenten*$_i$ *kennen sich*$_i$ *gut.* ≅ (a) 'The students$_i$ know themselves$_i$ well.'
≅ (b) 'The students$_i$ know each other$_i$ well.'

(12) *Die Leute*$_i$ *sind sich*$_i$ *gleichgültig* ≅ (a) 'The people$_i$ don't care for themselves$_i$.'
≅ (b) 'The people$_i$ don't care for each other$_i$.'

This option seems to be blocked, or at least not easily available, if the reflexive is the object of a proposition, as shown in (13) to (15). As a matter of fact, the reciprocal interpretation for objects of prepositions must be realized by *einander*, as is indicated in (13') to (15'):

(13) *Die Kinder*$_i$ *sprachen über sich*$_i$. ≅ (a) 'The kids$_i$ talked about themselves$_i$.'
≠ (b) 'The kids$_i$ talked about each other$_i$.'

(14) *Eure*$_i$ *Meinung über euch*$_i$ ≅ (a) 'Your$_i$ opinion about yourselves$_i$.'
≠ (b) 'Your$_i$ opinion about each other$_i$.'

(15) *Sie malten Bilder von sich.* ≅ (a) 'They$_i$ painted pictures of themselves$_i$.'
≠ (b) 'They$_i$ painted pictures of each other$_i$.'

(13') *Die Kinder*$_i$ *sprachen über einander*$_i$. = (13b)

(14') *Eure*$_i$ *Meinung über einander*$_i$ = (14b)

(15') *Sie*$_i$ *malten Bilder von einander*$_i$ = (15b)

These options and their constraints as well as the problems of reciprocal interpretation cannot be appropriately pursued here.

2. The morphology of reflexives

The German pronouns *mich, mir, dich, dir, uns,* und *euch,* can appear as reflexive or anaphoric pronouns, whereas *sich* is the only item that can only occur as reflexive. It is exclusively reflexive also in the sense that it indicates no other distinctions: case, gender, and number, systematically distinguished for non-reflexive pronouns, are all collapsed in one form. Case-syncretism furthermore eliminates the dative/accusative-distinction for the plural of first and second person. Schematically, we have:[5]

1st Person		2nd Person		3rd Person		
Singular	Plural	Singular	Plural	Masculine	Neuter	Feminine/Plural
ich	wir	du	ihr	er	es	sie
mich	uns	dich	euch	sich		
mir		dir				

If we assume a feature [+Reflexive] to indicate the reflexive use of a pronominal, then *sich* must be marked as [+Refl], while the optional reflexives would be marked as [αRefl], where α is either + or –, and the nominative case of all pronouns is (redundantly) marked [–Refl].

Assuming furthermore without detailed justification the following tentative features for lexical categories and for the specification of person, number, gender, and case in German, the relevant entries could be categorized as indicated in (16)–(18).

Category	Noun	Verb	Adjective	Preposition
Nominal	+	–	+	–
Dependent	–	–	+	+

Person	2nd	1st	3rd
Participant	+	+	–
Addressee	+	–	–

Number	Singular	Plural
Plurality	–	+

Gender	Masculine	Neuter	Feminine
Masculine	+	–	–
Feminine	–	–	+

Case	Nominative	Accusative	Dative	Genitive
Governed	–	+	+	–
Oblique	–	–	+	+

---

5 I did not include the genitive forms *meiner, deiner, seiner, ihrer, unsrer, eurer,* because it is unclear whether they can be used as reflexives at all. They can hardly be used as adnominal genitives, as shown in (i), but even for verbs their reflexive interpretation is dubious, cf. the contrast of (ii) vs.(iii):

(i) ??*Peters*$_i$ *Kritik* *meiner*$_k$ / *seiner*$_i$
Peter's$_i$ critique of.me$_k$ / himself$_i$

(ii) ??*Max*$_i$ *erinnerte* *sich*$_i$ *seiner*$_i$ *nicht.*
Max$_i$ remembered to.himself$_i$ himself$_i$ not

(iii) *Max*$_i$ *erinnerte* *sich*$_i$ *seiner*$_k$ *nicht.*
Max$_i$ remembered to.himself$_i$ him$_k$ not

(16)	/ich/	[+N;	+ Part,	–Addr,	–Plur,	–Gov,	–Obl,	–Refl]
	/mich/	[+N;	+ Part,	–Addr,	–Plur,	+Gov,	–Obl,	αRefl]
	/mir/	[+N;	+ Part,	–Addr,	–Plur,	+Gov,	+Obl,	αRefl]
	/wir/	[+N;	+ Part,	–Addr,	+Plur,	–Gov,	–Obl,	–Refl]
	/uns/	[+N;	+ Part,	–Addr,	+Plur,	+Gov,	βObl,	αRefl]
(17)	/du/	[+N;	+ Part,	+Addr,	–Plur,	–Gov,	–Obl,	–Refl]
	/dich/	[+N;	+ Part,	+Addr,	–Plur,	+Gov,	–Obl,	αRefl]
	/dir/	[+N;	+ Part,	+Addr,	–Plur,	+Gov,	+Obl,	αRefl]
	/ihr/	[+N;	+ Part,	+Addr,	+Plur,	–Gov,	–Obl,	–Refl]
	/euch/	[+N;	+ Part,	+Addr,	+Plur,	+Gov,	βObl,	αRefl]

(18)	/es/	[+N;	– Part,	–Addr,	–Fem,	–Masc,	–Plur,	αGov,	–Obl,	–Refl]
	/sich/	[+N;	– Part,	–Addr,	βFem,	γMasc,	δPlur,	+Gov,	αObl,	+Refl]

There is a number of interesting regularities in the array of these items. Thus, the "stems" /mi-/ and /di-/ of the first and second person singular and the "inflections" /-r/ and /-ch/ of dative and accusative forms deserve attention in this respect. Whether and in which way an account in terms of some sort of rules is indicated must be left aside here. Most of the information must be relegated to lexical entries in any case. The specifications in (16)–(18) are highly redundant, however, and must be reduced by principles of morphological redundancy. If for example [–F] is taken to be the unmarked value for [F], which normally need not be specified, then (16)–(18) reduce to (19)–(21):

(19)	/ich/	[+N;	+ Part,					]
	/mich/	[+N;	+ Part,			+Gov,		αRefl]
	/mir/	[+N;	+ Part,			+Gov,	+Obl,	αRefl]
	/wir/	[+N;	+ Part,		+Plur,			]
	/uns/	[+N;	+ Part,		+Plur,	+Gov,		αRefl]
(20)	/du/	[+N;		+Addr,				]
	/dich/	[+N;		+Addr,		+Gov,		αRefl]
	/dir/	[+N;		+Addr,		+Gov,	+Obl,	αRefl]
	/ihr/	[+N;		+Addr,	+Plur,			]
	/euch/	[+N;		+Addr,	+Plur,	+Gov,		αRefl]
(21)	/es/	[+N;				αGov,		– Refl]
	/sich/	[+N;				+Gov,		+ Refl]

Instead of simply supplying negative values for unspecified features, more intricate conditions might be appropriate or necessary. The point to be made here is simply that pronouns in general and reflexives in particular are categorized by sets of morpho-syntactic features in terms of which the agreement condition applying to them can be stated.

(22) The categorization Cat of a pronominal X_i must match the features $[F_n, \ldots, F_m]$ of the categorization of its antecedent Y_i.

The choice of [F$_n$,..., F$_m$] is determined by the language-particular morphological system, such that the overall condition (22) can be assumed to be part of UG, presumably a condition on the binding relation. It is worth noticing that (22) is not restricted to reflexives, but concerns the relation of antecedent-pronominal more generally, including ordinary cases like (23):

(23) *Mary$_i$ talked to the waiter$_k$ hastily, because she$_i$/he$_k$ was in a hurry.*

For German, the choice for [F$_n$, ..., F$_m$] includes [Part, Addr, Plur, Masc, Fem], i.e. the features for person, number, and gender. Case features must not be included, as they specify the argument position of a pronominal, rather than its binding relation. Also [Refl] cannot be included, although it directly defines the need of an antecedent and the conditions imposed on it.

Notice, incidentally, that gender as a characteristic category of antecedent selection becomes irrelevant for German reflexives, because only third person pronouns exhibit gender distinctions, which are neutralized under [+Refl] altogether. The effect of (22) is illustrated by cases like (24)–(26):

(24)	a.	*Du$_i$*	*kannst*	**sich$_i$/*	*dich$_i$ / *euch$_i$*	*im*	*Spiegel*	*sehen.*
		you$_i$	can	*themselves$_i$/	yourself$_i$/ you$_i$	in.the	mirror	see
	b.	*Du$_i$*	*kannst*	*mich$_k$ /*	*euch$_l$ / *dich$_k$*	*im*	*Spiegel*	*sehen.*
		you$_i$	can	me$_k$ /	you$_l$ / *you$_k$	in.the	mirror	see

(25)	a.	*Wir$_i$*	*möchten*	**sich$_i$ /*	*uns$_i$*	*bei*	*euch$_k$*	*bedanken.*
		we$_i$	want	*themselves$_i$	ourselves$_i$	at	you$_k$	thank
		'We want to thank you.'						
	b.	*Er$_i$*	*möchte*	*sich$_i$ /*	**ihn$_i$*	*bei*	*euch$_k$*	*bedanken.*
		he$_i$	wants	himself$_i$	*him$_i$	at	you$_k$	thank
		'He wants to thank you.'						

(26)	a.	*Euer$_i$*	*Ärger*	*über*	**sich$_i$*	*/ euch$_i$*	
		Your$_i$	vexation	at	*themselves$_i$	/ yourselves$_i$.	
	b.	*Ihr$_i$*	*werdet*		**sich$_i$*	*/ euch$_i$*	*lästig.*
		You$_i$	become		*themselves	/ yourselves$_i$	annoying
		'You are becoming a burden to yourselves$_i$.					

3. Syntactic conditions on reflexives

The morphological features considered so far play a role in the two syntactic relations reflexives characteristically enter: saturation of argument positions and binding by antecedents.

To begin with, reflexives must be licensed as complements of appropriate lexical heads, which assign a thematic role to proper arguments and impose case conditions on both proper and improper arguments. The difference with regard to thematic roles of proper, improper, and pseudo-reflexives has been illustrated by (1) to (3) above and can easily be recognized in the following cases:

(27) a. *Dann hat sie*$_i$ *sich*$_i^{Acc}$ / *dich*$_k^{Acc}$ *angerufen.*
then has she$_i$ herself$_i$ / you$_k$ called
'Then she called herself/you'
b. *Dann hat sie*$_i$ *mit sich*$_i^{Dat}$ / *dir*$_k^{Dat}$ *gestritten.*
then has she$_i$ with herself$_i$ / you$_k$ quarreled
'Then she quarreled with herself/you.'

(28) a. *Ich*$_i$ *habe mich*$_i^{Acc}$ *geirrt.*
I$_i$ have myself$_i$ erred
'I was mistaken.'
b. *Ich*$_i$ *habe mir*$_i^{Dat}$ *das anders vorgestellt.*
I$_i$ have myself$_i$ that differently imagined
'I imagined that differently.'

(29) *Es*$_i$ *dreht sich*$_i$ *um eine neue Wohnung.*
It$_i$ turns itself$_i$ around a new appartment
'It's about a new appartment.'

In (27a), the thematic role of *sich* is the recipient or patiens of *anrufen*, while in (27b) roughly the same role is jointly assigned by *streiten mit*, i.e. by the preposition combined with the verb. In (28), the reflexive doesn't have a separate thematic role at all, it simply refers to that of the subject. And in (29), neither the Reflexive nor its antecedent *es* realize a thematic role. The only thematic role in (29) is realized by the object of the preposition *um*.

However different they are with regard to thematic roles, reflexives must uniformly realize the case conditions associated with the argument positions they saturate. As (27) and (28) show, the accusative/dative-distinction must be realized, although the syncretism of *sich* does not make it explicit.

The next point to be noted is the fact that German reflexives are subject to the usual binding conditions in the sense introduced in Chomsky (1981). They include, as already mentioned, the requirement of morphological identity with respect to person and number.

(30) a. *Wir*$_i$ *haben *sich*$_i$ / *uns*$_i$ *gründlich gereinigt.*
we$_i$ have *themselves$_i$ / ourselves$_i$ carefully cleaned
'We cleaned ourselves carefully.'
b. *Die Gäste*$_i$ *haben *sie*$_i$ / *sich*$_i$ *gründlich gereinigt.*
the guests$_i$ have *them$_i$ / themselves$_i$ carefully cleaned
'The guests cleaned themselves carefully.'

(31) a. *Clara*$_i$ *hat sich*$_i$ *sich*$_i$ *mit Hut vorgestellt.*
Clara$_i$ has herself$_i$ herself$_i$ with.a hat imagined
'Clara imagined herself with a hat.'
b. *Clara*$_i$ *hat sich*$_i$ / *ihr*$_k$ *den Mantel angezogen.*
Clara$_i$ has herself$_i$ / her$_k$ the coat on.put
'Clara put (her) the coat on.'

(32) a. *Hans*$_i$ *war sich*$_i$ / *ihm*$_{k/*i}$ *verdächtig.*
Hans$_i$ was himself$_i$ / him$_{k/*I}$ suspicious
'Hans found himself/him suspicious.'
b. *Hans*$_i$ *war mit sich*$_i$ / *ihm*$_{k/*i}$ *zufrieden.*
Hans$_i$ was with himself$_i$ / him$_{k/*I}$ satisfied
'Hans was satisfied with himself.'

It must be noted that there are at least three types of constructions which allow for reflexives without overt antecedents within the domain of the head determining the functional complex in the sense of Chomsky (1995). This applies to both proper and improper reflexives, as shown by the (a)- and (b)-cases of (33) to (35), respectively. The three types of constructions include infinitives, impersonal passives, and imperatives:

(33) a. *Es ist gut, sich*$_i$ *warm anzuziehen.*
it is good oneself$_i$ warmly to.dress
'It is good to dress warmly.'
b. *Sich*$_i$ *ärgern nützt nichts.*
oneself to.upset helps nothing
'Being angry does not help.'

(34) a. *Hier wird sich*$_i$ *gründlich gereinigt*!
here is oneself$_i$ carefully cleaned
'One cleans oneself carefully here.'
b. *Jetzt wird sich*$_i$ *nicht unterhalten*!
now is oneself$_i$ not conversed
'No conversation!'

(35) a. *Rasier dich*$_i$ *vorsichtig.*
shave yourself$_i$ carefully
'Shave with caution.'
b. *Setzt euch*$_i$!
seat yourselves$_i$
'Sit down!'

While for imperatives an implicit subject can naturally be assumed, as evidenced by the person and number of the reflexive, this is not the case for infinitives and passives without a subject. Here only the unmarked reflexive *sich* is possible.

Nominalizations exclude reflexives, as non-genitive arguments are excluded for independent reasons, and the genitive *seiner* cannot properly serve as a reflexive (as mentioned in fn. 5). Hence (36b) cannot nominalize the reflexive version of (36a), while (37b) and (38b) illustrate the regular pattern, according to which improper reflexives are obligatorily dropped under nominalization:

(36) a. *Peter*$_i$ *verteidigt sich*$_i$ / *einen Freund*$_k$.
Peter$_i$ defends himself$_i$ / a friend$_k$
b. *Peters*$_i$ *Verteidigung* (**sich*$_i$ / ?*seiner selbst*$_i$) / *eines Freundes*$_k$
Peters defence (*of.himself / ?of.his self / of.a friend$_k$
'Peter's defence of himself/of a friend'

(37) a. *Der Fahrplan*$_i$ *hat sich*$_i$ *geändert.*
the schedule has itself changed
'The schedule has changed.'
b. *Die* (**sich*$_i$) *Änderung des Fahrplans*$_i$
the (*itself) change of.the schedule
'the change of the schedule'

(38) a. *Peter*$_i$ *weigert sich*$_i$ *teilzunehmen.*
Peter$_i$ refuses himself$_i$ to.participate
'Peter is refusing to participate.'
b. *Peters*$_i$ (**sich*$_i$) *Weigerung teilzunehmen*
Peter's (*himself) refusal to.participate

As objects of prepositions, however, reflexives can become complements or adjuncts of nouns, bound to the inherited subject, as shown in (39b).

(39) a. *Hannah*$_i$ *ärgert sich*$_i$ *über sich*$_i$.
Hannah$_i$ annoys herself$_i$ about herself$_i$
'Hannah is annoyed by herself.'
b. *Hannahs*$_i$ *Ärger über sich*$_i$
Hannah's$_i$ annoyance about herself$_i$
'Hannah's self-annoyance'

Finally, a short remark must be made about the linear ordering of reflexives, which is fairly free with respect to other arguments and adjuncts of the Verb governing the reflexive. Presumably, any position between major constituents depending on the Verb is allowed, cf. (40).

(40) a. *Die Kinder*$_i$ *haben sich*$_i$ *mit Freude auf die Feier vorbereitet.*
the kids have themselves$_i$ with pleasure for the party prepared
b. *Die Kinder*$_i$ *haben mit Freude sich*$_i$ *auf die Feier vorbereitet.*
the kids have with pleasure themselves$_i$ for the party prepared
c. *Die Kinder*$_i$ *haben mit Freude auf die Feier sich*$_i$ *vorbereitet.*
the kids have with pleasure for the party themselves$_i$ prepared
'The kids prepared for the party with pleasure.'

Besides various word orders, various types of clitization are possible with the pronoun *es*:

(41) a. … *weil es*$_i$ *sich*$_i$ *so nicht machen läßt.*
… because it$_i$ itself$_i$ that.way not make let
b. ?… *weil sich*$_i$ *es*$_i$ *so nicht machen läßt.*
… because itself$_i$ it$_i$ that.way not make let
c. … *weil es*$_i$ *so sich*$_i$ *nicht machen läßt.*
… because it$_i$ that.way itself$_i$ not make let
d. … *weil's*$_i$ *sich*$_i$ *so nicht machen läßt.*
… because'it(CL)$_i$ itself$_i$ that.way not make let

e. … *weil sich$_i$'s so nicht machen läßt.*
… because itself$_i$'it(CL)$_i$ that.way not make let
'… because it cannot be done that way'

In spite of this wide range of options, there seems to be a sufficiently clear preference for what comes close to the so-called Wackernagel-Position, viz. (40a) and (41a). A systematic characterization of overt positions of reflexives in terms of underlying and surface conditions is a intriguing problem of its own, which cannot be pursued here. I merely add a hint to a peculiar phenomenon in this respect, which has been called the "Adorno-Position". This position, immediately preceding the governing verb, is illustrated in (42b) and (42c):

(42) a. *der Kleinbürger, der sich mit der Macht identifiziert, die er*
the petit-bourgeois who himself with the power identifies that he
nicht hat
not has

b. *der Kleinbürger, der mit der Macht sich identifiziert, die er*
the petit-bourgeois who with the power himself identifies that he
nicht hat
not has
'the petit-bourgeois who identifies with the power that he doesn't have'

c … *nennt man Kultur, was immer die Herrschaft des Tauschwerts zu*
…calls one culture what ever the governance of.the exchg.value to
akzeptieren sich weigert, so…
accept itself refuses, then…
'…if culture is called whatever refuses to accept the control of the exchangeable value, then…'

This is by no means a complete account of the general and language-particular syntactic conditions on reflexives in German,[6] but it should be sufficient to discuss the semantic background and consequences to be taken up in the following sections.

---

6 Thus, reflexives and reciprocals are subject to different conditions on antecedents, depending on presumably semantic conditions of the governing heads, as illustrated by the following examples:

(i) *Die Kinder$_i$ legten die Bücher$_k$ neben sich$_i$.*
the kids$_i$ put the books$_k$ next.to themselves$_i$

(ii) *Die Kinder$_i$ legten die Bücher$_k$ neben einander$_k$.*
the kids$_i$ put the books$_k$ next.to each.other$_k$

(iii) *Sie$_i$ sprachen zu den Eltern$_k$ über sich$_i$.*
they$_i$ talked to the parents$_k$ about themselves$_i$

(iv) *Sie$_i$ sprachen zu den Eltern$_k$ über einander$_i$.*
they$_i$ talked to the parents$_k$ about each.other$_i$

4. Reflexives in proper argument positions

Reflexives appear, among others, in argument positions of transitive verbs, saturating a thematic role with corresponding case conditions. Thus, the same position is filled with a reflexive in (43b) and an anaphoric personal pronoun in (43a):

(43) a. *Die Schwester*$_i$ *reinigte sie*$_k$.
the nurse cleaned her
b. *Die Schwester*$_i$ *reinigte sich*$_i$.
the nurse cleaned herself

Ignoring details that are irrelevant in the present context, the minimal syntactic structure underlying these cases can be indicated as in (44):

(44) [$_V$ [$_D$ *die Schwester*] [$_V$ [$_D$ *sie/sich*] [$_V$ *reinig-*]]]

Suppose that a lexical entry distinguishes at least four types of information: the phonetic form PF, the morpho-syntactic categorization Cat, specified in terms of features sketched in section 2, the argument structure AS, and the semantic form SF, which is made up from variables and conceptual constants according to standard assumptions, characterizing the meaning of the entry:

(45) /rein-ig/ [+V, –N, ...] { $\underline{y}$, $\underline{x}$ } [[x ACT] [CAUSE [BECOME [CLEAN y]]]]
[+Gov]

PF Cat AS SF

The argument structure AS indicates which variables in SF are to be saturated by which complements under which conditions of case marking. Each position in AS can in fact be construed as a pair <$\underline{x}$, F>, where $\underline{x}$ is the relevant position in SF, and F indicates the c-selection condition associated with $\underline{x}$, i.e. predictable and idiosyncratic case-features in particular. Thus in (45), a standard causative verb with the agent x and the theme y, the object position of the theme is marked the feature [+Gov], which identifies the accusative. While this condition is predictable as the default object case and could hence be dropped from the lexical information, the dative required by a verb like *begegnen* (encounter) must be indicated as part of AS:[7]

(46) /be-gegn-/ [+V...] { $\underline{y}$, $\underline{x}$ } [x ENCOUNTER y]
[+Obl]

Technically, a position in AS functions as a lambda-operator, such that assigning a thematic role boils down to functional application. In other words, the SF of the appropriate syntactic argument replaces the variable in question, while the corresponding argument position is deleted. Thus we get (47) as the effect of combining the verb (45) with the pronoun *sie*$_k$ of (43a), using x_k as a provisional representation of the oversimplified SF of *sie*$_k$:

---

7 For further discussion of these assumptions about the structure of lexical information, see e.g. Bierwisch (1997).

(47) /sie$_k$ reinig/ { $\underline{x}$ } [[x ACT] [CAUSE [BECOME [CLEAN x_k]]]]

The next step in the interpretation of (43a) must now combine (47) with the DP *die Schwester$_i$*, whose SF might provisionally be abbreviated by [DEF x_i [NURSE x_i]]. Instead of treating this expression as some sort of definite description to be substituted for the variable x bound by the argument position $\underline{x}$ in (47), we have to construe it as an operator that binds the variable x_i which saturates the argument position $\underline{x}$ in (47). This is not a trivial assumption, and it must ultimately be justified as part of an overall account of definiteness, and more generally of referential and quantificational conditions of natural language. As these matters can profitably be separated from the problems of proper and improper reflexives, I will assume that the provisional assumptions made here can eventually be made precise as required.

With this proviso, we get (ignoring tense and other details) (48) as the interpretation of (43a):

(48) [[DEF x_i [NURSE x_i]] [[x_i ACT] [CAUSE [BECOME [CLEAN x_k]]]]]

Looking at the interpretation of (43b) and the reflexive it contains with this background, we can easily identify (49) as the required result, namely the same referential variable as both the agent and the patient or theme of the verb:

(49) [[DEF x_i [NURSE x_i]] [[x_i ACT] [CAUSE [BECOME [CLEAN x_i]]]]]

There are various ways in which the reflexive *sich$_i$* could yield this effect. The following proposal is motivated by the problems to be considered in the next section that deals with improper arguments. Suppose that the effect of reflexives is determined by the following principle:

(50) Let X_i with the binding index i be categorized by the feature [+Refl]. Then X_i saturates the argument position $\underline{x}$ by substituting z for x, where $\underline{z}$ is the argument position to which the antecedent of X_i is assigned.

(50) identifies the variable of the argument position filled by a reflexive with the variable to be replaced by its antecedent. To put it differently, the result of interpreting the reflexive is a second occurrence of the variable bound by the argument position of the antecedent. To illustrate the point, consider (43b), which combines (45) with the reflexive *sich$_i$* to yield (51), where the argument position $\underline{x}$ binds two occurrences of x.

(51) /sich$_i$ reinig/ { $\underline{x}$ } [[x ACT] [CAUSE [BECOME [CLEAN x]]]]

If this representation combines with *die Schwester$_i$*, we derive the desired (49) in strictly parallel fashion to (48). Intuitively, the semantic effect of reflexivization expressed by (50) is to bind a semantic position to a designated second occurrence.

Under this presumption, two closely related problems will have to be faced: reflexives as improper arguments and reflexives without antecedents.

5. Reflexives in improper argument positions

For absolute reflexive verbs like *sich schämen* (be ashamed), *sich verlaufen* (go astray)[8] with two positions in AS, but only one regular variable in SF, formally two options arise:

(52) a. /schäm/ { $\underline{y}$, $\underline{x}$ } [BE-ASHAMED x]
b. /schäm/ { $\underline{y}$, $\underline{x}$ } [BE-ASHAMED y]

Notice first, that $\underline{y}$ is the object-position, which must be saturated first. It differs from ordinary object positions as in *reinigen* above by obligatory association with the c-selectional condition [+Refl]. As we will see below, it is plausible to assume that the presence of this feature is predictable as the default for an argument structure containing an improper argument position – very much like the predictable case features requiring accusative for the object position.

The question to be decided is the following: Is the actual variable in SF bound to the subject or to the reflexive? One might intuitively be inclined to adopt (52a), i.e. to consider the subject as the proper argument position and the reflexive as the pseudo-position, which does not really participate in the semantic interpretation.

The previous considerations require the opposite choice, however. In fact, for intuitive and for purely formal reasons, absolute reflexive verbs must be represented according to (52b), where the variable in SF is bound to the reflexive in object position. The obvious derivation of cases like (53a) is shown in (53b), where the reflexive is interpreted according to the principle in (50), and (53c), where the subject position is saturated.

(53) a. ...*weil die Kinder*$_i$ *sich*$_i$ *schämen*
...because the kids of.themselves are.ashamed
b. /sich$_i$ schämen/ { $\underline{x}$ } [BE-ASHAMED x]
c. /die Kinder$_i$ sich$_i$ schämen/ [[DEF x$_i$ [KIDS x$_i$]] [BE-ASHAMED x$_i$]]

Now, the interesting point is the fact that this analysis automatically yields a plausible solution for the de-causative reflexive verbs of the type *sich setzen* (sit down), *sich ändern* (change), etc.

Suppose that the two variants of ergative verbs like *zerbrechen* (break) *schmelzen* (melt) are represented roughly as follows:

(54) a. /zerbrech/ { $\underline{y}$, $\underline{x}$ } [[x ACT] [CAUSE [BECOME [BROKEN y]]]]
b. /zerbrech/ { $\underline{y}$ } [BECOME [BROKEN y]]

[8] It might be noted that a large number of verbs are absolute reflexive on one particular reading, but open for non-reflexive use under alternative interpretation(s). Thus *verlaufen* ('lose one's way', 'disperse'...), *verschreiben* ('prescribe', 'make a slip of the pen'...), *versprechen* ('promise', 'make a slip of the tongue'...) and many others have different readings when used with and without reflexives. This phenomenon is still to be distinguished from the inchoative reflexives like *sich setzen* ('sit down'), which will be discussed below.

The causative verb (a) and the un-ergative, inchoative verb (b) can easily be collapsed into one complex entry, where the causative component – including the pertinent position in AS – is optional, marked by heavy parentheses:

(55) /zerbrech/ { $\underline{y}$, ($\underline{x}$ } [[x ACT] [CAUSE) [BECOME [BROKEN y]]]]

With this observation in mind, reflexive "ergative verbs" like (*sich*) *ändern* (change) can simply be represented as (56). The main difference with respect to (55) is the fact that the position bound to the agent of the causative verb is preserved in the AS of the inchoative:

(56) /änder/ { $\underline{y}$, $\underline{x}$ } [([x ACT] [CAUSE) [BECOME [DIFFERENT y]]]]

As in (55), two entries are collapsed in (56), as made explicit in (57):

(57) a. /änder/ { $\underline{y}$, $\underline{x}$ } [[x ACT] [CAUSE [BECOME [DIFFERENT y]]]]
b. /änder/ { $\underline{y}$, $\underline{x}$ } [BECOME [DIFFERENT y]]

The causative verb (57a) is a regular transitive entry with two proper argument positions. The interesting point is the inchoative verb (57b), which still has two argument positions, of which only the object position is bound to a variable in SF, however. This is exactly the configuration of absolute reflexive verbs illustrated in (52b).

Suppose now that we call a position in AS an improper position, if it does not bind a variable in SF, i.e. if it is semantically empty. With this proviso, the reflexive position in the AS of absolute reflexives like *sich ändern* can be warranted as follows:

(58) a. $\underline{x}$ is an improper position in AS if there is no variable x in SF.
b. $\underline{x}$ is associated with the feature [+Refl], if it directly precedes an improper position $\underline{y}$ in AS.

In addition to absolute and unergative reflexives, there is a small number of verbs with more than one improper position in AS, cf. (59).

(59) a. *Es handelt sich um Drucksachen.*
it concerns itself with printed.matter
'This is printed matter.'
b. *Es dreht sich um eine Genehmigung.*
it turns itself around a permission
'It concernes a permission.'
c. *Damit hat es sich.*
therewith has it itself
'It's enough now.'
d. *Es macht sich.*
it makes itself
'It's getting somewhere.'

In these cases, an absolute reflexive is combined with a pseudo-subject position as in weather verbs like *es schneit* (it is snowing), *es regent* (it's raining), *es ist kalt* (it's cold), etc. Improper positions of this sort require a condition for independent reasons, which guarantees that they must be saturated by the pronoun *es*, which is categorized by the features [–Masc, –Fem, –Plur].

We still have to account for (proper and improper) reflexives without an antecedent. There is no problem with imperatives: although reflexives don't have an overt antecedent in cases like (60), they must reasonably be assumed to have an underlying subject, as evidenced by the agreement conditions observed by the reflexive.

(60) a. *Benehmt euch / *dich/ *sich /*uns!*
behave yourselves / *you / *themselves /*us
b. *Setz dich / *mich / *sich!*
seat yourself / *myself/ *themselves
'Sit down!'

Examples like these corroborate the assumption that imperatives do have an underlying subject, which is fully present in SF and can also serve as antecedent of reflexives, to be checked for its morphological features, independently of its absence from PF.

What appears to be more intriguing is the question how infinitives like (61) are to be accommodated under the present analysis.

(61) a. *Sich aufzuregen ist nutzlos.*
oneself to.upset is useless
'Working oneself up is useless.'
b. *Es ist sinnlos, sich zu ärgern.*
it is meaningless oneself to vex
'It doesn't make sense to get angry.'

Two options might be taken into consideration: either infinitives are considered as verb phrases that are indeed subjectless, such that reflexives governed by the verb must be unbound. Or we assume some kind of an underlying subject – as for imperatives – which then may act as an antecedent. I will briefly look at these options in turn.

Suppose, for the sake of illustration, that *sich/jemanden ärgern* has a lexical entry like (62a), along the lines indicated in (56)/(57) for (*sich*) *ändern*. What we are interested in is the reflexive variant (62b)

(62) a. /ärger-/ { $\underline{y}$, $\underline{x}$ } [([x ACT] [CAUSE) [BE-IRRITATED y]]]
b. /ärger-/ { $\underline{y}$, $\underline{x}$ } [BE-IRRITATED y]
[+Refl]

The way in which the subject position $\underline{x}$ is to be dispensed with under the specific conditions of infinitivals, must be clarified for independent reasons. Suppose we indicate the suppressed subject position by $\langle \underline{x} \rangle$, such that the result would be a kind of defective argument structure with a suspended subject and a reflexive object that cannot be bound to an antecedent. Hence principle (50) cannot apply to the combination *sich ärger-* indicated in (63a), such that (63b) would be the resulting interpretation.

(63) a. /sich ärger/ { $\underline{y}$ $\langle \underline{x} \rangle$ } [BE-IRRITATED y]
[+Refl]
b. /sich ärger/ { $\langle \underline{x} \rangle$ } [BE-IRRITATED y]

The dispensed subject position must, of course, be ignored, and the free variable y is most naturally construed as having some kind of arbitrary reference. This result, although not completely implausible, requires certain ad-hoc stipulations and is therefore not very convincing.

The alternative option is based on the assumption that infinitivals without an overt subject do nevertheless have an underlying subject, realized by something like the PRO of Chomsky (1981), which does not show up in PF, but provides appropriate default properties which the reflexive can rely on. On this account, the reflexive *sich* in cases like (61) is bound by PRO as its antecedent. Hence (50) can apply, to the effect that the reflexive substitutes the variable y in the initial, lexical item (62b) by the variable x bound by the subject position in the usual way. As the result, we get (64), where the argument position $\underline{x}$ will eventually will be replaced by whatever the semantic contribution of PRO should be.

(64) /sich ärger/ { $\underline{x}$ } [BE-IRRITATED x]

This account is not only free of unmotivated stipulations, it also extends naturally to infinitivals the subject of which enters syntactic relations like Control or Exceptional Case Marking, as indicated in (65):

(65) a. *Eva$_i$ hat ihm$_k$ versprochen sich$_{i/*k}$ zu ändern.*
Eva$_i$ has him$_k$ promised herself$_i$/himself$_k$ to change
b. *Wir$_i$ haben ihn$_k$ sich$_k$ verbeugen sehen.*
we have him himself bow see
'We've seen him bow.'

This retreat to an invisible PRO as the reflexive's antecedent is not naturally available for impersonal passives like (34b) – *Jetzt wird sich nicht unterhalten* – , as the gist of passivization in German is the deletion of the subject position from the AS, as is obvious in cases like (66) or (67).

(66) a. *Karl hat das Buch gefunden.*
Karl has the book found
'Karl found the book.'
b. *Das Buch ist gefunden worden.*
the book is found been
'The book was found.'

(67) a. *Die Kinder schliefen.*
the kids slept
'The kids slept.'
b. *Es wurde geschlafen.*
it was slept
'People were sleeping.'

Differing from infinitives without subject, the subject of passives is not covert, but deleted and cannot be used as a hidden antecedent of a reflexive. An automatic consequence of the elimination of the subject from AS is furthermore the promotion of the direct object, in case there is one, to the subject position, as can be seen in (66). This usual promotion of the object creates a further conflict for reflexives with the basic position of the direct object, as a reflexive cannot function as a subject. As a matter of fact, passives with reflexives, although occasionally used, have a grammatically dubious status:

(68) a. *Man*$_i$ *hat sich*$_i$ *häufig rasiert.*
one has oneself often shaved
'People shaved often.'
b. **Es ist sich häufig rasiert worden*
it is oneself often shaved been
'There was frequent shaving going on.'

(69) a. *Heute ärgern sie*$_i$ *sich*$_i$.
today vex they themselves
'Today, they are angry.'
b. ?? *Heute wird sich geärgert.*
today is oneself vexed
'Today, people are angry.'

Hence the fact that the present analysis of reflexives does not provide a natural account for reflexive passives seems to me in plausible correspondence to the facts.

6. Middles and related constructions

It might finally be noted that (50), together with the general interpretation of reflexives, accounts for crucial properties of the German type of middle constructions as illustrated in (70):

(70) a.	*Sie liest das Buch.*	*Das Buch wird gelesen.*	*Das Buch liest sich* (*leicht*)
	'She reads the book.'	'The book is read.'	'The book reads easily.'
b.	*Hier schläft man.*	*Hier wird geschlafen.*	*Hier schläft es sich* (*gut*).
	'Here, people sleep.'	'Here, people sleep.'	'Here, one sleeps well.'

Like passives, middle constructions suppress the argument position of the subject. As a consequence, the direct object – if there is one, as in (70a) – moves up to the highest position. So far, middles seem to be parallel to passives. However, if there is no direct object, such that passives must be impersonal, i.e. without subject, middles introduce an empty subject that is obligatorily realized by the pronoun *es*. The actual peculiarity of German middles, however, is the addition of an improper position into AS, which leads to the characteristic obligatory presence of *sich*.

Given the analysis developed so far, the difference between a transitive verb and the corresponding middle is to be accounted for as shown in (71):

(71) a.	/les/	[Verb,…]	$\{ \underline{y}, \underline{x} \}$	[x READ y]
b.	/les/	[Verb,…]	$\{ \underline{y}, \underline{z} \}$	[x READ y]
c.	/les/	[Verb, +Pass…]	$\{ \underline{y} \}$	[x READ y]

(71a) is the standard transitive pattern with predictable case requirements for subject and object. In (71b), which has the same semantic content, the original subject-position $\underline{x}$ is dropped. Hence the original agent is not syntactically available any more. Now, while for passive – rather incompletely indicated in (71c) – the original object position $\underline{y}$ is the only position left in AS, such that it becomes the new subject position, the middle construction replaces it with an improper argument position $\underline{z}$.

The resulting configuration in AS is already familiar from inchoatives like *sich ändern*, etc. and the effect is completely parallel: The position $\underline{y}$ in (71b) is according to (58b) automatically associated with the feature [+Refl], the reflexive thus required replaces the variable y with the variable to be saturated by its antecedent according to (50), i.e. with z. This yields the desired result without any stipulation – except the assumption that middle formation replaces the subject in AS with an improper argument position.[9]

Looking finally at the middle construction of intransitives, exemplified in (70b), we notice that a regular intransitive (un-accusative) verb like *schlafen* or *atmen* as represented in (72a) is turned into an item with two improper positions in AS, which must be saturated by *sich* and *es*. Hence for verbs with only one proper position $\underline{x}$ in AS, middle formation does not only replace $\underline{x}$ by an improper position $\underline{z}$, but introduces an additional improper argument $\underline{y}$:

(72) a. /schlaf/ [Verb, ...] { $\underline{x}$ } [SCHLAF x]
b. /schlaf/ [Verb, ...] { $\underline{y}$, $\underline{z}$ } [SCHLAF x]

Whether and how the operation that changes (72a) into (72b) can be collapsed with that deriving (71b) from (71a), that is how middle formation can be reduced to one operation, is to be left open here. Once an intransitive verb is changed into a middle as indicated in (72b), however, all other consequences follow automatically: $\underline{y}$ is associated with the feature [+Refl] by (58b), $\underline{z}$ is restricted to *es* by conditions mentioned in connection to the cases in (59). The reflexive is bound to the antecedent *es*, without any semantic consequence, however, since it saturates an improper position, in which it cannot have any semantic effect. Finally, the pronoun *es* will then replace is semantically empty, such that the variable x in the SF of (72b) will remain a free variable with arbitrary, contextual interpretation.

7. In conclusion

Although there are many loose ends around the German reflexives, two fairly general points should be clear enough:

First, German reflexives serve a number of different constructions which differ from those of anaphors in other languages. These constructions include – besides standard

9 It must be noted here that this is by no means the complete story. The middle construction always requires in particular some sort of modal adverbial. This requirement must be accounted for by the operation creating the middle construction. I will put aside these issues here, as they do not directly affect the role of the reflexive.

and absolute reflexives – improper argument positions of various types, especially decausatives or unergatives and middle constructions. German reflexives furthermore allow for reciprocal interpretation under particular conditions.

Second, most of the effects of reflexives can be accounted for, if they are supposed to connect their referential option to that of their antecedent. In contrast to usual assumptions, this is done not in terms of identifying referential indices, but rather by unifying semantic variables underlying the argument positions saturated by reflexives and their antecedents.

References

Bierwisch, M. 1997. Lexical information from a minimalist point of view. In: C. Wilder, H.-M. Gärtner, and M. Bierwisch (eds.), *The Role of Economy Principles in Linguistic Theory,* 227–266. Berlin: Akademie-Verlag.

Chomsky, N. 1981. *Lectures on Government and Binding.* Dordrecht: Foris.

Chomsky, N. 1995. *The Minimalist Program.* Cambridge, Mass.: MIT Press.

The Acquisition of Japanese Word Structure

Sascha W. Felix[1]

1. Introduction

In this paper I wish to report on a number of experimental studies, some of them quite tentative, which focus on the acquisition of Japanese word structure by both monolingual and bilingual children. Possibly apart from semantic aspects concerning the organization of the mental lexicon, word structure and its acquisition do not seem to be particularly exciting areas of inquiry. After all, the vocabulary of a language is by and large finite and trivially children have to learn words individually so that there appears to be relatively little room for studying creative and rule-governed processes.

However, the Japanese lexicon has a number of very peculiar and almost unique properties which – to the best of my knowledge – it only shares with Korean. These properties are the result of historical accident in conjunction with a very specific geolinguistic configuration. The types of structures to be discussed are interesting not only because they are so rare, but also because they are extremely hostile to successful (oral) communication. Part of the motivation to study their acquisition is to find out how children cope with this hostility.

In general, linguists tend to agree that language structure is essentially a mechanism to map sound onto meaning. This is not only true of sentence structure, but also of individual words and morphemes. Thus a word is typically a structured sequence of sounds associated with a specific meaning or a limited range of closely related meanings. By way of example, the English word *table* maps a specific phonological representation, namely /teibl/, onto the reference to a (usually) four-legged object with a flat surface on which things may be placed.

The phonological representation of a word or morpheme may be subject to a very limited set of heavily constrained alternations, traditionally called allomorphs. Thus the English plural morpheme may surface as either unvoiced /s/ or voiced /z/ depending on the voice status of the preceding segment. More dramatic variations appear in such famous examples as *photograph* which may alternate between /f′outəgræf/, /fət′ogrəfər/, and /foutəgr′æfik/. Chomsky (1964) and Chomsky & Halle (1968) used this and similar examples in their early attacks on taxonomic phonemics showing that these alternations are rule-governed, i.e. they can be derived from a common underlying form by means of a set of context-sensitive phonological rules.

---

1 I am grateful to Suzuki Takeshi, Takagi Yoko, Saito Tomoko, Ogawa Ken, Bando Junichiro, Naganuma Masaaki, and Watanabe Takeshi for their invaluable assistance in finding informants and carrying out the experiments. Special thanks go to Tanaka Shin with whom I had many exciting discussions which inspired much of the work reported here. The usual disclaimers apply.

In other words, a morpheme can be viewed as a pair (π,μ), where π is a specific (underlying) phonological form and μ is the meaning of π. Phonological rules may change π within narrow limits to yield highly constrained alternate representations.

All this seems to be highly trivial in the sense that most languages appear to conform to this structural pattern. It thus seems difficult to imagine that it could be otherwise. However, in a language like Japanese this simple pattern does not hold. A Japanese word or morpheme regularly has more than one (underlying) phonological representation. Crucially, the relationship between these different phonological representations is not rule-governed, but largely arbitrary at the level of phonology. In other words, there is no phonological rule by which one of these representations can be derived from the other or by which all of these representations could be derived from a common underlying form (see also Lewin 1959, Miller 1967, McCawley 1968, Tsujimura 2002).

2. Some properties of Japanese word structure

Let us look at some examples.

The Japanese word for *day* is *hi*; the word meaning *rest*, *break*, *vacation* is *yasumi*. When these two words are combined to form a compound meaning *holiday*, *day off*, etc. the resulting phonological representation is *kyuujitsu*:

(1) *yasumi* + *hi* → *kyuu* + *jitsu*

Thus the phonological forms *yasumi* and *kyuu* represent one and the same word or morpheme meaning *rest* or *break*, while similarly *hi* and *jitsu* are alternate forms for the word meaning *day*.

The Japanese word for *man* is *otoko*, *woman* is *onna*, and *ko* is *child*. When we combine *child* with *man* or *woman* to denote *boy* and *girl* respectively, the result is as in (2).

(2a) *otoko* + *ko* → *dan* + *shi* (boy)
(2b) *onna* + *ko* → *jo* + *shi* (girl)

Again the phonological representations *otoko*, *onna*, and *ko* have the alternate forms *dan*, *jo*, and *shi* respectively.

For a final example let's look at the combination of *koroshi* (*killing*) and *hito* (*man*, *human being*) to yield a compound meaning *murder*:

(3) *koroshi* + *hito* → *satsu* + *jin*

These examples demonstrate that in Japanese a word has two independent phonological representations, where "independent" means that the two forms cannot be related to each other by means of a phonological rule.

However, the number of possible representations is by no means restricted to two. There are numerous cases in which three (or even more) representations belong to the same word. In (1) we saw that the word for *day* is either *hi* or *jitsu*. But there is a third form *nichi* which appears in the word for *whooping cough*:

(4) *hyaku + nichi + zeki*
hundred+days+cough
'hundred-days-cough'

The examples discussed so far are by no means rare or specific cases. On the contrary, virtually all Japanese words (with very few exceptions) have two or more independent phonological forms which are frequently called *readings* (for reasons that will become clear as we proceed).

Obviously, the question arises: what is the common denominator of these different phonological forms? How do *hi*, *jitsu*, and *nichi* relate to each other apart from the fact that they all mean *day*? The answer is fairly straightforward. The common denominator is the writing system, more specifically the pictographic characters originally imported from China (called *kanji* in Japanese). The three phonological forms *hi*, *jitsu*, and *nichi* are all written with the same character for the word *day*, namely 日.

To return to example (1), *yasumi* is written with the character 休, *hi* with the character 日. When both are combined the result is 休日, but this word is pronounced *kyuu+jitsu* rather than *yasumi+hi*. In the same way *satsu+jin* (*murder*) is written with the two characters 殺人, but when 殺 or 人 are used by themselves, they are pronounced as *koroshi* and *hito* respectively. In this sense it is said that a *kanji* (character) has different readings.

The reason for this phonological multi-level representation of Japanese words is historical accident. In the Japanese Middle Ages (roughly the second half of the first millennium) numerous groups of Buddhist monks traveled to China to study Chinese culture, art and political administration. These trips were extremely hazardous undertakings given the fact that nearly two thirds of them didn't even reach the Chinese shore due to heavy storms as well as bands of pirates of the Sea of Japan. Those who did arrive stayed in China for a fairly long period of time frequently exceeding 10 or even 20 years. During their lengthy stay they obviously learned both the Chinese language as well as the writing system. When they returned to Japan (and survived the trip back across the Sea of Japan), they were again confronted with a language that could be spoken, but not written. As a consequence, they standardly used Chinese to write and Japanese to speak in much the same way as European monks used Latin for writing, but the vernacular for speaking (for details see Reischauer 1955, Miller 1967, Samson 1958).

Since Chinese characters are heavily meaning-based, they could also be used for writing genuinely Japanese words so that during the following centuries a mixed vocabulary of Japanese and Chinese words arose, where in particular learned words were directly taken from Chinese in a phonological form that was rigorously adapted to Japanese phonological structure so that in particular tones were lost (see also Lewin 1959, Tateishi 1989, Itou & Mester 1993). Depending on the region of China in which the monks had resided and also the time period of their stay, Chinese words were introduced into Japanese in different pronunciations. This accounts for the fact that we find Chinese-based alternative forms such as *jitsu* and *nichi* both for *day*, in addition to the original Japanese word *hi*

In Japanese scholarly tradition the truly original Japanese pronunciation of a word is called its *kun* (訓)- reading, while the one adapted from Chinese is called the *on* (音)-

reading. Thus the character 日 meaning *day* has the *kun*-reading *hi* and the *on*-readings *nichi* and *jitsu*. Western linguistic literature frequently prefers the terminological distinction between *Japanese* (*kun*-reading) vs. *Sino-Japanese* (*on*-reading) which we will also adopt in this paper.[2]

A crucial property of Sino-Japanese readings and also compounds is their relatively low degree of transparency, i.e., it is frequently very difficult to determine their meaning from the phonological form alone without knowing the associated character(s). This low transparency becomes apparent in a wide variety of phenomena.

First, Sino-Japanese readings are extremely homophonous due to the fact that most of them are monosyllabic with the few disyllabic ones ending either in syllabic /n/ or in unvoiced /i/ or /u/ (cf. Felix 1975, Tateishi 1989). For example the *Gakken Kanjigen* (2002) character dictionary lists 61 meanings for *kyuu*, 6 meanings for *jitsu*, and 40 meanings for *dan*. But even compounds can be highly homophonous to the extent that it is virtually impossible to understand their meaning without seeing them either written or embedded in an appropriate and disambiguating context. Our previous example *kyuu-jitsu* is a fairly moderate case in that apart from *holiday* it has only one more meaning, namely *previous day* in which case it would be written 旧日 rather than 休日. But there are truly dramatic cases such as *koutou* for which the *Koujien* (1998) Dictionary lists 24 meanings. These are admittedly extremes, but roughly half a dozen meanings is by and large standard.

The fact that Sino-Japanese compounds are often semantically uninterpretable, unless the appropriate *kanji* are simultaneously available, is responsible for a number of peculiarities in Japanese life which are unknown in Western cultures. Thus TV news, political interviews and other high-level language presentations are regularly subtitled because seeing the associated *kanji* will greatly facilitate comprehension of complex topics and matters. Even in normal conversation speakers may make reference to *kanji* in the case of specific or rarely used terms. In other words, reference to writing is ubiquitous even in everyday Japanese spoken life.

Another phenomenon that strongly contributes to semantic ambiguity in Sino-Japanese compounds is the strong tendency towards restricting compounds to two morphemes in cases where adding up the original members of the compound would lead to more than two morphemes.

The following examples illustrate this point:

(5)	会社 + 長	→	社長	*会社長
	kaisha + *chou*	→	*shachou*	**kaishachou*
	company + boss	→	company	president

2 The question of when a given lexeme is pronounced in its Japanese or in its Sino-Japanese reading is a fairly complex matter. Many Japanese grammars and textbooks claim that, in general, complex learned compounds use the Sino-Japanese readings while simplex words use the Japanese reading. While this is true to some degree, there are also numerous "exceptions". We find both simplex Sino-Japanese words (e.g. 銃 *juu* - *pistol*, *gun*) as well mixed compounds (e.g. 両手 *ryou+te* - *both hands*). For details see also Itou & Mester (1993).

(6)	受＋試験	→	受験	*受試験
	ju + shiken	→	*juken*	**jushiken*
	take + examen	→	take-examen	
(7)	飛行機＋内	→	機内	*飛行機内
	hikouki + nai	→	*kinai*	**hikoukinai*
	airplane + inside	→	inside the	airplane/in-flight
(8)	学校＋生徒	→	学生	*学校生徒
	gakkou + seito	→	*gakusei*	**gakkouseito*
	school + student	→	(high)	school student
(9)	東京＋成田	→	京成	*東京成田
	toukyou + narita	→	*keisei*	**toukyounarita*
	Tokyo + Narita	→	Tokyo-Narita (express train)	

Example (9) illustrates a particularly dramatic case. The express bus connecting the international airport at Narita with central Tokyo is called *keisei basu* (京成バス), where *basu* means *bus* and *keisei* is Tokyo-Narita. Obviously the relationship between *tokyou-narita* and *keisei* is only understandable on the basis of the associated *kanji*. *Kei* is the alternative Sino-Japanese reading of *kyou* in *tokyou*, while *sei* is the Sino-Japanese reading of *nari* in *narita*. Though this case may be extreme, there are other examples of the same kind. Thus the bus from Tokyo to Osaka bears the name of *keihan basu* (京阪バス) where *kei* is again the *kyou* in *tokyou* and *han* is the Sino-Japanese reading of *saka* in Osaka (大阪).

Some compounds are both phonologically and semantically fully transparent as e.g. *chuushajou* (駐車場 – *parking lot*), where *chuu* is *park*, *sha* is *car*, *jou* is *place*. For other compounds the relationship between pronunciation and writing is fully idiosyncratic. The most famous case is probably *ebi* (*shrimp*) which is written 海老 where 海 means *sea* with the Sino-Japanese reading *kai*, and 老 means *old* with the reading *rou*.[3] Another often-cited case is *urusai* which is written 五月蝿 and means *noisy*, *bothersome*. 五月 is standardly read *gogatsu* and means the month of May (fifth month). 蝿 means *fly* (*insect*) with the Sino-Japanese reading *you*. The intuition behind this combination of characters is fairly clear: *bothersome* is like a fly in May. But again there is no systematic phonological relationship between the *kanji* combinations and the pronunciation.

What these examples are intended to demonstrate is that there is a very close and systematic relationship between word structure and word formation on the one hand and the *kanji* writing system on the other. Much of what happens at the level of individual words as well as compounding can only be understood with respect to the underlying rule system, if the associated *kanji* writing system is taken into account.

3 Traditional folklore has it that the writing derives from the observation that shrimps have the form of the back of an old person and at the same time come from out of the sea.

3. The acquisition problem

The logical acquisition problem that arises from the multi-level word structure described above is fairly obvious. While semantic interpretability of Sino-Japanese compound elements is closely connected to knowledge of the associated *kanji*, Japanese children start learning the writing system no earlier than at the age of six when entering school. Consequently, the question arises of how and when Japanese children acquire the basic principles of Sino-Japanese word formation as well as the different readings associated with a specific *kanji*.

Note that this acquisition problem has two logically independent aspects. One concerns the general principle of word structure, namely the fact that (virtually) each and every word has more than one pronunciation where the different pronunciations are not related to each other by phonological rules. The second aspect concerns the question of which specific pronunciations belong to one and the same word. In some intuitive sense these two aspects are reminiscent of similar problems in the acquisition of morphology. A child must learn first of all that e.g. in English nouns are consistently marked for plurality and verbs are marked for tense. In addition to that she must also learn which specific morphemes are used for plurals and past tense. This latter acquisition problem presupposes, of course, knowledge of the notions of plurality and tense, but *knowing that* does not automatically lead to *knowing how* (see also Clahsen 1991). Much the same applies to the acquisition of Japanese word structure.

Given these facts at least two hypotheses come to mind of how children may acquire the relevant word structures. One possibility would be that to the extent that children learn Sino-Japanese compounds at all they learn them as unanalyzed units. To return to our original example, a child might learn *kyuujitsu* with the proper meaning of *holiday* without realizing that *kyuu* is an alternative pronunciation of *yasumi* (*rest*) while *jitsu* corresponds to *hi* (*day*). In other words, *kyuujitsu* is totally unrelated to *yasumi* and *hi* in terms of the underlying word structure. Under this hypothesis it is not until the child enters school and starts to learn *kanji* that she comes to realize that *kyuu* and *yasumi* on the one hand and *jitsu* and *hi* on the other are alternative readings of exactly the same word. Similarly, the child might learn a word like *gyuuniku* (牛肉) meaning *beef* (*cow meat*) without realizing that *gyuu* is an alternative pronunciation of *ushi* (牛) meaning *cow* while *niku* (肉) is *meat*. That is, *gyuuniku* and *ushi* are treated in much the same way as *beef* and *cow*. Though the words are semantically related (they both refer to the same specific animal), there is no lexical relationship at the level of word structure.

Under a somewhat different hypothesis one might surmise that even prior to entering school the child does learn the basic principle of word structure, namely that one and the same word may have different, phonologically unrelated pronunciations without realizing, of course, that the link between these two pronunciations derives from the specific nature of the writing system to which she is only exposed after entering school. To return to the previous example, the child might learn that the word for *cow* may be rendered as either *ushi* or *gyuu* depending on the specific word formation context. That is, in the context of *niku* (*meat*) it has to be *gyuu+niku*, not **ushi+niku*, but outside of compounds one has to choose *ushi* rather than *gyuu*.

There are a number of a priori considerations favoring each of these two hypotheses. With respect to the first hypothesis one might expect children to largely resist the acquisition of Japanese multi-level word structure until they are literally forced to at the age of entering school. The reason might be that the multi-level structure of words is extremely rare in the languages of the world. To the best of my knowledge only Korean patterns like Japanese in the relevant domain essentially because only Korea and Japan share a common historical and linguistic heritage with respect to China. In contrast to other neighboring countries Japanese and Korean are typologically unrelated to Chinese and it is precisely the borrowing of words from a typologically unrelated language that led to the multi-level word structure we witness in current Japanese (and Korean). As a consequence, while principles of Universal Grammar must trivially be consistent with the multi-level word structure of Japanese (and Korean), the rareness of this type of word structure should require extremely strong positive evidence for the learning child to entertain the appropriate hypothesis and one might expect this evidence to be absent until formal teaching of *kanji* sets in at the beginning of the child's school career.

On the other hand there are some cases in which it is difficult to imagine how the child could possibly overlook the relevant generalization, thus favoring the second hypothesis. The following examples illustrate the relevant facts.

(10) 牛 + 肉 → 牛肉
ushi +*niku* → *gyuuniku*
cow +meat → beef

(11) 豚 + 肉 → 豚肉
buta +*niku* → *butaniku*
pig +meat → pork

(12) 魚 + 肉 → 魚肉
sakana +*niku* → *gyoniku*
fish +meat → fish (meat)

The word *niku* (*meat*) is of Sino-Japanese origin and is also used by itself, i.e. outside of compounds. In conjunction with *ushi* (*cow*) the resulting compound is *gyuuniku* (beef). Similarly *sakana* (*fish*) becomes Sino-Japanese *gyo* in the compound *gyoniku* (fish meat). However, the Japanese word *buta* (*pig*) remains *buta* in the compound *butaniku* (pork)[4].

Assuming that preschool children are exposed to food terms of the above type, the underlying regularity is fairly obvious so that it is difficult to imagine how children could miss it. Some words are pronounced in a different way when occurring in compounds while others are not. Intuitively it seems that we are not attributing too much sophistication to the child's mind if we assume that children will be able to become aware of this regularity.

However, a few examples such as those above are clearly not sufficient to discover the *general* principle of multi-level word structure in Japanese. Semantically similar, but

4 豚肉 can also be read as *tonniku*, where *ton* is the Sino-Japanese counterpart of *buta*. The more common pronunciation, however, is *butaniku*

phonologically unrelated variants of a word are not uncommon in the languages of the world. English *cow – beef*, French *aveugle – cécité*, Spanish *ir – voy* may be cases in point. What makes Japanese different from such cases is the fact that due to the aforementioned historical circumstances multi-level word structure is systematic in Japanese, while in English or French such cases are only occasional and accidental. This latter aspect is something the child cannot possibly learn from a few examples such as (10)–(12).

Since both hypotheses have some initial plausibility, it seems reasonable to test them empirically. In the following sections we will describe a number of tests which are meant to determine if children discover the multi-level principle of Japanese word structure prior to entering school and/or independently of knowledge of the underlying writing system.

4. Test subjects

The basic problem of finding an appropriate test population derives from the fact that the majority of Sino-Japanese compounds belong to the more learned and sophisticated vocabulary of the language and are thus not words that preschool children would typically use (cf. Itou & Mester 1993). Moreover, in many cases the meaning of compounds can be equally well expressed by syntactic constructions and the latter would be preferred when talking to children. For example, the word for *girl* could be either *onna no ko* (女の子 – *woman-of child*) or the compound *joshi* (女子 – *woman-child*). The compound would typically be used e.g. in newspaper texts, but rarely, if ever when talking to preschool children.

Since the crucial variable, however, is knowledge of *kanji* rather than school age *per se* we decided to include in our test population children beyond the age of six without any knowledge of the writing system. These were found among the Japanese-German bilingual children of either Japanese or mixed parents residing in the city of Düsseldorf/Germany. The criterion for selecting these test subjects was that they had to be fluent in Japanese (or both languages), but were unable to read or write Japanese.

The following table lists our test subjects.

No.	Name	Age	Residence	Parents	School	Writing
1	Ken (m)	5.3	Tokyo	J/J	None	None
2	Noriko (f)	5.4	Tokyo	J/J	None	None
3	Takeshi (m)	6.1	Tokyo	J/J	Elem.	Kana
4	Haruo (m)	5.9	Tokyo	J/J	None	None
5	Reiko (f)	13.5	Düsseldorf	J/J	High	Kana
6	Nobuo (m)	14.7	Düsseldorf	J/J	High	Kana
7	Yoko (f)	8.11	Düsseldorf	J/J	Elem.	None
8	Michael (m)	10.3	Düsseldorf	G/J	High	None
9	Sarah (f)	9.3	Düsseldorf	G/J	Elem.	Kana
10	Nanako (f)	9.4	Düsseldorf	G/J	Elem.	None
11	Ayumi (f)	10.11	Düsseldorf	G/J	High	None
12	Sabine (f)	14.1	Düsseldorf	G/J	High	None
13	Hanako (f)	11.5	Düsseldorf	J/G	High	Kana
14	Naomi (f)	7.8	Düsseldorf	J/G	Elem.	None

Children (1)–(4) are monolingual speakers of Japanese residing in Japan. Only Takeshi has just started elementary school and is thus beginning to learn the Japanese syllabic scripts (*kana*). The parents of these children are both Japanese.

Children (5)–(14) are bilingual speakers of Japanese and German. Three children have both a Japanese father and a Japanese mother. Five children have a German father married to a Japanese mother, and two children have a Japanese father married to a German mother. All of these children go to German schools, but consistently communicate in Japanese with their Japanese parent.

Düsseldorf was chosen primarily because the city has the largest Japanese community in Germany with roughly 8 000–9 000 people. The majority of Japanese citizens in Düsseldorf belongs to the business world and typically stays in Germany for a limited period of 2–3 years. In many cases they leave their families behind in Japan; in particular, fathers tend to prefer Japanese schooling for their children. These temporary residents form a fairly closed community with at best occasional contact to the German population. They visit their own clubs and restaurants and in most cases don't feel inclined to adapt to the German environment beyond the most inevitable requirements. Their knowledge of German ranges from extremely poor to moderately fluent. Accompanying wives typically speak very little or no German at all.

In addition to this group a substantial number of mixed couples resides in Düsseldorf. In the majority of cases these are Japanese women married to German husbands, who at some point during their professional career spent a limited period of time in Japan. The opposite combination, i.e., Japanese men married to German wives, is much rarer in Düsseldorf.[5]

The parents of children (5)–(7) are both Japanese, so that at home communication is entirely in Japanese. The fathers of children (5) and (6) do not belong to the typical

5 For reasons unclear beyond speculation, these Japanese husbands frequently refuse to speak Japanese with their children, so that for all practical purposes these are raised as monolingual speakers of German.

Japanese business world in Düsseldorf, but are rather both artists who have lived in Germany for more than ten years, without any short-term intention to return to Japan. Child (6) was born in Japan and came to Germany at the age of 2.5. Child (5) was born in Germany and has never lived in Japan outside vacation times.

Yoko (7) is the only child in our test population whose parents belong to the business world. The family came to Germany shortly after Yoko's fourth birthday. The father originally worked in a Japanese bank, but then decided to join a German bank where he is responsible for trade between Japan and Germany. His German is excellent. Quite unusual for a Japanese father, he decided to send his daughter to a German school and the family is determined to stay in Germany at least until Yoko has finished schooling. Yoko herself is bilingual. Her German is in no way different from that of a typical German 9-year old. The language at home is exclusively Japanese. She regularly associates both with German and Japanese playmates. For her summer vacation she is regularly sent to Japan to her grandparents.

Children (8)–(12) have a German father and a Japanese mother. The mothers speak consistently and almost exclusively Japanese with the children. The father of Nanako (10) is quite fluent in Japanese so that both German and Japanese are spoken within the family. The German fathers of the other children speak little or no Japanese, so that the children communicate in Japanese with their mothers, but in German with their fathers.

Children (13) and (14) have a Japanese father and a German mother. The mother of Naomi (14) has a moderate command of Japanese, so that both Japanese and German are used for communication within the family. Hanako's (13) mother is Austrian, speaking virtually no Japanese at all. As a consequence Hanako speaks German to her mother and mostly Japanese to her father. Whether or not Hanako is truly bilingual is somewhat dubious. According to her father she understands almost everything in Japanese, but both her vocabulary and the way of expressing herself seem to be much more restricted than that of a typical 12-year old speaker of Japanese.

5. Test material

Finding adequate test material for the entire population is not so easy, because – as mentioned before – Sino-Japanese compounds tend to belong to a speech style which is not normally that of a young child.

As a consequence the major challenge was to find an array of words which clearly belong to the vocabulary of even preschool children and still exhibit the crucial property of combining Sino-Japanese readings with identifiable Japanese readings. It appears that such an appropriate array can be found with the Japanese words for the days of the week which show a pattern similar to the one encountered in Western languages, i.e., the word for *day* is combined with various denominators which refer to objects perceivable in the real world. The following table lists the relevant data:

English	Kanji	Japanese	Kanji	English	On	Kun
Sunday	日曜日	nichi-youbi	日	sun	nichi	hi
Monday	月曜日	getsu-youbi	月	moon	getsu	tsuki
Tuesday	火曜日	ka-youbi	火	fire	ka	hi
Wednesday	水曜日	sui-youbi	水	water	sui	mizu
Thursday	木曜日	moku-youbi	木	tree	moku	ki
Friday	金曜日	kin-youbi	金	metal	kin	kane
Saturday	土曜日	do-youbi	土	earth	do	tsuchi

The Japanese word for *Sunday* is *nichiyoubi* (日曜日) where *nichi* means *sun* and *youbi* means *day*. The crucial character is 日 (*sun*) which has the Sino-Japanese (*on*) pronunciation *nichi* and the Japanese (*kun*) reading *hi*. The other days of the week pattern in a similar way.

Note that all these denominations appear in many compounds both with Japanese and Sino-Japanese readings which young children can be expected to know. E.g. *hi* (sun) occurs in *hizashi* (日差し, *sun ray*), *hi* (fire) in *hanabi* (花火, *flower fire = fireworks*), *moku* in *mokuzou* (木造, *wooden*), etc.

Initially we made sure that the children were familiar with the names for the days of the week. This turned out to be the case for all children with the following exceptions. (2), (8), and (11) did not present the names in the correct order. Interestingly, in all three cases it was always *Saturday* (*doyoubi*) that was misplaced. (13) could not remember the words *mokuyoubi* (Thursday) and *doyoubi* (Saturday); (4) could not remember *suiyoubi* (Wednesday).

During the actual testing the children were presented with 22 small cards showing pictures of various objects such as sun, moon, tree, etc. It was the task of the children to select the correct picture for each day of the week. For example, when given the word *getsuyoubi* (*Monday*), the child was expected to select the picture card showing the moon. If the child chose the correct picture, it was concluded that the child knew the meaning of the word.

In addition to the seven pictures showing the objects encoded in the days of the week, the following additional distractor objects were depicted on the remaining cards:

(13)	a walking man	a flower	an eye
	a house	a dragon	a fish
	a horse	a rock/stone	a lantern
	a star	a wheel	a temple
	a sitting child	a bridge	a mountain

By assigning the correct picture to the day presented, the child, of course, merely demonstrated that she knew the correct meaning of the name of the day. This assignment by itself is no proof yet that the child was aware of the fact that one and the same word can have different pronunciations. Therefore each child was presented the 22 cards in random order and was asked to name the objects shown. For example, when the child was given the word *getsuyoubi* (Monday), she was expected to point to the card with the moon. But when shown the card with the moon, the child should name the object as *tsuki*, not *getsu*. In other words, the child was expected to know that the word for *moon* is pronounced *tsuki* when occurring by itself, but *getsu* in the compound *getsuyoubi*.

With three exceptions all children knew the words for the objects on the card. (7), (9), and (13) didn't know the word for *dragon* (*ryou*); (7) answered *doubutsu* (動物; *animal*) instead.

We furthermore tested knowledge of the word for *eat* which is *taberu* with the Sino-Japanese pronunciation *shoku*. The test items are shown in the following list.

Word	Reading	English	Kanji	English	On	Kun
朝食	chou-shoku	breakfast	朝	morning	chou	asa
昼食	chuu-shoku	lunch	昼	noon	chuu	hiru
夕食	yuu-shoku	dinner	夕	evening		yuu
食事	shoku-ji	meal	事	thing	ji	koto
食卓	shoku-taku	meal table	卓	table	taku	
外食	gai-shoku	eating out	外	outside	gai	soto
和食	wa-shoku	Jap. meal	和	Japanese	wa	

In a pretest we checked again whether or not our test subjects were familiar with all of the above words. It turned out, however, that some words presented problems to our monolingual speakers. In particular, *choushoku* (*breakfast*) and *chuushoku* (*lunch*) were not always correctly understood by the monolingual children. This result is not all unexpected, since these two words belong to a higher level of expression and have alternative words which are more commonly used by children, namely *asa-gohan* (朝ご飯 – *morning meal*) and *hiru-gohan* (昼ご飯 – *noon meal*). Therefore we decided to exclude the monolingual children from this test.

Basically, the task of the children was to identify the pronunciation *shoku* with the Japanese verb *taberu*. As in the previous test the children were first shown pictures of the objects named or of some scenes involving the objects and were then asked to assign the appropriate card to the corresponding word. They were subsequently tested for their ability to identify *shoku* as an alternative pronunciation for *taberu*.

We administered a third test involving the word for *shop* which is *mise* (店) in Japanese with the Sino-Japanese reading *ten*. For reasons of space we cannot discuss this test here so that the data will be presented elsewhere.

6. Results

In this section we briefly present the results of our experiments and then discuss their significance in the subsequent section.

The following two tables (one for Sunday to Wednesday, one for Thursday to Saturday) present the results of the first experiment concerning the identification of word meanings and their associated pronunciations in the area of the days of the week:

Child	Sunday		Monday		Tuesday		Wednesday	
	on	kun	on	kun	on	kun	on	kun
	nichi	hi	getsu	tsuki	ka	hi	sui	mizu
1	yes	yes	yes	yes	yes	yes	yes	yes
2	yes	yes	yes	yes	yes	yes	yes	yes
3	yes	yes	yes	yes	no	yes	yes	yes
4	yes	yes	yes	yes	yes	yes	yes	yes
5	yes	yes	yes	yes	yes	yes	yes	yes
6	yes	yes	yes	yes	no	yes	yes	yes
7	yes	no	yes	yes	no	yes	yes	yes
8	yes	yes	yes	yes	yes	yes	yes	yes
9	yes	yes	yes	yes	yes	yes	yes	yes
10	yes	yes	yes	yes	yes	yes	yes	yes
11	yes	yes	yes	yes	yes	yes	yes	yes
12	yes	no	yes	yes	yes	yes	yes	yes
13	yes	no	yes	yes	no	yes	yes	yes
14	yes	no	yes	yes	no	yes	yes	yes

Child	Thursday		Friday		Saturday	
	on	kun	on	kun	on	kun
	moku	ki	kin	kane	do	tsuchi
1	no	yes	yes	yes	no	yes
2	yes	yes	yes	yes	no	no
3	yes	yes	yes	yes	yes	yes
4	no	yes	yes	yes	no	yes
5	yes	yes	yes	yes	yes	yes
6	yes	yes	yes	yes	yes	no
7	yes	yes	yes	yes	yes	yes
8	no	yes	yes	yes	no	no
9	no	yes	yes	yes	yes	yes
10	yes	yes	yes	yes	no	yes
11	yes	yes	yes	yes	yes	yes
12	yes	yes	yes	yes	no	no
13	no	yes	yes	yes	no	no
14	no	yes	yes	yes	no	yes

The results shown in the tables are fairly straightforward. With some exceptions, to be discussed below, the children were able to identify the meanings in the names of the days of the week and they were also able to present the correct pronunciation, if these words appear in isolation, i.e. outside a compound.

Monday, Wednesday, and Friday show a perfect score, i.e. all children gave correct answers in all cases. Sunday is near-perfect in a specific sense. Children (7), (12), (13), and (14) offered *taiyou* (太陽) for *sun* which is a correct alternative for the word *hi*. However, *taiyou* itself is a Sino-Japanese compound and thus cannot be the Japanese reading for the Sino-Japanese *nichi*. In principle, one could tell by its phonological structure that *taiyou* must be Sino-Japanese, but this knowledge is obviously beyond the competence of our test subjects. Therefore the answer was counted as incorrect.

Tuesday and Thursday show a very similar pattern of imperfection. Not unexpectedly, all children could name the Japanese words for *fire* (*hi*) and *tree* (*ki*), but five children could not associate the Sino-Japanese *ka* with *hi* and six children could not associate *moku* with *ki*. Children (8), (9) and (14) chose the picture with the eyes for Thursday (*moku-youbi*). This is interesting in that *moku* is also the Sino-Japanese reading for the word *me* meaning *eye* and occurs in a large number of compounds which also young children can be expected to know; e.g. *moumoku* (盲目 – *blind*) or *menmoku* (面目 – *face*,e.g. *lose one's face/save one's face*).

Saturday was the day name with the largest number of misses. Eight children could not identify the meaning of *do* in *doyoubi* and five children could not even name the word for *earth* when they were shown the picture. This might, however, be an artefact of our test material. The picture showed a pile of dirt on an earthy ground and possibly this picture was not appropriate enough to elicit the intended meaning.

The picture does not change very much if we look at the results of the second test which are summarized in the following table (X stands for the Sino-Japanese *shoku* (*eat*):

Ch	breakfast	lunch	dinner	meal	meal table	eat out	Jap. meal	eat
	chou+X	chuu+X	yuu+X	X+ji	X+taku	gai+X	wa+X	taberu
5	yes	yes	yes	yes	yes	yes	yes	yes
6	yes	yes	yes	yes	yes	yes	yes	yes
7	no	no	yes	yes	no	no	yes	yes
8	yes	yes	yes	yes	yes	yes	yes	yes
9	yes	yes	yes	yes	yes	yes	yes	yes
10	no	no	yes	yes	yes	no	yes	yes
11	yes	yes	yes	yes	yes	yes	yes	yes
12	yes	yes	yes	yes	yes	yes	yes	yes
13	no	no	yes	yes	no	no	yes	no
14	no	no	yes	yes	yes	no	yes	no

The table shows that all children were familiar with the words *yuushoku* (*dinner*), *shokuji* (*meal*), and *washoku* (*Japanese meal*). In contrast, *choushoku* (*breakfast*) and *chuushoku* (*lunch*) were unknown to four children. There is possibly a correlation with age in that these words were generally known only by the older children. However, due to the limited number of test subjects it is difficult to determine whether or not this was pure coincidence.

Similarly, four children failed on *gaishoku* (*eating out*) and two failed on *shokutaku* (*meal table*). In the former case this might be due to personal experience, but we could not detect any correlation with age in these cases.

The crucial column is, of course, the rightmost one for the standard Japanese verb *taberu* (*eat*). Note that the scores in this column do not indicate whether or not the children knew the word *taberu* which trivially they did. Rather the scores indicate whether or not the children were capable of relating the pronunciation *shoku* in the Sino-Japanese compounds with the verb *taberu*. As the table indicates, only two children failed to see this relation.

7. Discussion

The overall result of our experiments is fairly clear. Children are, in fact, able to acquire the basic principle of Japanese multi-level word structure without knowledge of the underlying writing system which is responsible for this structure. That is, children do not learn Sino-Japanese compounds exclusively as unanalyzed units, but are rather aware of the fact that these words are actually compounds and they are also able to relate parts of these compounds to the appropriate genuinely Japanese nouns and verbs. It thus turns out that it is the second of our hypotheses that seems to be true.

This result is, we believe, interesting for a number of reasons. As mentioned before, the particular property of Japanese word structure discussed in this paper appears to be extremely rare in the languages of world and is due only to a very specific historical coincidence in that a unilateral linguistic (and also cultural) transfer occurred between two communities speaking languages that are very different from one another typologically.

It goes without saying that this aspect of Japanese word structure, rare though it may be, must be a property specified for the class of humanly accessible languages, i.e. by Universal Grammar. If this is the case, then there is reason to believe that Universal Grammar also allows for structural properties that rarely occur at all. In other words, Universal Grammar is possibly much more flexible and thus much more abstract than standardly assumed, something I also argued for in a different context (Felix 2003).

Of course, one might be inclined to argue that the properties discussed in this paper have little if anything to do with Universal Grammar, simply because they are fairly superficial properties which, in principle, are easy to detect. The crucial test case for such a contention would be second language learners of Japanese. Following the work of Meisel (1997, 2004) and many others, one might claim that adult second language learners have little or no access to Universal Grammar and rely primarily on general cognitive abilities to acquire linguistic knowledge. If this is true, we would predict that adult second language learners should have extreme difficulties in discovering the multi-level structure of Japanese words if they are not simultaneously taught *kanji*. That is, there should be a fairly clear difference between adult second language learners with and without knowledge of the *kanji* system. There is some evidence (Felix, in prep.) suggesting that this is, in fact, the case.

There are also some obvious limitations to our experimental setup which require further, more specific research.

What we have shown is that the multi-level structure of Japanese words is acquired fairly early, i.e. considerably prior to exposure to the *kanji* writing system. But the question remains: how early? Do children acquire the structural property in question as soon as they are regularly exposed to Sino-Japanese compounds in their environment? If such is the case, then acquisition should indeed occur very early, since Sino-Japanese compounds are ubiquitous in Japanese speech.

Alternatively, one might assume that Sino-Japanese compounds are first and consistently learned as unanalyzed units and are recognized as compounds only after some unknown time interval. There is some indication that this might, in fact, be the case. Notice that e.g. children (7) and (10) were familiar with some, but not all compounds with *shoku* (*eat*). Nevertheless, for those compounds they did know they could identify *shoku* with the standard verb for *eat*, i.e. *taberu*. In contrast, child (14) was unable to relate *shoku* to *taberu*, even though she knew the correct meaning of some *shoku*-compounds. If we assume that compounds are first learned as unanalyzed units, then we need to ask which factors are responsible for revising the original analysis. That is, how does the child discover that the original unit-analysis was wrong and needs to be replaced by a compound-analysis?

There is a third alternative that needs to be considered. What our experiments showed was that children are able to associate Japanese with Sino-Japanese readings. That is, they could somehow relate *taberu* to *shoku*, *hi* to *nichi*, *sui* to *mizu*, etc. But the question remains: what exactly is the relation that children assume before they are exposed to writing? Even though the child knows that both *taberu* and *shoku* have something to do with *eating*, this knowledge cannot necessarily be equated with the knowledge that *taberu* and *shoku* are alternative pronunciations of one and the same word. The child might simply assume that *taberu* and *shoku* are near-synonyms or semantically very closely related different words, somewhat on a par with English *food* and *meal*. In this case, however, we would expect the child to use *shoku* also outside of compounds as a simplex word. Obviously, only longitudinal studies can clarify this issue.

Even though a number of important questions remain, our experiments have shown that children acquire the multi-level structure of Japanese words fairly early and independently of any exposure to the underlying writing system. How these levels are related to each other in the child's mind and how exactly children proceed to discover this rare property must be clarified in future studies.

References

Bhatia T. & Ritchie, W. (eds.) 2004. *The Handbook of Bilingualism (Blackwell Handbooks in Linguistics)*. Oxford: Blackwell.

Chomsky, N. 1964. *Current Issues in Linguistic Theory*. The Hague: Mouton.

Chomsky, N. & Halle, M. 1968. *The Sound Pattern of English*. New York: Harper & Row.

Clahsen, H. 1991. Constraints on parameter setting: a grammatical analysis of some acquisition stages in German child language. *Language Acquisition* 1, 361–391.

Felix, S. 1975. A phonological grammar of Japanese compound morphemes. *Folia Linguistica* 7, 135–148.
Felix S. 2003. Theta parametrization: predicate-argument structures in Japanese. *Linguistische Berichte* 194: 131–152
Felix, S. in prep. *Multi-level Word Structure in Japanese Second Language Acquisition*. University of Passau.
Gakken Kanjigen (学研漢字源) 2002. Epwing, Tokyo.
Itou, J. & Mester, A. 1993. *Japanese Phonology: Constraint Domains and Structure Preservation*. University of California, Santa Cruz.
Koujien (広辞苑) 1998. 5th edition. Iwanami Shoten, Tokyo.
Lewin, B. 1959. *Abriss der japanischen Grammatik auf der Grundlage der klassischen Schriftsprache*. Wiesbaden: Otto Harrassowitz
McCawley, J. 1968. *The Phonological Component of a Grammar of Japanese*. The Hague: Mouton.
Meisel, J. 1997. The acquisition of the syntax of negation in French and German: contrasting first and second language development. *Second Language Research* 13, 227–263.
Meisel, J. 2004. The bilingual child. In: Bhatia & Ritchie, 91–113.
Miller, R. 1967. *The Japanese Language*. Chicago: Chicago University Press.
Reischauer, E.O. 1955. *Ennin's travels in T'ang China*. New York: The Ronald Press Company.
Samson, G. 1958, *A History of Japan to 1334*. London: Cresset Press.
Tateishi, K. 1989. Phonology of Sino-Japanese morphemes. *University of Massachusetts Occasional Papers* 13, Amherst, Mass.
Tsujimura, N. (ed.) 2002. *The Handbook of Japanese Linguistics*. Malden: Blackwell.
Vance T. 1987. *An Introduction to Japanese Phonology*. Albany: SUNY.

Observing Non-Finite Verbs: Some 3V Phenomena in German-Dutch

Tilman N. Höhle

1. Non-finite forms

The system of non-finite verb forms of Standard German and Standard Dutch is rather poor. It comprises just three forms: the bare infinitive, the *zu*-infinitive, and the participle. While the *zu*-infinitive is just the bare infinitive with the prefix *zu* (or *te* etc.) attached to it, the form of the participle is somewhat more subtle.

It has the 'weak' suffix *-t* (or *-d*), with the stem usually unmodified; or it has the 'strong' suffix *-en* (often developed into *-e*), usually with ablaut of the stem vowel. In most parts of Low German and in Frisian, this is all there is. In part of southern Low German (Ostfälisch), the participle in addition bears the prefix *e-*. In High German and most of Dutch, it instead bears the prefix *ge-* according to rather peculiar rules.

The fundamental condition in Standard German (and many High German dialects), as proposed in Curme (1922: 257), is whether or not the first syllable of the stem bears the main word accent. A prefix never bears the main accent of the verb.[1] Thus, the participles in (1) have their main accent on the vowel marked in bold, and they never bear the participial *ge-*:

(1) be-sch**ä**digt ('damaged'), er-l**i**tten ('suffered'), ge-h**o**rcht ('obeyed'), über-s**e**tzt ('translated')

When the verb has no prefix, its first syllable may escape main accent because of a verbal suffix that attracts stress (*-ier*, *-isier*), as in (2i); or for any other (often unknown) reason, as in (ii):

(2) (i) halb-**ie**rt ('halved'), gödel-is**ie**rt ('treated in Gödel's manner')
(ii) berl**i**nert ('talked Berlinian'), mi**au**t ('mewed'), schmar**o**tzt ('sponged'), stib**i**tzt ('filched')

Absence of main accent on the first syllable is sufficient for the absence of participial *ge-*. It is also almost necessary. In Standard German, just one verb lacks *ge-* idiosyncratically, i.e., immediately before the accent (viz., the 'passive auxiliary' *werd-*). In many dialects, more verbs idiosyncratically do without the prefix.

In some dialects, the accent condition is insufficient. E.g., in Sondershausen (3i) and in Hämmern (near Sonneberg, (ii)) the prefix *gə-* (or *jə-*) is present in cases like (2):

1 Prefixes thus differ sharply from particles such as *aus* in (3i). (Cf. Müller 2002 for a study of particles.)

(3) (i) sə hɛt åləs kli:ç u:s jəbosåʊnt (Döring 1912: 36)
'sie hat alles gleich aus pos**a**unṯ' ('she trumpeted everything at once')
(ii) eəʃd hodə siç gəbålwi:əd (Sperschneider 1959: 74)
'erst hat er sich balb**ie**rṯ' ('first he shaved')

In such dialects, *ge-* is absent just before a verbal prefix, as in (1) (and idiosyncratically with certain verbs). This is also the rule in Standard Dutch.

The rules governing the form of participles are formally interesting upon explicit analysis.[2] But to the extent that the system of non-finite forms has attracted attention, it is mostly restricted to the 'substitute infinitive' exemplified in (4), which all members of the German-Dutch linguistic continuum are supposed to have.[3]

(4) (i) sie soll ihn haben schnarchen hören
'she is said to have heard him snore'
(ii) wenn ich ihn doch hätte schnarchen hören!
'if I only had heard him snore!'

The construction is puzzling in that the 'perfect auxiliary' *hab-* ordinarily requires its verb complement to be a participle, not an infinitive, cf. (5), (7); and the position of *hab-* is counter to the ordinary rule of leftward selection in the Standard German verb complex[4], cf. (6), (7).

(5) (i) sie soll ihn schnarchen gehört haben
(ii) wenn ich ihn doch schnarchen gehört hätte!

(6) (i) *sie soll ihn haben schnarchen gehört
(ii) *wenn ich ihn doch hätte schnarchen gehört

(7) (i) *sie soll ihn schnarchen hören haben
(ii) *wenn ich ihn doch schnarchen hören hätte

Thus, four questions pose themselves:

(8) (i) How can a different form appear where a participle is required?
(ii) Why is it, of all forms, the bare infinitive that replaces the participle?

---

2 By a 'formally explicit analysis' I mean a theory expressed in a model-theoretic framework such as (R)SRL; cf. Richter (2000). For steps towards grammatical research on German-Dutch within this framework, cf., e.g., Meurers (2000), Reinhard (2001), Bouma (2003), and references therein.

3 Except for Yiddish and (major parts of) Low German. Cf. Schmid (2005) for some comparative information.

4 I take a *verb complex* (*vc*) to be either *pure* or *impure*. The head of a *vc* is a verb or a *pure vc*. Assuming strictly binary branching, the non-head in a *pure vc* is a verb or a particle or a *pure vc*. In an *impure vc* it is any other verbal projection, and it follows the head, e.g., in Standard German (i), West Flemish (ii), and (19iii), (21iii) below. (This can look confusingly similar to clause or conjunct initial preposed (finite or non-finite) verbs; cf. Höhle (1997).)

(i) … daß dir hätten die Knie zittern können
'… that your knees could have trembled'
(ii) … da ze Valère Marie zag an Jan geven (Haegeman 1994: 510)
'… that Valère saw Marie give them to Jan'

(iii) Why is the participle only replaced when it (a) is selected by a particular perfect auxiliary and (b) selects another verb in a non-participial form by itself?[5]
(iv) Why cannot *hab-* occur in the ordinary word order here?

There is an extensive literature that seeks (or presupposes) causal connections between answers to these questions (primarily, questions (ii) and (iv)); cf. Schmid (2005) and references therein; also contributions in Seuren and Kempen (2003).

It is hardly possible, though, to maintain any truly causal connection between (iv) and the rest in view of the fact that there are speakers who find (7) quite unobjectionable (cf., e.g., Patocka 1997: 279). Although this fact does not make (7) any more palatable to Standard speakers, it would seem to exclude any explanatory account that goes beyond descriptive correlations.

A similar caveat applies to question (ii). There is a large Middle German area where it is not the infinitive that substitutes for the participle. We will also see that not only participles can be replaced (question (i)) and that there are remarkable displacement phenomena that surface when three (or more) verbs are related such that V_1 selects V_2 and V_2 selects V_3 (question (iii)).[6] Thus, the substitute infinitive is just one special case in the family of what I call '3V phenomena'; a case whose very simplicity leads astray many attempts to understand it plainly in terms of what strikes the eye. The observations reported on below should thus be conducive to any more reliable research into the non-finite system.

2. Substitutes

Trebs (1899) reports on the dialect of Oberschwöditz (part of Trebnitz, NW of Zeitz). The participle is formed just as in Standard German. In place of the Standard's substitute infinitive, this dialect has special forms not used elsewhere in the language. These are the data:[7]

(9) (i) ij håwe musd gi:e
'ich habe müssen gehen' ('I had to go')
(ii) ij håwəse kund sa:e
'ich habe sie können sehen' ('I could see them')
(iii) ij håmn wuld drafe
'ich habe ihn wollen treffen' ('I wanted to meet him')
(iv) de håsd suld kume
'du hast sollen kommen' ('you were supposed to come')

---

5 In some variants of German the third verb can be replaced by some pronominal, as in *das hättest du nicht müssen* ('you wouldn't have had to do that'). I leave that aside. Still, it is clear from cases like this that the causal link from (8iiib) to (8ii) found by Kurrelmeyer (1910: 167ff.) for the early times has got lost.

6 For general information on this area, Spangenberg (1989) and Rowley (1989) can be consulted. Spangenberg (1993a) has accumulated much detailed information. The *Thüringisches Wörterbuch* (ThürWb) can also be useful.

7 The relevant data from Trebs (1899: 7, 20ff.) are reproduced here in full. Weise (1906) took up and commented on part of the data; with this exception, they have been ignored in the literature.

(v) de håsd darfd driŋke
'du hast dürfen trinken' ('you were allowed to drink')
(vi) de håsd muchd lo:fe
'du hast mögen laufen' ('you wanted to run')
(vii) e håd larnd få:re
'er hat lernen fahren' ('he learned to ride')
(viii) e hådn måchd gefri:re
'er hat ihn machen gefrieren' ('he made him freeze')
(ix) se hå:dn hi:rd siŋe
'sie hat ihn hören singen' ('she heard him sing')
(x) mər hunn halfd drå:e
'wir haben ihm helfen tragen' ('we helped him to carry [s.t.]')
(xi) mər hun[n] lå:sd le:e
'wir haben ihn lassen liegen' ('we left him lying')
(xii) hudərnij sa:d lo:fe
'habt ihr ihn nicht sehen laufen' ('haven't you seen him run')
(xiii) hå:dərsche nij he:sd size
'habt ihr sie nicht heißen sitzen' ('haven't you told them to sit')
(xiv) se hun du:d schi:wundscharje
'sie haben tun schieben und schergen (stoßen)'
('they helped in every way')
(xv) se hunn waisd danze
'sie haben ihm weisen tanzen' ('they taught him to dance')

These forms are built, it seems, by suffixing *-d* to the bare stem. This suffix looks like the weak participial suffix; but the participle bears the prefix *ge-*. Sometimes the participle's vowel is different, e.g., with *dürf-* in (v) (gedorfd); and several verbs of course have a strong participle, e.g., *seh-* in (xii) (gesa:n), *heiß-* in (xiii) (gehe:sən), *tu-* in (xiv) (gedå:n), and *weis-* in (xv) (gewisən). Exploiting classical terminology, I will call such forms "supine".[8]

In this dialect, then, a supine, rather than an infinitive, substitutes for a participle. But there is more. When the perfect auxiliary has subjunctive form, the supine changes likewise:

(10) (i) ij hedəs misd wise
'ich hätte es müssen wissen' ('I should have known it')
(ii) ij hedn kend drafe
'ich hätte ihn können treffen' ('I could have met him')

Cf. (9i, ii). The same would happen with these forms:

(11) *weld* for *wuld* in (iii); *seld* for *suld* in (iv); *mejd* for *muchd* in (vi); *lesd* for *lå:sd* in (xi); *di:d* for *du:d* in (xiv)

8 No term is established in the literature. Weldner (1991) calls the form "participial infinitive".

The suffix *-d* is here attached not to the stem but to something similar to forms of the preterite subjunctive. I will call this the "complex supine", in distinction to the "simple" supine in (9).

The situation in Kranichfeld (SW of Weimar) is partially similar to Oberschwöditz. There is a (substitute) simple supine with *brauch-*[9], *müss-*, *soll-*, as in (12i–iii):[10]

(12) (i) eç hɛdsn neç braʊxd sə ga:n (Schachtschabel 1910: 85)
'ich hätte es ihm nicht brauchen.SS zu geben.G'
('I wouldn't have had to give it to him')

(ii) a håds mʊsd måxə (ib., 84)
'er hat es müssen.SS machen.IN' ('he had to do it')

(iii) a håd hɛn sʊld gɪ: (ib., 85)
'er hat hin sollen.SS gehen.IN' ('he was supposed to go there')

(iv) a hådsç hɛn lo:sd få:rə (ib., 86)
'er hat sich hin lassen.SS fahren.IN' ('he had himself carried there')

(v) a hådsç hɛn lo:sə få:rə (ib.)

Supines of *dürf-* (dɔrfd), *könn-* (kund), *mög-* (muxd), *woll-* (wuld) occur under the same conditions. Occasionally, also *lass-* appears as a supine, as in (iv). But unlike Oberschwöditz, Kranichfeld has also the substitute infinitive as seen in (v). This is usual with *lass-*, and it is the only choice with *fühl-* ('feel'), *helf-* ('help'), *heiß-* ('order'), *hör-* ('hear'), *lern-* ('learn'), *seh-* ('see'), *spür-* ('feel').

In Ruhla (Regel 1868: 116ff.) there are just simple supines of *dürf-* (durft), *könn-* (konnt), *mög-/möcht-* (moijt), *müss-* (mutt), *soll-* (sollt), *woll-* (wollt); otherwise, the substitute infinitive is found. Still, there is something extra: *konnt*, *sollt* and *wollt* freely alternate with *t*-less forms *konn*, *soll*, *woll* (which are different from the infinitives *könn*, *söll*, *wöll*).

Sondershausen has simple and complex supines of the same verbs: *dürf-* (dærft), *könn-* (kunt, kænt), *mög-/möcht-* (moxt, meçt), *müss-* (mʊt, mɪt), *soll-* (solt, sɪlt), *woll-* (wolt). Their distribution is much less clear than in Oberschwöditz. It appears they sometimes alternate freely with each other and with the bare (substitute) infinitive. Thus, one of Döring's (1912) examples is this:

(13) mə hædn mʊt / mɪt / mɪsə nu:(s)ʃmɪ:ßə (Döring 1912: 37)
'wir hätten ihn müssen.SS/CS/IN rausschmeißen.IN'
('we would have had to expel him')

There are at least some supines in Bleicherode near Nordhausen (cf. ThürWb s.v. *sollen*), but I have not found any north of this. Southward, we find supines in Obermaßfeld (near Meiningen), but only with *müss-* (mößt) and *mög-* (mö:gt):

---

9 As a participle, *brauch-* from (12i) follows the V_3 it selects: *a håd neç sə komŋ gəbraʊxd* ('he didn't have to come').

10 The position in (12iii–v) of the particle *hɛn* (in a verb complex with a branching head) is typical of part of Thuringia. (This is another similarity to Standard Dutch.) See the map in Maurer (1926) for the regional distribution; and cf. Weldner (1991: 210f.) for one local dialect.
From now on, the verb form is often indicated on the Standard German gloss: CS for complex supine; G for gerund; IN for infinitive; P for participle; SS for simple supine; ST for bare stem.

(14) (i) mi honn lang mößt wært (Spieß 1873: 56)
'wir haben lange müssen.SS warten.IN' ('we had to wait long')
(ii) mä jång håt e pfærr mö:gt gewæ:r; å:ber mä fra: håt'ß nét wöll hå: (ib.)
'mein Sohn hat ein Pfarrer mögen.SS werden.IN, aber meine Frau hat's nicht wollen.IN haben.IN'
('my son wanted to be a pastor, but my wife didn't want it')

All other verbs form the substitute infinitive, such as *woll-* (wöll) in (ii). In nearby Wasungen, only *müss-* has a supine (mösd), as in (15). These are the southernmost supines I have seen.

(15) ich hu:ən do:edrü: so: se:ər mösd lach (Reichardt 1914: 205)
'ich habe darüber so sehr müssen.SS lachen.IN'
('I had to laugh about it so much')

Roughly, then, the supine is found throughout the Thuringian area, excepting the most northern and the southern fringe, with large differences in detail from one local dialect to the next.

But it is also found outside this area. Thus, Hanke (1913: 65, 69) reports on the simple supine of *soll-* (sult) and *könn-* (kunt) in a part of Silesia. Graebisch (1907), covering a larger part of Silesia, has found in addition *müss-* (mußt), *mög-/möcht-* (mucht) and *dürf-* (durft), but also *brauch-* (braucht). In eastern Bavarian, supines of *müss-* (miaßt), *möcht-* (mecht), *könn-* (kunnt, kennt), *woll-* (woit), *soll-* (sollt) are found in scattered places (Patocka 1997: 260ff., 264).

Looking westward, I have not found any supine west of the Fulda region, with the exception of two regions near the border to French. Labouvie (1938: 105) reports on substitute forms of *brauch-*, *dürf-*, *könn-*, *mög-/möcht-*, *müss-*, *soll-*, *woll-* in Dillingen on the Saar river similar to forms of the preterite subjunctive with the strong participial suffix *-en* attached to them, e.g., (16):

(16) (i) du hättest das nicht bräuchten zu verraten (p. 132)
'you wouldn't have had to give that away'
(ii) … daß ich nicht habe dürften kommen (p. 123)
'... that I wasn't allowed to come'
(iii) ich hätte gern möchten zu Fuß gehen (p. 112)
'I would have liked to go on foot'

These forms are in no way sensitive to the mood (subjunctive or indicative) of the perfect auxiliary, even though they are formally a variant of the complex supine.

In part of Alsace, by contrast, there are complex supines that regularly participate in the subjunctive system; cf. Philipp (1987: 135f.). One variant of these systems is found in Strasbourg:[11]

(17) (i) van's vɛʀməʀ ksɪ: va:ʀ, hatʃ ɪm hɔft khɛntə ʃpɪ:lə̣
'if it had been warmer, you could have played in the yard'
(Philipp and Bothorel-Witz 1989: 327)

11 Thanks to Marthe Philipp for giving me (17ii, iii) in personal communication.

(ii) ich hätt wotte gehn
'I would have liked to go'
(iii) ich hätt mieste uffpasse
'I should have paid attention'

These forms are similar to finite preterite subjunctive forms. They occur in place of the substitute infinitive when the perfect auxiliary is in subjunctive form. In addition to *könn-*, *woll-* and *müss-* in (17), also *dürf-* (tɛʀftə), *soll-* (sɔtə) and *möcht-* (mɛçtə) are in use.

Returning to the supine mainland, we look into Altenburg.[12] The dialect has simple supines of, at least, *hör-* (hiert), *könn-* (kunnt), *lass-* (lußt), *lern-* (larnt), *mög-/möcht-* (mucht), *müss-* (mußt), *soll-* (sullt), *woll-* (wullt). As usual, they are used in 3V configurations when selected by (indicative or subjunctive) perfect *hab-*. It is thus surprising that with *sollt*, the selecting verb can also be the 'future auxiliary' *werd-*, which ordinarily requires a bare infinitive:

(18) uf su en Ristwogen warn Se doch nich iernd Pfannkuchen sullt hule (Daube 1895: 56)
'auf so einem Rüstwagen werden Sie doch nicht Pfannkuchen sollen.SS holen.IN'
('you are surely not supposed to carry any pancakes on such a truck')

With some verbs – at least *müss-* (muß), *helf-* (half), and *hör-* (hier) – an infinitive is replaced by a form that might either be a *t*-less supine (reminiscent of forms in Ruhla) or a bare stem:

(19) (i) su schien's, all sal se all alle Jungfer sich dorchs Lam muß schlo (Daube 1897: 47)
'so schien's, als sollte sie als alte Jungfer sich durchs Leben müssen.ST schlagen.IN'
('so it seemed as though she should have to spend her life as an old spinster')
(ii) wulln se mir die Sochen widder half uf'n Buckel hiebe? (Daube 1895: 61)
'wollen Sie mir die Sachen wieder helfen.ST auf den Rücken heben.IN'
('do you wish to help me get the stuff on my back?')
(iii) mer hette kunnt ene Schteckenahle uf de Are hier folle (Daube 1897: 29)
'man hätte können.SS eine Stecknadel auf die Erde hören.ST fallen.IN'
('one could have heard a needle fall on the ground')

The bare stem appears as a substitute for the participle and the infinitive at least with *lass-* (loß):

---

[12] My report is based on an inspection of the extant volumes of Daube's 'reader'. Statements in Pasch (1878: 80) and Weise (1900: 154f.) are largely verified by it, but Weise (1906: 195ff.) only partially so.
The infinitive has a Schwa suffix when the stem ends in a consonant; otherwise, it has no suffix. In Daube, *(d)er* (as in (21i) below) is always 'you (pl.)', never 'he', contrary to Weise (1900: 14).

(20) (i) ich hob mer en Zwarnsfonn loß gabe (Daube 1905: 56)
'I had them give me a thread'
(ii) ich will en loß ufmorschiere (Daube1905: 55)
'I want to deploy him'

The regular infinitive does appear on V_2, though, whenever it selects V_3 to the left, e.g.:

(21) (i) wos der nich glei widder ward ausfrasse kunne (Daube 1897: 20)
'was Ihr nicht gleich wieder werdet ausfressen.IN (auslöffeln) können.IN'
('which you will not be able to compensate for immediately')
(ii) mer mußte'n gieh loße (Daube 1897: 16)
'man mußte ihn gehen.IN lassen.IN'
('one had to let him go')
(iii) doß mer hette de Fliegen ibber de Fansterscheim kunnt loofe hiere (Daube 1905: 21)
'daß man hätte die Fliegen über die Fensterscheiben können.SS laufen.IN hören.IN'
('that one could have heard the fleas run across the windows')

Judging from Daube, Altenburg is thus like Oberschwöditz in that it shows no substitute infinitive; instead it has a supine and something like a substitute stem. Some of these forms, however, are not only induced by perfect *have*, but also by 'future' and 'modal' auxiliaries, replacing the infinitive (question (8i)).[13]

In passing, we note that there is a region where a substitute infinitive on *lass-* is not only induced by perfect *have* (question (8iiia)). Constellations such as (22) with the 'passive auxiliary' *werd-* are fully acceptable (and the only choice) in Barchfeld (east of Bad Salzungen):[14]

(22) dɪ sy:ʁ meʊɪç wiʁd duʁç ə sagçə duʁç lɔs lœʏf (W)
'die saure Milch wird durch ein Säckchen durch lassen.IN laufen.IN'
('the sour milk is passed through a sackcloth')

Many similar examples can be found in the *Koch- und Backrezepte* from Steinbach-Hallenberg (SE of Schmalkalden).

The Barchfeld dialect also provides a constellation that is somewhat similar to (18) above:

---

13 The dialect of Rudolstadt as exemplified in Sommer (1906) is broadly similar (as Weise 1900: VI notes). It differs, e.g., in that there are substitute infinitives with most verbs (next to simple supines and some bare stems).

14 My information on Barchfeld comes from three sources: the monograph by Heinrich Weldner (1991), which is by far the finest work of its kind; a series of booklets that contain dialect texts and various informations (Weldner [1994] through 2000, available with the Verein Heimatgeschichte Barchfeld); and most importantly, Weldner has been overwhelmingly generous in sharing his native speaker judgement and his linguistic insight with me in an extended correspondence. I am also grateful for his detailed comments on an earlier version of these pages. Examples from these personal communications are marked "W". (The supines induced by *möcht-* in (23) and *brauch-* in (41) below complement the distributional statement in Weldner (1991: 132f.).)

(23) (i) ɪç meçd aɪ diʁfd gədɑ:nts (W)
'ich möchte auch dürfen.CS tanzen.IN'
('I want to be allowed to dance, too')
(ii) ɪç meçd ned med dɔbɛɪ gəsɛɪ (W)
'ich möchte nicht müssen.CS dabei sein.IN'
('I don't want to be forced to be present')

With *möcht-* (meçd), one could expect the infinitives *diʁf* in (i) and *mɪs* in (ii); instead, we find the complex supines *diʁfd* and *med.* But we also see here a strange prefix *gə* on the infinitives *da:nts* and *sɛɪ.* This is part of a larger system of non-finite verb forms.

3. Further non-finite forms

In addition to the bare infinitive, Frisian has another non-finite form. It looks like the infinitive with a suffix *-n* attached to it. In honour of a long tradition, this is often called the "gerund". Frisian has no *te*-infinitive; rather, *te* is prefixed to the gerund. The bare gerund is selected by sensory verbs such as *see*, *hear*, *feel*, etc. A few verbs, among which *stand*, *sit* and *lie* are prominent, appear as gerunds when selected by *stay* or *go*.

There is but little information on the gerund in Dutch; cf. Marle (1994) and references therein. Low German has two areas with a *zu*-gerund in *-ene*. In a large Swabian area the *zu*-gerund is conspicuous in that it ends in *-et* etc. The more typical gerund in *-e(n)* is found on the border to Italian and in eastern Low German,[15] and also, fortunately, throughout the Thuringian area, with much variation in detail. E.g., Oberschwöditz, Kranichfeld, Bad Frankenhausen, and Sondershausen have just the *zu*-gerund. Thus, *brauch-* ('need') in (12i) above selects the *zu*-gerund *sə ga:n* ('to go'), where *-n* is the gerundial suffix. In Altenburg the gerund is also selected by *bleib-* ('stay') and *hab-* ('have'). In Hämmern near Sonneberg, many more verbs select the gerund: *helf-* ('help'), *hör-* ('hear'), *seh-* ('see'), *spür-* ('feel'), *mach-* ('make'); *bleib-* ('stay'), *hab-* ('have'), and *lass-* ('let'); *werd-* ('future aux') and *tu-* ('do'); e.g., from Sperschneider (1959):

(24) (i) miə blaem dra:dn (p. 40)
'wir bleiben stehen.G'
('we stay standing')
(ii) ʃdarm wen iç ne döid (p. 75)
'sterben.G wenn ich nur täte'
('if I only would die')

In (i), the gerund *dra:dn* ('stand', with suffix *-n*) is selected by *blaem* ('stay'); in (ii), the gerund *ʃdarm* ('die', with stem-final *b* and suffix *-n* developed into *m*) is selected by *döid* ('do').

---

15 See *Deutscher Sprachatlas*, map 54 (*auszutrinken*) for the German situation at large, and *Sprachatlas der deutschen Schweiz*, vol. III (pp. 1 and 2) for details in Switzerland.

Regel (1868) reports that in Ruhla the gerund can be selected by *hear* and *see*, but only when selection is to the left, as in (25); otherwise the bare infinitive is required, as in (26):

(25) bamme den hirz låufen senn (p. 102)
'wenn wir den Hirsch laufen.G sehen'
('when we see the deer run')

(26) (i) mäi såchen den hirz låuf (ib.)
'wir sahen den Hirsch laufen.IN'
('we saw the deer run')
(ii) dæ:r hä:t d'n dü:fel sä: barwes låuf (p. 127)
'der hat den Teufel sehen.IN barfuß laufen.IN'
('he saw the devil run barefoot')

In Steinach (north of Sonneberg), the gerund is selected by a (large) subset of the selectors in Hämmern. There is a delightful additional caprice, though. In a small area that used to include Sonneberg, Steinach, and a few small villages north of these (Spechtsbrunn, Siegmundsburg, Altenfeld, Neustadt, Gabel), the gerund can have the special prefix *bə-*. This *be*-gerund occurs if and only if it is the verb complement of *bleib-* ('stay'). The examples that can be found in the literature are reproduced in (27)–(29):

(27) (i) blei bətraatən (Schleicher 1894: 71)
'bleib stehen.G !' ('stand!')
(ii) blei bəsitzən (ib., 62)
'bleib sitzen.G !' ('stay sitting!')
(iii) əs bleit bəschtenna[16] (ib.)
'es bleibt stehen' ('it stays standing')

(28) (i) iç blai b(ə)sidsn (Luthardt 1963: 289)
'ich bleibe sitzen.G' ('I remain seated')
(ii) a is bəli:çŋ gəbli:m (ib.)
'er ist liegen.G geblieben.P' ('he stayed lying')
(iii) iç sol b(ə)sidsn blai (ib., 290)
'ich soll sitzen bleiben' ('I am supposed to remain seated')

(29) … bəʃdagŋ gəbli:m … (Sperschneider 1959: 95)
'(… daß ich) stecken.G geblieben.P (wäre)'
('(… that I would have) got stuck')

In (23) above, we were intrigued by a prefix *ge-* appearing on the infinitives of *sei-* and *tanz-*. This prefix is similar to the *be-* of the *be*-gerund in that it is bound to nonfinite verbs in particular contexts. We had an instance of it even earlier in (14ii). There, the form *gewæ:r* is built by attaching *ge-* to the bare infinitive of *werd-* ('become'). This is the rule for the *ge*-infinitive throughout.[17] Just like the participial prefix, this *ge-*

---

16 The form *(bə)schtenna* is special; see the exposition concerning (57) below.

17 I am aware of one exception. In a region near Fulda, the bare infinitive of stems ending in *r* takes *-n*. According to Noack (1938: 49) the *-n* is dropped when *ge-* ist prefixed, as in *εʌ wil fɔaʌn ε:wəʌ εʌ kon*

obeys the rule(s) we discussed with (1)–(3) above. (But unlike the participle, *ge*-infinitives do not idiosyncratically do without *ge*-, it seems.)

The regional distribution of the *ge*-infinitive partially overlaps that of the supine in the Thuringian area. The northern border of the prefix is north of Nordhausen (cf. Haushalter 1884 and Rudolph 1924/25); the border in the south is somewhere near Bad Mergentheim (cf. Wolf 1998). In the west, Fulda is within the area, but the border is probably not far from it. In the east, the border is not far from Sonneberg, which is inside (cf. Rosenkranz 1938, map 9a). Rudolstadt and Kranichfeld are outside, but a story about Rottenbach near Rudolstadt shows several instances of the prefix (Firmenich 1846: 169). Gamstädt near Erfurt is within the area (Spangenberg 1993b: 33), and so is Bad Frankenhausen. There is, thus, a small northern and a substantial southern region that has *ge*- but no supine. Similarly, there is a large eastern region that has the supine but no *ge*-. But in the rest, both phenomena occur, roughly from Nordhausen to Wasungen and from Fulda to Erfurt.

The basic cooccurrence condition for the *ge*-infinitive is that it is selected by *könn*-, as in (30i) from Stützerbach near Ilmenau and in (ii) from Sondershausen:[18]

(30) (i) dos muß mer ne laß, gearbt konn er (ThürWb s.v. *lassen*)
'das muß man ihm lassen.IN, arbeiten.IN kann er'
('… work, he can')
(ii) ıç khån mıç nıç mı: sʊ sı:rə ob jəʃtråptsı:rə (Döring 1912: 36)
'ich kann mich nicht mehr so sehr ab strapazieren.IN'
('I can no longer exert myself that much')

(Recall (3) for the accent condition.) In Sondershausen, *könn*- is the only selector for the *ge*-infinitive. But through most of the *ge*-area, *mög-/möcht*- also selects it, as in (14ii). In some places additional verbs can select *ge*-. E.g., Frank (1898: 41) illustrates *dürf*- and *müss*- for Bad Frankenhausen, Haushalter (1884: 14) cites *woll*- and *lass*- as selectors from a publication in the region of Nordhausen. The text in Firmenich (1846: 169) concerning Rottenbach has *woll*- and *soll*-.

4. Displacement

Leaving substitutes aside, we now have, in some places, six non-finite forms: the bare infinitive, the *ge*-infinitive, the bare gerund, the *zu*-gerund, the *be*-gerund, the participle. We illustrate some of this with Sonneberg data from Schleicher (1894: 62f.).

Comparing the forms in (31), we see that *müss*- selects the bare infinitive, *könn*- the *ge*-infinitive, and *werd*- the bare gerund:

---

ned gəfɔaʌ ('he wants to drive, but he cannot drive'). – For information on the Fulda region, see also Weber (1959), Wegera (1977), Wild (1991).

18 Heiligenstadt is unusual in that the *ge*-infinitive (selected by *könn*-) appears to be optional; cf. Firmenich (1846: 199f.). According to Rudolph (1924/25: 263), only epistemic *könn*- takes the *ge*-infinitive in Rottleberode (near Nordhausen).

(31) (i) döös mußmər sough
'das muß man sagen.IN' ('one must say so')
(ii) mər kåå gəsough
'man kann sagen.IN' ('one can say')
(iii) ich waarsch soughən
'ich werde's sagen.G' ('I will say it')

In (32), the gerund (müßən) is selected by *werd-*, and it selects the bare infinitive (reiß, måch). The freedom of word order that there is does not affect these relations.

(32) (i) ich waarsch runtər müßən reiß
'ich werde's runter müssen.G reißen.IN' ('I probably have to tear it down')
(ii) mər warns måch müßən
'wir werden's machen.IN müssen.G' ('we probably have to do it')

The situation in (33) is structurally identical. The gerund (künna) is selected by *werd-*, and it selects the *ge*-infinitive (gəreiß, gəmåch).

(33) (i) ich waarsch runtər künna gəreiß
('I am probably able to tear it down')
(ii) ich waarsch schä gəmåch künna
('I am probably able to do it')

Finally, the substitute infinitive *höör* in (34) selects the gerund (singa):

(34) ich houna höör singa
'ich habe ihn hören.IN singen.G' ('I heard him sing')

Thus, everything is just as one would expect.[19]

However, this is atypical. Other dialects for which sufficient data are available never accord to these plain expectations, even in close vicinity to Sonneberg. They rather display a rich array of additional 3V phenomena.

In Kleinschmalkalden (north of Schmalkalden, east of Barchfeld and Bad Salzungen), *müss-* again selects the bare infinitive, and *werd-* selects the gerund. But a constellation similar to (32) looks different here:

(35) (i) iç wæʃ müd əråb dun (Dellit 1913: 168)
'ich werde's müssen.CS herab tun.G' ('I probably have to put it down')
(ii) mə wæn müd glün (ib., 143)
'wir werden müssen.CS klagen.G' ('we probably have to go to law')

---

19 With one exception: in (i), (ii), *help* would be expected to be a participle/gerund but appears as an infinitive. This is likely related to an anomaly with *help* in Wasungen (Reichardt 1914: 207) that is well-known from Swabian (cf. Heilmann 1999: 61ff.). (Cf. also Schmeller (1821: 380f.) for a similar anomaly with *go* in Bavarian.)

(i) ich hou half gəarbət
'ich habe helfen.IN gearbeitet.P' ('I helped working')
(ii) ich wüür half arbətən
'ich würde helfen.IN arbeiten.G' ('I would help work')

In a 3V configuration, the gerund selected by *werd-* is regularly replaced by the (complex) supine, hence the form *müd* ('must'). This is just like the substitution we have seen for Altenburg in (18). Remarkable is the fact that *glün* and *dun* are, by their suffix *-n*, unmistakably gerunds, even though *müd* would require them to be infinitives. The gerundial form that *werd-* requires is thus displaced to a verb where it plainly does not belong.

We can say that in the Sonneberg data (32)f. the formal requirement of a verb V_n is satisfied immediately by its verb complement V_{n+1}. In (35), in contrast, immediate satisfaction of V_1's requirement is excluded when the complement V_2 selects a verb V_3 by itself. Apparently, satisfaction of V_1's requirement is mediated to V_3, while V_2 appears in a substitute non-finite form.[20]

The differing behaviour of Sonneberg and Kleinschmalkalden with respect to displaced (mediated) selectional requirements is reminiscent of a well-known difference between Dutch and Standard German. Examples such as (36i) with a displaced *zu*-infinitive used to be a topic of traditional German syntax ever since they were observed (and criticized) by Grimm (1837: 949):

(36) (i) ich glaube es haben tun zu können
('I believe I was able to do it')
(ii) *ich glaube es zu haben tun können
(iii) ik geloof het te hebben kunnen doen

(Cf., e.g., Curme 1922: 259.) The majority of Standard speakers feel more or less uneasy with this construction. But the deeper problem is that the construction in (ii), which conforms to our grammatical expectations, is absolutely impossible.[21] This, as Merkes (1895: 66f.) notes, is all the more remarkable as its counterpart in Dutch (iii) is quite unobjectionable (and the only choice).

It is not known what this sharp difference between Dutch and German is induced by. The same question would seem to arise for the Thuringian dialects that have displaced gerundial forms and the Sonneberg dialect that does not.

With respect to constructions such as (36), our dialects are even farther away from Dutch (while their word order properties are conspicuously similar). The displaced *zu* is perfectly natural in, e.g., Rudolstadt, Barchfeld, and Steinbach-Hallenberg[22]:

(37) (i) da braucht mersch Pitschen nur lasse zu merken (Sommer 1906, vol. 1:70)
('one just needs to make Pitsch notice it')
(ii) bɑn ə sɪç ned fʊn ən bʀyçd lɔs ʊ:n tsə ʃny:tsə (W)
'wenn er sich nicht von ihm braucht lassen.IN an zu schnauzen.G'
('if he doesn't need to be snapped at by him')

---

20 This formulation insinuates that the displacement of form is more than just an 'edge phenomenon' but involves the selectional system. Cf. Bader (1995) and references therein for the former, and our final section for relevant data. It seems natural that questions (8i, iii) for substitution find a similar answer. These considerations, though, are really beyond the pretensions of this report.

21 Cf. Meurers (2000) for thorough development of this point.

22 My information on Steinbach-Hallenberg stems from the data in Anita Steube's (1995) article and from her tutorial correspondence, for which I am deeply grateful. Data from the latter source are marked "S". Some dialect texts can be found in the booklets *Sellemoa* and *Koch- und Backrezepte*.

(iii) hä bruchd sich ned fu:nən bəschaid lås zu sœ:wə (S)
'er braucht sich nicht von ihm Bescheid lassen.IN zu sagen.G'
('he doesn't need to take advice from him')

We expect to find the *zu*-gerund required by *brauch-* on *lass-*, but *lass-* is an infinitive, and in place of the infinitive that *lass-* requires we find the *zu*-gerund on *merk-*, *schnauz-* (tsə ʃny:tsə), and *sag-* (zu sœ:wə).[23] This corresponds exactly to constructions well-known from Bern (Hodler 1969, Bader 1995), Gurin (Comrie and Frauenfelder 1992), and Zürich (Cooper 1995).

Returning to the Kleinschmalkalden data, we see a variation of (35) in (38i):

(38) (i) ə wyəds ned wöłd hun (Dellit 1913: 143)
'er wird's nicht wollen.CS haben.G' ('he won't want to have it')
(ii) ə wyəd(s) ʃund lås maxə (ib., 137)
'er wird's schon lassen.IN machen.G' ('he will have it done')

In (i), *woll-* appears as a supine (wöłd), and its verb complement is a gerund (hun) rather than the infinitive that *woll-* ordinarily requires. In (ii), the gerund expected on *lass-* is not replaced by a supine but by an infinitive (lås); and the complement again appears as a gerund (maxə), rather than the expected infinitive.

The same kind of displacement can be seen with the *ge*-infinitive:

(39) (i) ə kon ən iu lås gəkom (ib., 137)
'er konnte ihn ja lassen.IN kommen.IN' ('he could let him come')
(ii) kåsd mə heləf gəʃri: (ib., 168)
'kannst mir helfen.IN schreiben.IN' ('[you] can help me to write')

Here, *lass-* and *helf-* appear as infinitives, even though *könn-* ordinarily requires a *ge*-infinitive. Instead, their complement (*komm-* or *schreib-*) appears with *ge-*.

What happens, then, when a gerund selector (e.g., *werd-*) and a *ge*-infinitive selector (e.g., *könn-*) come together? There was no problem in (33), but now displacing the form would require that one verb be marked for both gerund and *ge*-infinitive. What we actually find is this:

(40) ə wyəds ne(d) könd əråb gəris (ib., 137)
'er wird's nicht können.CS herab reißen.IN'
('he probably isn't able to tear it down')

Thus, *könn-* does find its *ge*-infinitive (gəris), but *werd-* does not find any gerund.[24]

---

23 Altenburg, on the other hand, prefers to drop the *zu*-gerund in this constellation:
(i) dar braucht sich's bluß emol vun dan loß derstreeche (Daube 1897: 38)
'der braucht sich's bloß mal von dem lassen.ST erklären.IN'
('he only needs to let him explain it to him')

24 This is so in all dialects with displacement that I have seen, with one exception. Speakers of Steinbach-Hallenberg are split. There are some who strongly prefer the form *gespräche* in (i) and (ii) to *gespräch*, the latter being the *ge*-infinitive required by the supine *könnt*, while the former in addition bears the gerundial *-e*. Cf. Steube (1995: 432). Others have the judgements reversed.

It thus seems that V_1's requirement is mediated to V_3 if V_2 selects a bare infinitive, as in (35), (37)–(39). This much seems to be necessary and also sufficient for displacement to occur.

The situation in Bad Salzungen is largely identical (Hertel 1888). Looking back to the Barchfeld examples (23), we now recognize the displaced *ge*-infinitive, e.g., in 'ɪç meçd aɪ diʁfd gədɑ:nts': *dürf-* (diʁfd) requires an infinitive; the *ge*-infinitive (gədɑ:nts) is required by *möcht-*. However, in the rare case where additional configurations of verbs can be observed, we find that the displacement of forms can actually be more involved. In (41), *könn-* (kend) requires its verb complement *les-* to be a *ge*-infinitive; but *les-* instead appears as the *zu*-gerund required by *brauch-*. Thus, the *zu*-gerund required by V_1 is mediated to V_3 even when V_2 selects a *ge*-infinitive.

(41) dɔs de dɔs ned bʀyçsd kend tsə la:sə (W)
'daß du das nicht brauchst können.CS zu lesen.G'
('that you need not be able to read that')

As a further dimension of variation, displacing the gerund need not be obligatory. In Wasungen (Reichardt 1914) and in Ruhla, it is optional:

(42) (i) mäi wæ:rens no:ch villmå: mütt hürrn (Regel 1868: 117)
'wir werden's noch oft müssen.IN hören.G'
('we will have to hear that often')
(ii) hæ: würds net wöll hü:r (ib., 116)
'er wird's nicht wollen.IN hören.IN'
('he probably doesn't want to hear it')

(43) sü: wæ:rns ü:r mütt gånn *or* gå: (ib., 117)
'sie werden's ihr müssen.IN geben.G/IN'
('they probably have to give it to her')

In Steinach, displacement of the gerund does not seem to occur (as in (55) below). This also applies to Coburg (Hermann 1957) and Hämmern (Sperschneider 1959).

For a more systematic picture of one local dialect, we look into Barchfeld. First, the *ge*-infinitive. The *ge-* prefixes appear in accordance with (1)f. (with some verbs doing idiosyncratically without *ge-* in the participle; Weldner 1991: 113). Therefore, no *ge-* is to be expected in (44):

(44) (i) də kɔsd mɪç əmɔ: fedəmeçəʊ (Weldner 1991: 217)
'du kannst mich mal vettermicheln.IN'
('leave me alone')

(i) ich wüür dås net könnt gespräche / gespräch, ban ich's net seiwer häd gesie (S)
'ich würde das nicht können sagen, wenn ich's nicht selber hätte gesehen'
('I couldn't say that if I hadn't seen it myself')
(ii) ich war das fileicht niimåls richdich könnt gespräche / gespräch (S)
'ich werde das vielleicht niemals richtig können sagen'
('perhaps I will never be able to say that clearly')

I am grateful to Anita Steube and Eberhard Jäger for discussing this issue with me.

(ii) mə kond sıç ned lɔs bədi:n (W)
'man konnte sich nicht lassen.IN bedienen.IN'
('one couldn't enjoy service')

In (i), the verb *vettermichel-* has its word accent on the third syllable, hence there is no *ge-*. In (ii), the *ge*-infinitive must be displaced toward the last verb *bedien-*. This verb's word accent is on the second syllable (because of the prefix *be-*); hence again no *ge-* here.

In (45), an interesting constellation arises:

(45) ə meçd lıwɚ kend gəaʁwəd (W)
'er möchte lieber können.CS arbeiten.IN'
('he would prefer to be able to work')

Both *möcht-* (meçd) and *könn-* (kend) select a *ge*-infinitive. In this constellation we expect the *ge*-infinitive required by the first verb to be displaced to the last verb. Thus, the single verb *gəaʁwəd* apparently satisfies the selectional requirements of two different verbs.

This kind of situation is well known from Swiss German. Thus, in the Bernese example (46), the *zu*-infinitive *z'häuffe* ('to help') is required by both *schiint* ('seems') and *probiere* ('try'):

(46) dr Hans schiint sine Fründe probiere z'häuffe (Bader 1995: 22)
'der Hans scheint seinen Freunden versuchen zu helfen'
('Hans appears to try to help his friends')

See also Comrie and Frauenfelder (1992: 1058) and Cooper (1995: 187ff., 193).

It seems possible that (47) follows the same pattern:

(47) (i) ə wiʁd ned we:sd tsəʀa:çd tsə komə (W)
'er wird nicht wissen.CS zurecht zu kommen.G'
('he probably doesn't know how to get along')
(ii) sı wiʁd dɔs ned bʁyçd tsə dʊn (W)
'sie wird das nicht brauchen.CS zu tun.G'
('she probably doesn't need to do that')

In (i), *wiss-* (we:sd) requires a *zu*-gerund. The bare gerund that *werd-* requires should be displaced, and the *zu*-gerund *tsə komə* conceivably satisfies both requirements. Similarly in (ii): the *zu*-gerund *tsə dʊn* required by *brauch-* (bʁyçd) conceivably also satisfies *werd-*. Alternatively, (47) might belong to the class of (48) where the gerund from *werd-* must be unrealized in the way discussed with (40) above because of the *ge*-infinitive required by *könn-*.

(48) (i) dɔ: wiʁʃdə ned kend nɑ: gəsɔ:ı (Weldner 1991: 132)
'da wirst du nicht können.CS nein sagen.IN'
('there you probably cannot say no')
(ii) də wiʁʃd hɔ: kend gəkom (W)
'du wirst haben.IN können.CS kommen.IN'
('you were probably able to come')

(The complex supine *kend* is induced by *werd-* in (i) but by the infinitive *hab-* in (ii).)

Displacement of the *ge*-infinitive is usually obligatory, but there is a class of configurations where it is dispreferred or even impossible, viz., with *hear* and *see*. In short examples such as (49) the *ge*-infinitive seems possible, but the bare infinitive is preferred:

(49) (i) ɪç meçd sə hɪːʁ seŋ *or* gəseŋ (W)
'ich möchte sie hören.IN singen.IN'
('I want to hear her sing')
(ii) də kɔsd ən sɛː lœʏf *or* gəlœʏf (W)
'du kannst ihn sehen.IN laufen.IN'
('you can see him run')

With more complex examples such as (50), the *ge*-infinitive is felt to be deviant:[25]

(50) (i) ɪç meçd ən gaʁn hɪːʁ bosœʏnə blɔːs (W)
'ich möchte ihn gern hören.IN Posaune blasen.IN'
('I want to hear him play the trombone')
(ii) mə kondə dn hɑns ɪmɚ in goʁdə sɛː aʁwəd (W)
'wir konnten den Hans immer im Garten sehen.IN arbeiten.IN'
('we could always see Hans work in the garden')

The displacement of the gerund is dispreferred under slightly more inclusive conditions. With *lass-*, displacement seems possible, but not obligatory:

(51) (i) ɪç waːʁ ən lɔs ʀʊf *or* ʀʊfə (W)
'ich werde ihn lassen.IN rufen.IN/G'
('I will have [s.o.] call him')
(ii) mə waʁn ən ned soː lɔs fuʁd geː *or* gen (W)
'wir werden ihn nicht so lassen.IN fort gehen.IN/G'
('we won't let him leave like this')

With *hear* and *see*, the bare infinitive is strongly preferred even in simple examples such as (52):

(52) (i) sɪ wiʁd ən sɛː kom (W)
'sie wird ihn sehen.IN kommen.IN'
('she will see him come')
(ii) ə wiʁd sə hɪːʁ seŋ (W)
'er wird sie hören.IN singen.IN'
('he will hear her sing')

The periphrastic *do* (which is not particularly popular in Barchfeld) ordinarily requires a gerund, as in (53i), but this is left unrealized with *heiß-* ('tell'), as in (ii):

25 The data in Steinbach-Hallenberg are similar; cf. Steube (1995: 433).

(53) (i) ıç dɛd lıwɚ aʁwədə (Weldner 1991: 158)
'ich täte lieber arbeiten.G'
('I would prefer to work')
(ii) ə dɛd mıç hɛıs fuʁd ge: (W)
'er täte mich heißen.IN fort gehen.IN'
('he would tell me to leave')

5. Word order

We have noted that details of selection and substitution sometimes correlate with the order of verbs in the verb complex. Thus, in Standard German (and various dialects, including, e.g., Zaans (Hoekstra 1994)), the order V_1 < (i.e., before) V_2 correlates with the substitute infinitive on V_2. In Kranichfeld and Altenburg, $V_2 < V_3$ correlates with the substitute supine and stem on V_2. In Ruhla, $V_n < V_{n+1}$ correlates with the infinitive (instead of the gerund) on V_{n+1}. We now turn to observations on displacement.

In Steinach (near Sonneberg), *werd-* requires the gerund, as in (54). But in (55), with the order $V_2 < V_3$, immediate satisfaction is excluded, and displacement is avoided as well. Nearby Hämmern is similar (Sperschneider 1959: 32).

(54) (i) iç wa:ʃ måxŋ (Luthardt 1963: 290)
'ich werde's machen.G'
('I will do it')
(ii) a weʌds ne: gəmåx khünʌ (ib.)
'er wird's nicht machen.IN können.G'
('he is probably not able to do it')

(55) (i) iç wa:ʃ råu müs rais (ib.)
'ich werde's runter müssen.IN reißen.IN'
('I probably have to tear it down')
(ii) a weʌd ne: döf låx (ib., 370)
'er wird nicht dürfen.IN lachen.IN'
('he is probably not allowed to laugh')

In Barchfeld, the order of verbs is less variable, selection to the right being strictly adhered to. (Cf. Weldner 1991: 199f.) There are two main exceptions: participles and special gerunds.[26]

Participles precede the auxiliary they are selected by (except, optionally, if the latter is in finite form). We thus have the relative order seen in (56):

---

26 As a third exception, *könn-* can follow its *ge*-infinitive, but only when finite (while in (54ii) it is a gerund):

(i) dɔs ə ned gəaʁwəd kend (Weldner 1991: 158)
'daß er nicht arbeiten könnte' ('that he couldn't work')

(ii) gück emå, bi dær äkel noch gehepf kunn (Weldner 1999: 57)
'guck mal, wie der Ekel noch springen kann' ('look how that bloke still can jump')

(56) (i) bene nåch dn lætzte Krieg vertreewe sein wurn (Weldner 1997: 99)
'welche nach dem letzten Krieg vertrieben.P sind worden.P'
('who were expelled after the last war')
(ii) ə wiʁd gʀɑ:ŋg gəwɑ:sd sɛɪn (Weldner 1991: 165)
'er wird krank gewesen.P sein.G'
('he probably was ill')
(iii) ɛɪ had dɔs ʃond kond gəmɔxd gəhɔ: / ʃond gəmɔxd kond gəhɔ: (ib., 200)
'ihr hättet das schon können.SS gemacht.P haben.IN'
('you could have done that already')

The gerund (sɛɪn) in (ii) and the *ge*-infinitive (gəhɔ:) in (iii) immediately satisfy their selector's requirements, even though they select a verb V_3 by themselves. A similar constellation was seen in (54ii). This thus requires our sketch of the displacement rule (discussed with (35) above) to be modified: it must be order sensitive such that the immediate satisfaction by V_2 of V_1's formal selectional requirement is excluded only if $V_2 < V_3$. (This might also apply to (36i); but it does not apply to the substitution case (4).)

The situation with special gerunds is similar. The ordinary gerunds of *steh-* ('stand') and *lieg-* ('lie') are *ʃden* and *lɛɪn*; but there are also special long forms *ʃdenə* and *lɛɪnə*. They must be selected by *bleib-* ('stay') or *hab-*, *lass-*, *seh-* (cf. Weldner 1991: 113).[27] These verbs must follow the gerund they select, as in (57). (Recall also (25).)

(57) (i) iç wa:ʁ setsə blin (W)
'ich werde sitzen.G bleiben.G'
('I will remain seated')
(ii) iç meçd lɛɪnə gəbli: (W)
'ich möchte liegen.G bleiben.IN'
('I want to remain lying')
(iii) mə hʊnən aɪnfɑx ʃdenə gəlɔsə (Weldner 1991: 113)
'wir haben ihn einfach stehen.G gelassen.P'
('we simply left him standing [there]')

They thus differ sharply from *werd-*, which must precede the gerund it selects:

(58) ɪç glaɪ dɔs ɚ bɪ: gəwe:nlɪç əʀem wiʁd ʃden / gəwe:nlɪç wiʁd əʀem ʃden (W)
'ich glaube, daß er wie gewöhnlich herum wird stehen.G'
('I believe that he will hang about as usual')

This is why *werd-* is standardly involved in the displacement of gerunds while *bleib-* etc. are not.

There is a rare constellation in (59) and (60) (from Steinach and Wasungen) that is, in a sense, complementary to (56iii) and (57ii):

27 With the long gerund and two supines (but no *be*-gerund), Barchfeld thus has eight non-finite verb forms. The few bare gerunds found in Sommer (1906) (Rudolstadt) are all selected by *bleib-*. Otherwise, the verbs that in Barchfeld take the special gerund select a Thuringian adjective in *-nig*, e.g.:

(i) da blieb 'r liegnig / stihnig / sötznig / stecknig / ausnig (Sommer 1906, vol. 1: 31, 39, 46, 66, 397)
('there he stayed lying / standing / sitting / stuck / out')

(59) (i) du hesds jå li:wə los khün gəsai (Luthardt 1963: 370)
'du hättest's ja lieber lassen.IN können.IN sein.IN'
('you should have preferred to abstain from it')
(ii) iç håusnʌ los khün gəmåx (ib.)
'ich habe's ihn lassen.IN können.IN machen.IN'
('I could make him do it')

(60) ich wü:ərən ned hå: laəs kön gəruf (Reichardt 1914: 207)
'ich würde ihn nicht haben.IN lassen.IN können.IN rufen.IN'
('I wouldn't have been able to have [s.o.] call him')

What is unusual here is the order in the verb complex: $V_2 < V_1 < V_3$, where V_2 is *lass-* and V_1 is *könn-*. The varying direction of selection does not seem to disturb the usual displacement relations: the last verb (V_3) is selected to the right by *lass-*; this is sufficient for the *ge*-infinitive (required by *könn-*) to be mediated to V_3, it appears.

References

Bader, T. 1995. Missing and misplaced *z'* in Bernese Swiss German. In: Z. Penner (ed.), *Topics in Swiss German Syntax*, 19–28. Bern: Peter Lang.

Bouma, G. 2003. Verb clusters and the scope of adjuncts in Dutch. In: Seuren and Kempen (eds.) (2003), 5–42.

Comrie, B. and U. Frauenfelder. 1992. The verbal complex in Gurinerdeutsch. *Linguistics* 30, 1031–1064.

Cooper, K. E. 1995. Topics in Zurich German syntax. *Groninger Arbeiten zur Germanistischen Lingustik* 38.

Curme, G. O. [1922]. *A Grammar of the German Language*. 2nd. rev. ed. New York: Frederick Ungar.

[Daube, E.] Sporgel. 1895. *Noch Feierohmds. E Lasebuch in Altenborjscher Mundort. Zweetes Heft.* Altenburg.

[Daube, E.] Sporgel. 1897. *Noch Feierohmds. E Lasebuch in Altenborjscher Mundort. Drittes Heft.* Altenburg.

Daube, E. (Sporgel). 1905. *Noch Feierohmds. E Lasebuch in Altenborjscher Mundort. Viertes Heft.* Altenburg.

Daube, E. (Sporgel). 1908. *Noch Feierohmds. E Lasebuch in Altenborjscher Mundort. Fünftes Heft.* Altenburg.

Dellit, O. 1913. *Die Mundart von Kleinschmalkalden. (Laut- und Formenlehre, Syntax und Wortschatz)*. Marburg.

Deutscher Sprachatlas. Marburg: Elwert [1927ff.] (cf. www.diwa.info).

Döring, E. 1912. Beiträge zu einer Laut- und Wortlehre der Sondershäuser Mundart. (Beilage zum Programme des Fürstlichen Gymnasiums mit Realschule zu Sondershausen, Ostern 1912.)

Firmenich, J. M. (ed.). 1846. *Germaniens Völkerstimmen, Sammlung der deutschen Mundarten in Dichtungen, Sagen, Mährchen, Volksliedern u.s.w.* Zweiter Band. Berlin.

Frank, J. 1898. *Die Frankenhäuser Mundart.* Phil. diss. Leipzig. Halle.

Graebisch, F. 1907. I. Ich habe gehen müssen und Verwandtes. II. Ich habe ihn singen hören und Verwandtes. *Zeitschrift für Deutsche Mundarten* [2], 181–185.
Grimm, J. 1837. *Deutsche Grammatik.* Vierter Theil. Göttingen.
Haegeman, L. 1994. Verb raising as verb projection raising: some empirical problems. *Linguistic Inquiry* 25, 509–522.
Hanke, L. 1913. *Die Wortstellung im Schlesischen.* Breslau.
Haushalter, B. 1884. *Die Mundarten des Harzgebietes.* Halle.
Heilmann, A. 1999. *Die VP im Schwäbischen.* Phil. diss., University of Stuttgart.
Hermann, E. 1957. *Die Coburger Mundart.* Aus dem Nachlass des Verfassers hg. v. Adolf Siegel. Coburg.
Hertel, L. 1888. *Die Salzunger Mundart.* Meiningen.
Hodler, W. 1969. *Berndeutsche Syntax.* Bern: Francke.
Höhle, T. N. 1997. Vorangestellte Verben und Komplementierer sind eine natürliche Klasse. In: C. Dürscheid, K.-H. Ramers, and M. Schwarz (eds.), *Sprache im Fokus. Festschrift für Heinz Vater zum 65. Geburtstag*, 107–120. Tübingen: Niemeyer.
Hoekstra, E. 1994. Woordvolgorde en het Infinitivus-pro-Participio Effect in het Zaans. *Taal en Tongval* 46, 132–141.
Koch- und Backrezepte aus der Steinbach-Hallenberger Küche. Hg. v. d. Mundartgruppe im Verein "Heimatliches Brauchtum" Steinbach-Hallenberg. o.O. [2005]. [distributed by Touristinformation Steinbach-Hallenberg]
Kurrelmeyer, W. 1910. Über die Entstehung der Konstruktion "Ich habe sagen hören". *Zeitschrift für Deutsche Wortforschung* 12, 157–173.
Labouvie, E. 1938. *Studien zur Syntax der Mundart von Dillingen an der Saar.* Marburg.
Luthardt, E. 1963. *Mundart und Volkstümliches aus Steinach, Thüringerwald und dialektgeographische Untersuchungen im Landkreis Sonneberg, im Amtsgerichtsbezirk Eisfeld, Landkreis Hildburghausen und in Scheibe im Amtsgerichtsbezirk Oberweißbach, Landkreis Rudolstadt.* Phil. diss. (masch.) Hamburg.
Marle, J. van. 1994. Oppervlakte-gelijkvormigheid als conditionerende factor bij taalverandering. Iets over de resten van het gerundium in het Middelnederlands en de Nederlandse streektalen. *Taal en Tongval* 46, 14–32.
Maurer, F. 1926. *Untersuchungen über die deutsche Verbstellung in ihrer geschichtlichen Entwicklung.* Heidelberg.
Merkes, P. W. 1895. *Der neuhochdeutsche Infinitiv als Teil einer umschriebenen Zeitform. Historisch-grammatische Betrachtungen.* Phil. diss. Göttingen. Leipzig.
Meurers, W. D. 2000. Lexical generalizations in the syntax of German non-finite constructions. (*Arbeitspapiere des Sonderforschungsbereichs 340.* Bericht Nr. 145.) Universität Tübingen.
Müller, S. 2002. *Complex Predicates. Verbal Complexes, Resultative Constructions, and Particle Verbs in German.* Stanford, Cal.: CSLI.
Noack, F. 1938. *Die Mundart der Landschaft um Fulda.* Marburg.
Pasch, E. 1878. *Das Altenburger Bauerndeutsch, eine sprachliche Studie.* Altenburg.
Patocka, F. 1997. *Satzgliedstellung in den bairischen Dialekten Österreichs.* Frankfurt/M.: Peter Lang.

Philipp, M. 1987. Les verbes modaux en alsacien. In: C. Buridant (ed.), *Romanistique – germanistique. Une confrontation*, 133–137. Strasbourg: Association des Publications près les Universités de Strasbourg.

Philipp, M. and A. Bothorel-Witz. 1989. Low Alemannic. In: C. V. J. Russ (ed.), *The dialects of Modern German. A linguistic survey*, 313–336. Stanford, Cal.: Stanford University Press.

Regel, K. 1868. *Die Ruhlaer Mundart.* Weimar.

Reichardt, E. 1914. Die Wasunger Mundart, 2. Teil. (= *Schriften des Vereins für Sachsen-Meiningische Geschichte und Landeskunde.* 71. Heft.) Hildburghausen.

Reinhard, S. 2001. *Deverbale Komposita an der Morphologie-Syntax-Semantik-Schnittstelle: ein HPSG-Ansatz.* Phil. diss., Universität Tübingen.

Richter, F. 2000. *A Mathematical Formalism for Linguistic Theories with an Application in Head-driven Phrase Structure Grammar.* Phil. diss., Universität Tübingen.

Rosenkranz, H. 1938. *Mundart und Siedlung im Gebiet der obern Saale und des nördlichen Frankenwalds*. Jena.

Rowley, A. R. 1989. East Franconian. In: C. V. J. Russ (ed.), *The dialects of Modern German. A linguistic survey*, 394–416. Stanford, Cal.: Stanford University Press.

Rudolph, F. 1924/25. Dialektgeographie des Honsteinischen. *Teuthonista* 1, 193–200, 257–285. [map after p. 320]

Schachtschabel, O. 1910. *Die Mundart von Kranichfeld in Thüringen.* Phil. diss., Straßburg.

Schleicher, A. 1894. *Volkstümliches aus Sonneberg im Meininger oberlande.* Zweite [unveränderte] Aufl. Sonneberg. (1st ed. 1858.).

Schmeller, J. A. 1821. *Die Mundarten Bayerns grammatisch dargestellt.* München.

Schmid, T. 2005. *Infinitival syntax. Infinitivus Pro Participio as a Repair Strategy*. Amsterdam: John Benjamins.

"Sellemoa önnern alle Schloos" 2. Heft – "Vun Waschett, Huis u Hoof". Hg. v. d. Mundartgruppe im Verein "Heimatliches Brauchtum" Steinbach-Hallenberg. o.O. o.J.

Seuren, P. A.M. and G. Kempen (eds.). 2003. *Verb Constructions in German and Dutch.* Amsterdam: John Benjamins.

Sommer, A. 1906. *Bilder und Klänge aus Rudolstadt in Volksmundart.* Gesammt-Ausgabe, 1. Band und 2. Band. 17. Aufl. Rudolstadt.

Spangenberg, K. 1989. Thuringian. In: C. V. J. Russ (ed.), *The dialects of Modern German. A linguistic survey*, 265–289. Stanford, Cal.: Stanford University Press.

Spangenberg, K. 1993a. *Laut- und Formeninventar thüringischer Dialekte.* Beiband zum Thüringischen Wörterbuch. Berlin: Akademie Verlag.

Spangenberg, K. 1993b. Zur Erforschung der Umgangssprache in Thüringen. In: W. Lösch (ed.), *Beiträge zur Dialektforschung in Thüringen. Heinz Rosenkranz zum 80. Geburtstag*, 17–33. Jena.

Sperschneider, H. 1959. *Studien zur Syntax der Mundarten im östlichen Thüringer Wald.* Marburg.

Spieß, B. 1873. *Die Fränkisch-Hennebergische Mundart.* Wien.

Sprachatlas der deutschen Schweiz. Bern: Francke [1962ff.].

Steube, A. 1995. Flexibler Umgang mit funktionalen Kategorien: Unterstützung aus einer Mundartgrammatik. In: G. Lerchner, M. Schröder, and U. Fix (eds.),

Chronologische, areale und situative Varietäten des Deutschen in der Sprachhistoriographie. Festschrift für Rudolf Große, 421–434. Frankfurt/M.: Peter Lang.

Thüringisches Wörterbuch. Berlin: Akademie Verlag [1965ff.].

Trebs, E. 1899. Beiträge zur osterländischen Mundart. (Beilage zum Programm des Gymnasiums zu Fürstenwalde a. Spr. Ostern 1899. Progr.-Nr. 75.) Fürstenwalde.

Weber, E. 1959. *Beiträge zur Dialektgeographie des südlichen Werra-Fuldaraumes*. Tübingen.

Wegera, K.-P. 1977. *Kontrastive Grammatik: Osthessisch-Standardsprache. Eine Untersuchung zu mundartbedingten Sprachschwierigkeiten von Schülern am Beispiel des 'Fuldaer Landes'*. Marburg.

Weise, O. 1900. *Syntax der Altenburger Mundart*. Leipzig.

Weise, O. 1906. Ich habe gehen müssen und Verwandtes. – Ich habe ihn singen hören und Verwandtes. *Zeitschrift für Deutsche Mundarten* [1], 193–196–198.

Weldner, H. 1991. *Die Mundart von Barchfeld an der Werra*. Stuttgart: Steiner.

Weldner, H. [1994]. *Onse klai Waelt. Heimat zwischen Rhön und Thüringer Wald*. o.O.

Weldner, H. 1995. *Von Land und Leuten, Liedern und Lauten*. o.O.

Weldner, H. 1997. *Eindrücke und Ausdrücke*. o.O.

Weldner, H. 1999. *Erlebtes, Erlesenes und Erhörtes*. o.O.

Weldner, H. 2000. *Jakob Ludwig Güths "Bärfelser Ditsch" (rezensiert und kommentiert)*. o.O.

Wild, K. 1991. Zur Satzgliedstellung in den "Fuldaer" deutschen Dialekten Südungarns. *Zeitschrift für Dialektologie und Linguistik* 58, 24–43.

Wolf, N. R. 1998. Zum verbalen "Präfix" *ge-* in Dialekten Unterfrankens. In: W. Bauer and H. Scheuringer (eds.), *Beharrsamkeit und Wandel. Festschrift für Herbert Tatzreiter zum 60. Geburtstag*, 337–345. Wien: Praesens Verlag.

A Note on the Pronoun System and the Predicate Marker in Tok Pisin*

Manfred Krifka

1. The personal pronoun system of Tok Pisin

The personal pronoun system of Tok Pisin, spoken in Papua New Guinea, and of the closely related Melanesian Pidgin languages Bislama in Vanuatu and Pijin in Solomon Islands, is both one of the most complex and most perspicuous pronoun system in any human language. According to traditional descriptions (see e.g. Mihalic 1971, Verhaar 1995, and most recently Smith, 2002, p. 76) it features a distinction between first, second and third person, an inclusive/exclusive distinction, and four numbers: singular, dual, trial and plural.

(1) Personal Pronouns in Tok Pisin

Person \ Number		Singular	Dual	Trial	Plural
First	exclusive	*mi*	*mitupela*	*mitripela*	*mipela*
	inclusive		*yumitupela*	*yumitripela*	*yumi*
Second		*yu*	*yutupela*	*yutripela*	*yupela*
Third		*em*	*(em)tupela*	*(em)tripela*	*ol*

Occasionally we even find forms that are to be analyzed as quadrals, like *mifopela*, and quintals, like *mifaipela*, cf. Verhaar (1995).

The motivation of these forms by the lexifier language English is obvious:

(2)

mi	*me*	*tu*	*two*	*ol*	*all*
yu	*you*	*tri*	*three*	*pela*	*fellow*
em	*him*				

The morpheme *pela* is used in a variety of functions. In addition to personal pronouns, it occurs with many preposed adjectives (*bikpela taun* 'big town'), demonstratives (*dispela toktok* 'these words'), quantifiers (*wanpela meri* 'a certain woman', *sampela man* 'some men'), and numerals (*fo*, *fopela* '40'), in particular multiplicative numerals (*fopela ten* '40'), see Verhaar (1995). Besides the forms given in table (1) there are a number of variants, especially in creolized versions of the language; for example, we find *mitupla*, *mitla*, *mila* in addition to *mitupela.* There are similar variations in Bislama and Pijin.

* Thanks to the participants of my class in 2005 on Melanesian Pidgin languages, and to an anonymous referee.

The structure of the pronoun system of Tok Pisin is very similar to the system of Austronesian substrate languages, in particular Tolai, presumably the most influential one of these languages (cf. Mosel, 1980, Keesing, 1988):

(3) Personal Pronouns in Tolai

Person \ Number		Singular	Dual	Trial	Plural
First	exclusive	*iau*	*amir*	*amital*	*avet*
	inclusive		*dor*	*datal*	*dat*
Second		*u*	*amur*	*amutal*	*avat*
Third		*ia*	*dir*	*dital*	*diat*

While the coding of the person distinctions in Tolai is not as transparent as in Tok Pisin, the coding of number shows obvious relations to the number words of this language: *ura* 'two', *utul* 'three' and also *ivat* 'four', perhaps evidence of a quadral that was generalized to a plural.

The pronoun system of Tolai (and also Tok Pisin) is typical of Austronesian languages in having an inclusive/exclusive distinction, a rich number system, and no gender distinction. Papuan languages are different; in particular, they rarely exhibit an exclusive/inclusive distinction. This has influences on the variety of Tok Pisin spoken in areas with Papuan substrate, as the corresponding distinction may be less entrenched there.

Now the regularity of the pronoun system as tabulated in (1) breaks down at two points: First, we do not find the form **empela*, or *olpela*, for third person plural; instead we find *ol* (the form *olpela* exists as an adjective, meaning 'old'). Second, we do not find the form **yumipela* for first person plural inclusive; instead we find the simpler form *yumi*. To be sure, the form *yumipela* is reported as an occasional variant of *yumi* in Mandang province. cf. Romaine (1992). Also, Mühlhäusler (1985) notes that this form occurs as a 1st person plural form in varieties without an inclusive/exclusive differentiation in areas in which local languages lack this distinction. But these forms are certainly exceptions.

The lack of **yumipela* might be explained by pointing out that *yumi*, 'you'+'me' is inherently non-singular already, hence *–pela*, apparently a non-singular marker with personal pronouns, would be redundant. This is essentially the line of argumentation I will propose. However, notice that *–pela* does occur in forms like *mitupela*, where it would be equally redundant; we do not find forms like **mitu*. Also, it would be redundant in existing forms like *yumitupela*. Furthermore, if *yumi* just combines the meanings of *yu* and *mi*, it should be restricted to refer to the speaker and the addressee, a group of two persons, while in fact in may well refer to larger groups.

I would like to propose that the irregularities of the pronominal system can be reduced considerably if we apply, first, an alternative view of the inclusive/exclusive distinction, and second, a new perspective on the pragmatics of number.

2. Minimal/Augmented person systems

European languages in general lack an inclusive/exclusive distinction, and languages with such a distinction were not easy to incorporate into the system developed, originally, for classical European languages. One problem posed by a number of such languages can be illustrated with the traditional analysis of the pronoun system of Ilocano, a Philippine language, as proposed by Bloomfield (1942):

(4) Ilocano pronominal system, traditional

		Singular	Dual	Plural
1st Person	inclusive	*ko*	*ta*	*tayo*
	exclusive			*mi*
2nd Person		*mo*		*yo*
3rd Person		*na*		*da*

The problem of this analysis is that there are a number of gaps in the paradigm. In particular, it assumes a dual that just occurs for 1st person inclusive, and not for the other persons. In fact, there is no other evidence for the category dual in the language. Thomas (1955) showed that a more congenial analysis is possible. The following table gives the essence of it; it differs from the original analysis of Thomas insofar as it subsumes his singular/plural distinction under his categories of "simple" and "plus".

(5) Ilocano pronominal system, alternative analysis:

	simple (minimal)	plus (augmented)
Speaker	*ko*	*mi*
Hearer	*mo*	*yo*
Speaker-Hearer	*ta*	*tayo*
neither	*na*	*da*

In this system, the category Dual does not exist. Rather, there is an additional person, Speaker-Hearer, consisting of the combination of first and second person. All four forms exist in a simple form, which later has been called "minimal", and a plus form, which later has been called "augmented"; this latter category identifies forms that refer to non-minimal groups that satisfy the person descriptions. What was analyzed as dual now appears as a simple first+second person, referring to one speaker and one hearer. The augmented form *tayo* then refers to a group of speakers and hearers consisting of at least three persons, one speaker and two hearers.

Minimal/augmented systems have been identified beyond the languages of the Philippines. see Cysouw (2003) for discussion. In particular, the Australian languages of Arnhem land show complex versions of this distinction. McKay (1978) discusses the case of Rembarrnga, which has a "Unit Augmented" form referring to groups of the size minimum + 1, and an "Augmented" form referring to larger groups.

3. Blocking in number distinctions

The second idea that I would like to draw on is a somewhat different view of grammatical number. Standardly, singular stands for reference to single entities, and plural for reference to groups consisting of more than one entity. If a language has a dual, this category identifies reference to groups consisting of two entities. But this view was already challenged by de Saussure, guided by the structuralist insight that one cannot determine the "value" of an expression of a language in isolation. In lecture notes of June 30, 1911, he was reported as saying (cf. de Saussure, 1993):

> The value of a German or Latin plural is not the value of a Sanskrit plural. But the meaning, if you like, is the same. In Sanskrit, there is a dual. Anyone who assigns the same value to the Sanskrit plural as to the Latin plural is mistaken because I cannot use the Sanskrit plural in all the cases where I use the Latin plural. Why is that? The value depends on something outside.

In other words, the presence of a dual blocks the application of the plural for reference to groups consisting of two members. Even for Sanskrit, we can assume that the plural refers to groups of two or more entities; it just has no chance of getting applied to groups consisting of two entities because the dual is a better choice in this case. The idea that more narrow rules should be applied before more general rules is of course much older; it is central to the organization of grammatical knowledge in the work of Pāṇini (cf. Kiparsky 1994). The application of this principle in semantics is grounded in the theory of implicatures by H. P. Grice (cf. Grice, 1975), in particular in the Maxim of Quantity, first submaxim, which requires the hearer to maximize information. In the case at hand, this means that whenever the more specific dual is applicable, it has to be used instead of the less specific plural. Consequently, if the plural was used, the addressee can conclude that the dual was not applicable, and hence that the speaker indeed referred to a group consisting of at least three entities.

This argument can be applied to the relation between singular and plural as well. In Krifka (1989) I have argued that bare plurals like *children* apply also to single entities, but that their use in this case is blocked by the presence of the more specific singular form, *a child*. Evidence for the wider meaning of bare plurals like *children*, in contrast to expressions like *more than one child*, comes from data like the following:

(6) A: *Do you have children?*
B: *Yes, one. / *No, (only) one.*

(7) A. *Do you have more than one child?*
B: * *Yes, one. / No, (only) one.*

If the plural *children* were indeed restricted to groups of two or more children, as with *more than one child*, we should expect that question (6A) cannot be answered by *yes, one* by someone that has only one child, but can be answered by *No, (only) one* – just the opposite of what we find.

Sauerland, Anderson and Yatsushiro (2005) discuss further evidence for a meaning of plurals that includes reference to single individuals. For example, a sentence like (8a) is easily understood as including boys that have only one sister, whereas (8b) suggests, or even presupposes, that the boys talked about have exactly one sister.

(8) a. *Every boy should invite his sisters.*
b. *Every boy should invite his sister.*

Sauerland et.al. (2005) also discuss evidence that plural pronouns may include reference to single objects. This is the source of plural politeness forms such as *Vous* in French or *Ihr/Sie* in German, and also historically *you* vs. *thou* in English. Also, this explains why, in current English usage, gender-unspecific reference is possible with plural pronouns, such as in the following case:

(9) a. *Everyone should show their passport.*
b. *Someone left their umbrella.*

By this reasoning, what is traditionally called plural should better be called "neutral number", as the corresponding forms are not restricted to plural entities. Consequently, English should be analyzed as a language that has a singular and a neutral number.

It is important to notice that number neutrality is a semantic notion that is different from morphological underspecificiation, as e.g. with *sheep*, which is underspecified as to number:

(10) a. *The sheep is in the pen.*
b. *The sheep are in the pen.*

With morphological underspecification, we can assume feature-value pairs like [NUM –], where "–" can be specified as sg or pl, according to context. Number neutrality, on the other hand, is a semantic notion that applies to English plurals and mass nouns as well, as they do not impose any restriction as to the size of the entities to which they refer. This can be illustrated as follows. Let CH be the set of all entities that consist of one or more children. Let #(x) be a measure function that yields the number of atomic entities an individual x consists of. Then we have the following meanings for *a child* and *children*:

(11) $[\![child]\!]$: $\lambda x\, [\#(x) = 1 \wedge x \in CH]$
$[\![children]\!]$: $\lambda x\, [x \in CH]$

Notice that $[\![child]\!]$ is a proper subset of $[\![children]\!]$. This is a systematic property; all noun phrases with the feature [NUM sg] will denote a proper subset of noun phrases of the feature [NUM pl]. (See Krifka 1995, 2004 for a more refined view of the semantics of singular nouns, plural nouns and mass nouns). In order to distinguish these semantic concepts from morphological notions, I will use the feature [1 Num] for nouns restricted in their application to single entities, as e.g. in singulars, [2 Num] for nouns restricted to groups consisting of two entities, as e.g. in Sanskrit plurals, and [–Num] for nouns that are unrestricted as to the number of entities they refer to, as e.g. in English and Sanskrit plurals.

One question that has not been investigated so far is whether it is always the plural form that is semantically more general. We have argued that this is the case in English. But it might very well be that in other languages, it is the the singular that is more general, referring to single individuals or groups, and that reference to groups is typically blocked by a plural that is restricted to groups of two or more individuals. For example, question (6) in Bahasa Indonesia is *Sudah ada anak*, lit. 'Already exist child?', even though there exists a plural form, *anak-anak* 'children'. The plural form is not obligatorily used when referring to groups of more than one individual, cf. Chung (2000). Consequently, we should assume that Bahasa Indonesia has a true plural form and a number neutral form: *anak* is [– Num], and *anak anak* is [≥2 Num].

4. A new analysis of Tok Pisin pronouns

With the concept of minimal/augmented person systems and of blocking effects in the realm of number in mind, I propose the following analysis of the Tok Pisin personal pronouns:

(12) Tok Pisin pronoun system, revised analysis

Number / Person	[– Num] (number-neutral)	[≥2 Num] (plural)	[2 Num] (dual)	[3 Num] (trial)
[+Sp –Addr]	*mi*	*mipela*	*mitupela*	*mitripela*
[–Sp +Addr]	*yu*	*yupela*	*yutupela*	*yutripela*
[+Sp +Addr]	*yumi*	—	*yumitupela*	*yumitripela*
[–Sp –Addr]	*em*	*ol*	*(em)tupela*	*(em)tripela*

The category of Person is captured by two binary semantic features, [± Sp] and [± Addr]. Here, [+ Sp] is interpreted as saying that the reference includes the speaker(s), and [– Sp] as saying that the reference does not include the speaker(s); similarly, [+ Addr] says that the reference includes the Addressee, and [– Addr] that the reference does not include the Addressee. In this case, the category Inclusive is rendered by [+Sp +Addr], inspired by Thomas' proposal for Ilocano.

Notice that even under this analysis, there is an apparent gap in the paradigm, the missing form **yumipela*. Reorganizing the description by introducing two binary features for persons did not simplify the overall structure, as it did in the case of Ilocano. But it allows for a different address for the form *yumi*, which is not analyzed as a plural anymore, but as a number-neutral form, just like *mi* and *yu*.

We assume furthermore that more specific forms block general forms. This means that reference to a group containing the speaker cannot be expressed by *mi*, but must use *mipela*. The form *mipela* again is blocked by *mitupela* if the group contains two individuals, and by *mitripela* if it contains three. Hence the typical use of *mipela* is pragmatically restricted to groups of four or more individuals, and the typical use of *mi* is restricted to single individuals. This can be depicted as follows; dotted braces express the literal meaning of the expressions, and full lines express their range of use after blocking has been taken into account.

(13)

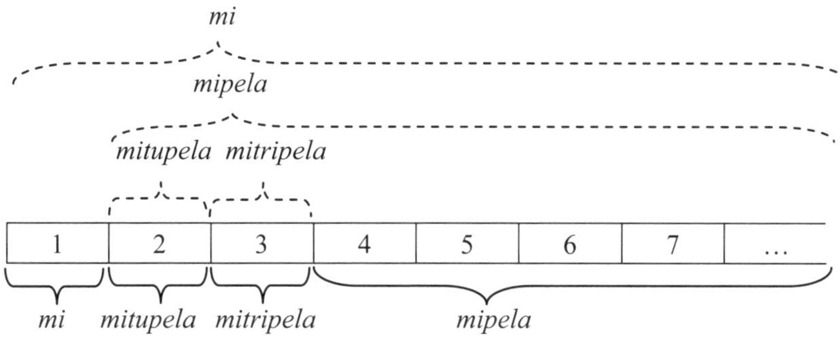

This analysis assumes that what is normally analyzed as singular in Tok Pisin is, in fact, a number-neutral form, and that plurals are indeed plurals, and not number-neutral forms as suggested for English above. There is independent evidence for that. For example, the nominal plural marker *ol* does not occur with number words or plural quantifiers, and it often does not occur with bare nouns with plural reference (cf. Verhaar, 1995, 346ff.). The proper way to ask whether the addressee has children is *Yu gat pikinini?*, not *Yu gat ol pikinini?* In the following example, we find the singular (= number-neutral) form *pikinini* 'child' instead of *ol pikinini* 'children', even though the context, according to Verhaar, is such that a group of children is referred to.

(14)	*Dispela*	*i*	*no*	*inap*	*help-im*	*yu*	*wantaim*	*pikinini*	*bilong*	*yu.*
	this	PM	NEG	enough	help-TR	2SG	with	child	POSS	you

'This is of no use to you and your children.'

The case of *yu, yupela, yutupela* and *yutripela* is exactly parallel. Now consider the case of inclusive person, [+Sp +Addr].

(15)

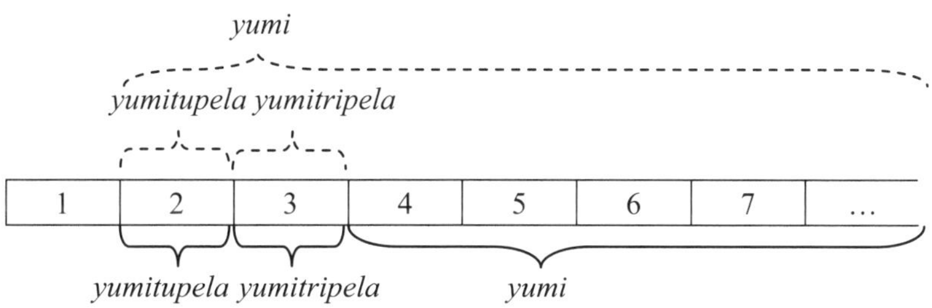

The number-neutral form is *yumi*, which is exactly parallel to the number-neutral forms *mi* and *yu*. As the person specification is [+Sp +Addr], reference to a single individual is excluded in this case by the person specification (presumably, even soliloquies cannot be analyzed in a way that speakers address themselves). Blocked by *yumitupela* and *yumitripela*, the form *yumi* will preferably be used for reference to groups containing speakers and hearers that consist of four or more individuals.

We can now explain why the form **yumipela* is missing as follows: The form *yumi* is a number-neutral form, just as *mi* and *yu*. But contrary to *mi* and *yu*, it cannot be applied to single persons, due to the person specification [+Sp, +Addr], which can be satisfied only when referring to two or more persons. Hence the only area in which *yumi* is applicable at all is for reference to two or more persons. We do not need to mark this by explicit plural marking as in the case of *mipela* and *yupela*.

We have seen in a number of examples that blocking is not always operative. So we should not be astonished, for example, to find examples of reference to a group of two people containing the speaker with *mipela*, in addition to *mitupela*. Indeed, Smith (2002) observes that forms like *mitupela* and *mitripela* occur quite rarely – too rarely, in fact, to assume that they cannot refer to groups of only two or thee persons. And we do find uses of *mipela* in texts that strongly suggest reference to just two persons, as in the following example, a text passage of an interview.

(16) *Orait ol salim mi na wanpela Tolai mitupela i go sikarapim ol trai kokonas na kukim oil. Mipela mekim dispela wok i go na bihain ol i askim husait i save long painim pis.*
'So they sent me and one Tolai person away, we two [*mitupela*] should make coconut oil. We [*mipela*] did this work and went (back), later they asked who knows how to fish.'
Source: http://ajrp.awm.gov.au/ajrp/remember.nsf/pages/NT00002C86

If the analysis proposed here is right, we could, in principle, even find cases in which the blocking of the ≥2 reading is not operative for the forms *mi*, *yu* and *em*. That is, there could be uses of these pronouns in which they refer to groups. I am not aware of cases like that, which might point to the fact that plural forms *mitupela*, *yutupela* and *ol* are more central to the grammar than dual and trial forms. Cases of total blocking are, of course, nothing special in grammar. For example, we find that the presence of irregular forms like *men* leads to ungrammaticality of regular forms like **mans*.

But there is at least one small piece of independent evidence that what traditionally has been analyzed as singular is, in fact, a number-neutral form. In the third person, we find dual and trial forms *emtupela* and *emtripela* (besides the more frequent forms *tupela* and *tripela*); forms like **oltupela* do not occur. These forms can be explained in a straightforward way if *em* is a number-neutral form, and *tupela* and *tripela* just add as number information [2 Num] and [3 Num], respectively. In this analysis, the only remaining irregular form in the paradigm is *ol*, where we would expect **empela* as the regular form.

5. Personal pronouns and the predicate marker

An important feature of Tok Pisin grammar is the predicate marker *i* that precedes main verbs. It is often thought to be derived from English *he* as a resumptive pronoun, but it also has a precursor in Austronesian languages (cf. Mihalic 1971, Mosel 1980, and Keesing 1988). The use of *i* is not restricted to third person singular, as its likely origin suggests, but it is not used everywhere. As a general rule, it is used for all persons and numbers except for *mi* [+Sp, -Addr], *yu* [–Sp, +Addr] and *yumi* [+Sp, +Addr]. Some examples:

(17) a. *Mi amamas.* 'I am happy.'
b. *Yu amamas.* 'You [sg] are happy.'
c. *Yumi amamas.* 'We [incl.] are happy.'
d. *Hem i amamas.* 'He/she is happy.'
e. *Mipela i amamas.* 'We [incl.] are happy.'
f. *Yupela i amamas.* 'You are happy.'
g. *Ol i amamas.* 'They are happy.'
h. *Yumitupela i amamas.* 'We two [incl.] are happy.'
etc.

To be sure, this is just an approximation of the use of the predicate marker. First, it also occurs with subjects *mi, yu* and *yumi* in case there are intervening constituents, as in *mi wanpela i amamas* 'I alone am happy', cf. Verhaar (1995), p. 70ff. Second, there is a tendency that *i* does not occur in the presence of other preverbal particles and tense, mood and aspect markers, most likely because they occupy the syntactic slot (presumably I^0) in which *i* occurs, cf. Smith (2002), p. 119ff. Furthermore, *i* can be dropped, especially in Highland regions with Papuan substrate, cf. Smith (2002), p. 115ff. We will concentrate here on the basic facts given in (17).

(18) Tok Pisin predicate marker

Number / Person	[– Num] number-neutral		[≥2 Num] plural		[2 Num] dual		[3 Num] trial	
[+Sp –Addr]	*mi*	∅	*mipela*	*i*	*mitupela*	*i*	*mitripela*	*i*
[–Sp +Addr]	*yu*	∅	*yupela*	*i*	*yutupela*	*i*	*yutripela*	*i*
[+Sp +Addr]	*yumi*	∅	—		*yumitupela*	*i*	*yumitripela*	*i*
[–Sp –Addr]	*em*	*i*	*ol*	*i*	*(em)tupela*	*i*	*(em)tripela*	*i*

Verhaar (1995) offers the following explanation for the distribution of the predicate marker. He points out that *i* is used whenever reference to an entity is included that is not given by the speech situation itself (p. 72f.); he distinguishes between "permanent files" referring to speaker and addressee (sometimes called "local persons"), and non-permanent files that are created by introducing entities in conversation. This explains why *i* is not used for *mi*, *yu* and *yumi*, as these forms refer to participants of the speech situation only. The form *i* is used for third person, but also for first and second person non-singular forms, as they also contain reference to entities not given by the speech situation, under the assumption that the speech situation identifies exactly one speaker and one addressee. Forms like *mipela* contain reference to a group that contains the speaker, where the delineation of this group is not given by the speech situation itself. In a sense, *mipela* is a mix of 1st person and 3rd person, not a plural of 1st person. Similarly, *yupela* contains reference to a group that contains the addressee, and again the identity of this group is determined by the larger context, and not only the speech situation itself.

One problem of this explanation is that we do not find the predicate marker with *yumi*, even though this form can definitely contain reference to a group greater than just the immediate speaker and the immediate addressee. Another problem is that we do find the predicate maker with *yumitupela*, which is restricted to immediate speaker and immediate addressee. Verhaar assumes that *pela* forms in a sense have a noun-like character, which is the reason why we find the predicate marker with them. But notice that if this were the case, then we would not need to explain the distribution of *i* by reference to explicit or implicit third persons in the first place.

A better explanation for the distribution of the predicate marker can be developed under the assumption of two markedness hierarchies:

(19) a. Participant hierarchies: {[+Sp], [+Addr]} > [–Sp, –Addr]
b. Number hierarchy: [– Num] > [+ Num]

Here, {[+Sp], [+Addr]} stands for the three feature combinations [+Sp, –Addr], [–Sp, +Addr] and [+Sp, +Addr]. And [+ Num] stands for all features in which the number is specified, that is, [≥2 Num], [2 Num] and [3 Num]. A more fine-grained scale would also state the hierarchy [≥2 Num] > [2 Num] > [3 Num], but this is irrelevant for our purposes.

The participant hierarchy states that forms that include reference to the speaker or to the addressee are unmarked with respect to forms that do not. This is a well-known preference scale for subjects, cf. e.g. Aissen (1999). It can be seen at work in languages with split ergativity in which the split is conditioned by 1st/2nd person vs. 3rd person. The best known example of this type of language is the Australian language Dyirbal, which has no case marking (= nominative case) for 1st and 2nd person subjects, and ergative case marking for other subjects. The following example is constructed after Dixon (1972).

(20) a. ***ŋana-∅*** *nyura-na* *bura-n.*
1PL-NOM 2PL-ACC see-NONFUT
'We saw you.'
b. ***yabu-ŋgu*** *nyura-na* *bura-n.*
mother-ERG 2PL-ACC see-NONFUT
'Mother saw you.'

Similarly, languages with inverse person marking typically distinguish between sentences with 1st and 2nd person subjects and 3rd person objects, with direct verb forms, and sentences with 1st and 2nd person objects and 3rd person subjects, which trigger inverse marking on the verb. The following example illustrates this for 1st person exclusive and 3rd person plural in Plains Cree, after Dahlstrom, 1986.

(21) a. *e:-wa:pam-a:-ya:hk-ik.*
DET-see-DIR-1PL.EXCL-3PL (CONJ)
'We (excl.) see them.'
b. *e:-wa:pam-iko-ya:hk-ik.*
DET-see-INV-1PL.EXCL-3PL (CONJ)
'They see us (excl.)'

The motivation for the participant scale is the distribution of persons in natural discourse: NPs with reference to the speaker or the addressee occur more often in subject position than in object position. By principles of coding economy, this can be captured by simplifying expressions for first or second person subjects (cf. Du Bois 1987, Bresnan et al. 2001, Zeevat and Jäger 2002).

The second hierarchy, the number hierarchy, does not need much motivation. It simply states that more general expressions are unmarked, compared to more general ones. It might be in conflict with intuitions in the face of previous claims concerns markedness in English, where we analyzed plural forms as number-neutral, and singular forms as specified for the number one. But notice that, as far as number of NPs is concerned, the markedness relations are less clear than it appears at first sight. While one can argue that *apple* is unmarked with respect to *apples*, notice that *apples* is a full NP, whereas *apple* is not; we need to fill the determiner position, as e.g. in *an apple*. This NP is arguably at least as complex as the NP *apples*.

When we align the two markedness scales in (19), we get the following combined scale:

(22) Aligned Participant and Number Hierarchy

$$\underbrace{\left\{\begin{array}{l}{[+\text{Sp}, -\text{Num}]}\\ {[+\text{Addr}, -\text{Num}]}\end{array}\right\}}_{\text{no predicate marker}} > \underbrace{\left\{\begin{array}{l}{[+\text{Sp}, +\text{Num}]}\\ {[+\text{Addr}, +\text{Num}]}\\ {[-\text{Sp}, -\text{Addr}, -\text{Num}]}\end{array}\right\} > [-\text{Sp}, -\text{Addr}, +\text{Num}]}_{\text{predicate marker}}$$

This aligned hierarchy is to be interpreted as: If the language makes a distinction in the marking of subjects, then subjects higher on the hierarchy should be unmarked, and subjects lower on the hierarchy should be marked. Tok Pisin does make a distinction: Subjects corresponding to the first feature combinations, namely *mi* [+Sp, –Addr, –Num], *yu* [–Addr, +Sp, –Num] and *yumi* [+Sp, +Addr, –Num] do not have the predicate marker, whereas subjects corresponding to all other feature combinations do require it.

The aligned hierarchy in (22) makes the prediction that *i* should be most firmly entrenched for the forms [–Sp, –Addr, +Num] forms. Considering the considerable variations in the use of *i* mentioned above, it would be worthwhile to investigate whether the factors inherent in the hierarchy can be identified in statistical tendencies in text corpora. For example, Mihalic (1971: 23) suggests that the predicate marker may be dropped after *mipela* and *yupela* (no information for dual and trials is given), which suggests that the participant hierarchy is more important than the number hierarchy, leading to the following refined scale:

(23)

$$\underbrace{\left\{\begin{array}{l}{[+\text{Sp}, -\text{Num}]}\\ {[+\text{Addr}, -\text{Num}]}\end{array}\right\}}_{\text{no predicate marker}} > \underbrace{\left\{\begin{array}{l}{[+\text{Sp}, +\text{Num}]}\\ {[+\text{Addr}, +\text{Num}]}\end{array}\right\}}_{\text{optional pred marker}} > \underbrace{[-\text{Sp}, -\text{Addr}, -\text{Num}] > [-\text{Sp}, -\text{Addr}, +\text{Num}]}_{\text{obligatory predicate marker}}$$

6. Conclusion

In this article I have tried to show that a curious irregularity in traditional analyses of the personal pronoun system of Tok Pisin – the lack of the form **yumipela* – can be explained within a new perspective on this system. The basic ingredients were, first, a componential analysis of the person category, using the features [± Sp] and [± Addr]. This made it possible to analyze First Person Inclusive as a separate person category. The second step was a blocking analysis of number. In particular, singular was analyzed as number-neutral, not imposing any restrictions, whereas plural was analyzed as restricting reference to groups of two or more individuals, and dual and trial were analyzed as restricting reference to groups of two and three individuals, respectively. I have also shown that this reanalysis of the person and number system opens a way to explain the distribution of the predicate maker *i* in a more coherent way. In this analysis, the predicate marker is used for subjects that are low on the scale that results when the markedness scales for person and number are combined, which fits to the general patterns of coding economy.

References

Aissen, J. L. 1999. Markedness and subject choice. *Natural Language and Linguistic Theory* 17, 673–711.

Bloomfield, L. 1942. Outline of Ilocano syntax. *Language* 18, 193–200.

Bresnan, J., S. Dingare, and C. D. Manning. 2001 Soft constraints mirror hard constraints: Voice and person in English and Lummi. In: M. Butt and H. King (eds.), *Proceedings of the LFG '01 Conference*. Stanford: CSLI Online Publications.

Chung, S. 2000. On reference to kinds in Indonesian. *Natural Language Semantics* 8, 157–171.

Cysouw, M. 2003. *The Paradigmatic Structure of Person Marking*. Oxford: Oxford University Press.

Dahlstrom, A. L. 1986. *Plains Cree Morphosyntax*. Doctoral dissertation, University of California at Berkeley.

Dixon, R. M. W. 1972. *The Dyirbal Language of North Queensland*. Cambridge: Cambridge University Press.

Du Bois, J. 1987. The discourse basis of ergativity. *Language* 63, 805–855.

Grice, P. 1975. Logic and conversation. In: P. Cole and J. L. Morgan (eds.), *Logic and Conversation*, 41–58. New York: Academic Press.

Keesing, R. M. 1988. *Melanesian Pidgin and the Oceanic Substrate*. Stanford: Stanford University Press.

Krifka, M. 1989. *Nominalreferenz und Zeitkonstitution. Zur Semantik von Massentermen, Pluraltermen und Aspektklassen*. München: Wilhelm Fink.

Krifka, M. 1995. Common nouns: a contrastive analysis of Chinese and English. In: G. N. Carlson and F. J. Pelletier (eds.), *The Generic Book*, 398–411. Chicago, London: The University of Chicago Press.

Krifka, M. 2004. Bare NPs: kind-referring, indefinites, both, or neither. In: R. B. Young and Y. Zhou (eds.), *Proceedings of SALT XIII*, Cornell: CLC Publications.

Kiparsky, P. 1994. Paninian linguistics. In R. E. Asher (ed.), *The Encyclopedia of Language and Linguistics*, 2918–2922. Oxford: Pergamon Press.

Mihalic, F. 1971. *The Jacaranda Dictionary and Grammar of Melanesian Pidgin*. Milton, Qld.: The Jacaranda Press.

McKay, G. R. 1978. Pronominal person and number categories in Rembarrnga and Djeebbana. *Oceanic Linguistics* 17, 27–37.

Mosel, U. 1980. *Tolai and Tok Pisin: The Influence of the Substratum on the Development of New Guinea Pidgin*. Canberra: Australian National University.

Mühlhäusler, P. 1985. Syntax of Tok Pisin. In: S. A. Wurm and P. Mühlhäusler (eds.), *Syntax of Tok Pisin*, 341–412. Canberra, Australian National University.

Romaine, S. 1992. *Language, Education and Development: Urban and Rural Tok Pisin in Papua New Guinea*. Oxford: Oxford University Press.

Sauerland, U., J. Andersen, and K. Yatsushiro. 2005. The plural involves comparison. In: S. Kepser and M. Reis (eds.), *Linguistic Evidence – Empirical, Theoretical, and Computational Perspectives* Berlin: Mouton de Gruyter.

de Saussure, F. 1993. *Third Course of Lectures on General Linguistics (1910-11)*. Oxford/New York: Pergamon Press.

Smith, G. P. 2002. *Growing up with Tok Pisin. Contact, Creolization and Change in Papua New Guinea's National Language*. London: Battlebridge Publications.

Thomas, D. D. 1955. Three analyses of the Ilocano pronoun system. *Word* 11, 204–208.

Verhaar, J. W. M. 1995. *Towards a Reference Grammar of Tok Pisin. An Experiment in Corpus Linguistics*. Honolulu: University of Hawaii Press.

Zeevat, H. and G. Jäger. 2002. A reinterpretation of syntactic alignment. In: D. de Jongh, H. Zeevat, and M. Nilsenova (eds.), *Proceedings of the 3rd and 4th International Symposium on Language, Logic and Computation.* Amsterdam: ILLC.

Pro-Drop and Impoverishment[*]

Gereon Müller

1. Introduction

It is often assumed that some notion of morphological richness plays a central role in the theory of pro-drop: In languages with sufficiently rich verbal ϕ-feature (person, number, gender) agreement morphology, pronominal arguments can (and, in some contexts, must) remain without phonological realization; in languages without such a rich verbal agreement morphology, pronominal arguments must be overtly realized. Focusing on subject pro-drop here and in what follows, this difference can be illustrated with evidence from Spanish and Italian on the one hand (languages with rich subject agreement morphology), and English and German on the other (languages without rich subject agreement morphology); see (1a,b) vs. (1c,d). Following Chomsky (1982), Rizzi (1986), and Grewendorf (1989), among many others, I assume that the pro-drop phenomenon does not involve post-syntactic deletion (cf. Perlmutter 1971), but an empty category *pro*. Such a non-overt pronoun *pro* is merged in the canonical position for subjects (in Spec of vP or within the VP, depending on its status as an external or internal argument), and undergoes Agree with T in the languages under consideration here (which rely on a nominative/accusative system of argument encoding), thereby ensuring nominative case and subject agreement; cf. Chomsky (1995); Chomsky (2001).[1]

---

* This paper is dedicated to Günther Grewendorf, who opened up a new world for me in my second semester when he came to Frankfurt University in 1984, and who got me (and Bernhard Rohrbacher) interested in pro-drop early on; see Müller and Rohrbacher (1989). For helpful comments of various kinds, I would like to thank Gisbert Fanselow, Jochen Geilfuß-Wolfgang, Fabian Heck, Gunnar Hrafn Hrafnbjargarson, Uwe Junghanns, Denisa Lenertova, Marga Reis, Ian Roberts, Bernhard Rohrbacher, Wolfgang Sternefeld, Juan Uriagereka, and Dieter Wunderlich, as well as audiences in Cambridge (UK), Leipzig, Tübingen, Potsdam, and Philadelphia. Last but not least, I am grateful to Patrick Brandt and Eric Fuß for their written comments.

1 See Grimshaw and Samek-Lodovici (1998) for an elaboration of the information-structural restrictions on overt vs. non-overt realization of subject argument pronouns in pro-drop languages. Non-pro drop languages may exhibit phenomena which at first sight resemble pro-drop constructions; compare, e.g., diary and cookbook contexts in English, or the topic drop phenomenon in German. I will disregard such phenomena throughout this paper. Furthermore, I will not have anything to say about constructions involving meteorological predicates, impersonal passives, subject inversion, and the like, where some languages may need overt expletive or 'quasi-argumental' pronouns, and others do not. It has been argued that these phenomenona may also involve *pro* (as argued, e.g., by Rizzi 1986 for Italian, Platzack 1987 for Icelandic, and Grewendorf 1989; Grewendorf 1990 for German), but there is also evidence against such a view; see, e.g., Haider (1988), Rohrbacher (1999). I will confine myself to clear cases of argumental pro-drop in what follows, and would like to leave open the question of whether *pro* can also show up in expletive or quasi-argumental contexts. Finally, I assume here that *pro* does not have to raise to SpecT in order to satisfy some EPP property, which brings the analysis closer to 'I-subject' approaches as in

(1) a. [$_{\text{TP}}$ *Hemos* [$_{\text{vP}}$ pro *trabajado todo el día*]].
have-3.PL worked all the day
b. [$_{\text{TP}}$ *Ha* [$_{\text{vP}}$ pro *cantato*]].
has-3.SG sung
c. **I think* [$_{\text{TP}}$ *have* [$_{\text{vP}}$ pro *worked all day*]].
d. **Ich denke, dass* [$_{\text{TP}}$ [$_{\text{vP}}$ pro *geschlafen habe*]].
I think that slept have-1.SG

Even though the hypothesis that morphological richness is involved in the licensing of argumental *pro* seems to be a natural one, and is widely accepted, it has proven extremely difficult to pin down. In fact, it seems that the notion of morphological richness relevant here is left somewhat vague in most of the relevant literature. Jaeggli and Safir (1989: 29-30) assume that pro-drop is possible in languages with morphologically uniform inflectional paradigms, where an inflectional paradigm counts as uniform iff it "has either only underived inflectional forms or only derived inflectional forms": The former option covers pro-drop languages like Spanish and Italian; the latter option covers pro-drop languages like Japanese or Chinese, which lack verbal ϕ-feature agreement completely; non-pro drop languages like English and French, which have bare-stem inflectional forms in their verbal paradigms, are correctly excluded. As noted by Rohrbacher (1999: ch. 5), this approach ignores the fact that base-stem forms can act as fully distinctive members of morphological paradigms (and are accordingly often analyzed as involving null morphemes in morphology that are accorded the same formal status as other, overt inflection markers). Rohrbacher proposes a slightly more elaborate approach according to which a language can have pro-drop if "in at least one number of one tense, the person features [1] and [2] are distinctively marked" (pp. 130, 247). Notwithstanding empirical differences between these approaches, it seems clear that they all do not straightforwardly exclude languages like German or Icelandic (which distinguish 1. and 2. person in paradigms) as pro-drop languages; but both languages lack subject argument *pro*. Furthermore, these approaches crucially presuppose a concept of inflectional paradigm that corresponds to the traditional notion adopted in reference grammars, but that is incompatible with recent developments in theoretical morphology. Here, paradigms are often viewed as epiphenomena, i.e., descriptive generalizations that principles of grammar cannot refer to by definition (e.g., this holds for all of the work carried out within Distributed Morphology; see Halle and Marantz 1993; Halle and Marantz 1994, Bobaljik 2002b, among many others); or they are viewed as abstract grammatical objects that bear little resemblance to the traditional reference grammar notion (compare, e.g., the notions of paradigm in Williams 1994, Wunderlich 1996, and Wiese 1999).

In view of this state of affairs, the main goal of the present paper is to argue for a concept of morphological richness underlying the theory of pro-drop that is based on recent morphological research and correctly derives the cross-linguistic distribution of subject argument *pro*. The central claim that I want to make here is that morphological

Borer (1986), Alexiadou and Anagnostopoulou (1998); however, not much depends on this in the present context.

richness should be captured not by looking at and counting distinctive forms of traditional paradigms (which I assume to be mere epiphenomena), but rather by invoking an abstract property of morphological inventories – more specifically, by relying on the concept of impoverishment developed in Distributed Morphology. For concreteness, I would like to suggest that *pro* cannot be licensed by T if T is subject to an impoverishment operation that leads to a neutralization of ϕ-features.

The argument consists of three steps. In section 2, I look at the morphological system of verb inflection in German. I develop a Distributed Morphology approach that makes crucial use of the classical morphological techniques of decomposition of person features and underspecification of inflection markers with respect to these features. In addition, it employs two mechanisms that only exist in Distributed Morphology: fission and impoverishment. Fission permits a *subanalysis* of markers. Even more importantly in the present context, impoverishment will be shown to derive *system-wide patterns of syncretism* (where syncretism is conceived of as systematic homonymy of inflection markers). In section 3, I make a first attempt to derive the absence of pro-drop from the presence of impoverishment, and I discuss the problems that arise with this view under standard conceptions of the morphology/syntax interface, where impoverishment is post-syntactic and *pro*-licensing is not. To solve this dilemma, a pre-syntactic concept of impoverishment is called for. In section 4, I argue that there is independent evidence for this: Given minimalist assumptions, post-syntactic vocabulary insertion as it is standardly assumed in Distributed Morphology should be replaced by pre-syntactic probe-driven Merge. In line with this, post-syntactic impoverishment should be reformulated as pre-syntactic impoverishment. In section 5, I present and justify the main claim of this paper, viz., that ϕ-feature impoverishment blocks the licensing of *pro*. This approach is then tested against evidence from Icelandic, Modern Irish, and Russian. Section 6 draws a conclusion.

2. German verb inflection

2.1 Background assumptions

Let me begin by sketching some basic assumptions of Distributed Morphology (see Halle and Marantz 1993; Halle and Marantz 1994, Harley and Noyer 2003). A core concept is that of *late vocabulary insertion*: Abstract functional morphemes like v and T (cf. Chomsky 1995) contain fully specified bundles of morpho-syntactic features in syntax; however, they do not yet contain phonological material.[2] The actual realization of these functional morphemes is brought about by post-syntactic insertion of vocabulary items – i.e., inflection markers – that pair phonological and (often underspecified) morpho-syntactic features. Affixal inflection markers presuppose complex head formation with functional and lexical morphemes. A simple assumption (and one that will do

2 It is unclear whether the same holds for lexical categories like V or N. Following Chomsky (2001: 11), I will here assume that it does not.

at least for present purposes) is that such complex heads are created by head movement in the syntax. This is shown schematically for V, v, and T in (2) (linear order irrelevant).

(2) $[_{TP} \ldots [T [_{v} V\ v\ ]\ T\ ]\ [_{vP} \ldots t_{v} [_{VP} \ldots t_{V} \ldots]]]$

The insertion of vocabulary items into functional heads takes place in accordance with the Subset Principle, which can be defined as in (3) (see Halle 1997, among many others).

(3) *Subset Principle*:
A vocabulary item *V* is inserted into a functional morpheme *M* iff (i) and (ii) hold:
(i) The morpho-syntactic features of *V* are a subset of the morpho-syntactic features of *M*.
(ii) *V* is the most specific vocabulary item that satisfies (i).

The Subset Principle contains two clauses: The first one ensures that an inflection marker can only be inserted if it is *compatible* with the morpho-syntatic feature specification of the syntactic context, and the second one states that to be inserted, an inflection marker must be the most *specific* marker among those that are compatible (see also Wunderlich 1996). Specificity can be defined as follows (see Lumsden 1992, Noyer 1992, Wiese 1999 for versions of this concept of specificity).

(4) *Specificity of vocabulary items*:
A vocabulary item V_i is more specific than a vocabulary item V_j iff there is a class of features F such that (i) and (ii) hold.
(i) V_i bears more features belonging to F than V_j does.
(ii) There is no higher-ranked class of features F′ such that V_i and V_j have a different number of features in F′.

Thus, to determine the degree of specificity of a vocabulary item, the rank of its morpho-syntactic features on an independently available feature hierarchy must be determined; as in Optimality Theory (see Prince and Smolensky 2004), quality is more important than quantitiy. The (partial) feature hierarchy that I assume here is given in (5). It follows that if there are two inflection markers that are compatible with a given syntactic context, but one bears, say, a tense feature, and the other one does not, then the Subset Principle will force insertion of the former vocabulary item.

(5) Tense > Number > Person

A further important assumption is that morpho-syntactic specifications of syntactic contexts (i.e., of functional morphemes, in the cases at hand) can be affected by deletion operations that take place after syntax proper (so they do not affect syntactic operations like agreement), but before morphological vocabulary insertion (see Bonet 1991, Halle and Marantz 1993; Halle and Marantz 1994, Bobaljik 2002b, Frampton 2002). Such *impoverishment* rules neutralize differences between syntactic contexts in morphology; this may effect a "retreat to the general case" because more specific morphological markers may not fit anymore into impoverished contexts, and will therefore lose out to less specific markers that still do.

Finally, I adopt the following notion of *fission* (see Noyer 1992, Frampton 2002, Müller 2005; the notion of fission in Halle and Marantz 1993 is very different).

(6) *Fission*:
If insertion of a vocabulary item V with the set of morpho-syntactic features β takes place into a fissioned morpheme M with the set of morpho-syntactic features α, then α is split up into β and α–β, such that (a) and (b) hold:
a. α–β is available for further vocabulary insertion.
b. β is not available for further vocabulary insertion.

Thus, insertion of the most specific vocabulary item into a fissioned functional morpheme leaves those features of the functional head which are not matched by features of this (underspecified) vocabulary item (and thereby discharged) available for further vocabulary insertion, and so on, until there is no inflection marker left that can be inserted in accordance with the Subset Principle. This way, a subanalysis of complex inflection markers becomes possible, and agglutinative-like structures can emerge in morphological systems of the Indo-European type. I assume that the functional morpheme T is subject to fission in German (and that v is not overtly realized).

2.2 Analysis

Let me now develop an account of the system of verb inflection in German (see Müller 2006 for a somewhat more comprehensive exposition). Paradigms of the three main inflection classes (weak, strong, and suppletive inflection) are given in (7).

(7) a. *Weak conjugation* *glauben* ('believe') b. *Strong conjugation* *rufen* ('call') c. *Suppletive conjugation* *sein* ('be')

	Present	Past	Present	Past	Present	Past
1.SG	glaub-e	glaub-te	ruf-e	rief	bin	war
2.SG	glaub-st	glaub-te-st	ruf-st	rief-st	bi-st	war-st
3.SG	glaub-t	glaub-te	ruf-t	rief	is-t	war
1.PL	glaub-en	glaub-te-n	ruf-en	rief-en	sind	war-en
2.PL	glaub-t	glaub-te-t	ruf-t	rief-t	seid	war-t
3.PL	glaub-en	glaub-te-n	ruf-en	rief-en	sind	war-en

These paradigms exhibit four instances of syncretism that I take to be non-accidental, and in need of an explanation. First, the inflection markers for 1.SG.PAST and 3.SG.PAST contexts are identical in all paradigms. Second, the inflection markers for 1.PL and 3.PL contexts are identical in all tenses and paradigms. Whereas these two instances of syncretism are straightforward and evidently systematic, the remaining two may at first sight look less obvious: The marker for 3.SG.PRES is *t*, and the very same markers shows up in 2.PL.PRES contexts in the two main inflection classes in (7a,b). Finally, the marker for 2.SG contexts (*st*) is identical throughout to the marker for 3.SG.PRES contexts (*t*), except for the initial *s*.

Of these four cases of syncretism, only the first two have been accounted for in the literature (see Wiese 1994, Wunderlich 1996, Eisenberg 2000, Frampton 2002); and only Frampton derives them as the general, system-wide properties that they arguably are (rather than as the result of individual inflection marker specifications of an arbitrary nature). To the best of my knowledge, the remaining two syncretisms have not yet been systematically addressed.[3]

The analysis I would like to propose is based on three building blocks: First, standard morpho-syntactic features are decomposed into combinations of more abstract primitive features. A cross-classification of these features yields the standard morpho-syntactic categories; and underspecification of inflection markers with respect to these features captures natural classes of categorizations and thereby accounts for cases of syncretism (see Bierwisch 1967, extending earlier work by Jakobson 1962). More specifically, I assume that the person features 1., 2., 3., and 1.INCL(USIVE) are to be decomposed into combinations of the more primitive features [±1], [±2], as shown in (8) (see Noyer 1992, Wiese 1994, Wunderlich 1996, Frampton 2002). It follows that (i.a.) 1. and 3. form a natural class (characterized by the underspecified information [–2]), as do 2. and 3. (characterized by [–1]). Consequently, we expect syncretism to affect 1./2. and 2./3. categorizations, but not 1./2. or 1.INCL/3. categorizations.[4]

(8) *Decomposition of person features*:
  a. 1 = [+1,–2]
  b. 1.INCL = [+1,+2]
  c. 2 = [–1,+2]
  d. 3 = [–1,–2]

Second, fission ensures that *st*, which at first sight looks like a single inflection marker without internal structure, can emerge as the combination of two vocabulary items /s/ and /t/, inserted in that order. Third, impoverishment also refers to natural classes of persons and can thus derive system-wide syncretism patterns. For concreteness, suppose that the two impoverishment rules in (9) apply to T morphemes in German (also see Frampton 2002).

---

3. Compare, however, the remarks in Bierwisch (1961: 62-66) on the third syncretism involving marker homonymy in 3.SG.PRES and 2.PL.PRES contexts. Also relevant are the empirical results reported in Fanselow and Frisch (2005), according to which DP coordinations with *oder* ('or') that involve agreement with the same finite verb and exhibit this syncretism are judged grammatical by speakers (see (i-a)), on a par with coordination examples that involve an instance of syncretism that is uncontroversially systematic (see (i-b)), but in contrast to minimally different coordination examples that do not involve a syncretism in the verb form (see (i-c)).

(i) a. *Er oder ihr wohnt in Frankfurt.*
  he or you.PL live-3.SG.PRES/2.PL.PRES in Frankfurt
 b. *Wir oder sie wohnen in Frankfurt.*
  we or they live-1.PL.PRES/3.PL.PRES in Frankfurt
 c.* *Ich oder du wohne/wohnst in Frankfurt.*
  I or you.SG live-1.SG.PRES/live.2.SG.PRES in Frankfurt

4 Of course, 1.INCL is a categorization that is not actually used in Indo-European languages; but the feature system is designed to be more general.

(9) *Impoverishment rules for German verb inflection*:
a. [±1] → Ø/[–2,–pl,+past]__
b. [±1] → Ø/[–2,+pl]__

Both impoverishment rules neutralize ϕ-feature distinctions: rule (9a) ensures that it is a defining property of the system of verb inflection in German that 1. and 3. SG.PAST contexts cannot have a different morphological realization, whatever the inflection class otherwise looks like; this is so because the sole feature that differentiates 1. and 3. person is deleted ([±1]). Similarly, rule (9b) has the effect of systematically neutralizing the syntactic difference between 1. and 3. person in the plural in general. Hence, the only domains where 1. and 3. person can be (and regularly are) distinguished are SG.PRES contexts.[5]

Given these assumptions, the set of inflection markers for T heads in (10) accounts for all instances of syncretism in German verb inflection (abstracting away from suppletion here and in what follows).[6]

(10) *Vocabulary items*:
a. /te/ ↔ [+past,–strong]
b. /s/ ↔ [+2,–pl]
c. /n/ ↔ [–2,+pl]
d. /t/ ↔ [–1]
e. /(e)/ ↔ [ ]

The most specific marker is the past tense marker for the weak inflection class, /te/. It is inserted wherever it is compatible with the fully specified syntactic feature context in T (cf. (3)), discharging the features [+past] and [–strong] in accordance with (6). The two next specific vocabulary items are /s/ and /n/ (since they bear number features, unlike the remaining items): /s/ is a singular marker for 2. person, and /n/ is a plural marker for non-2. person – an option made possible by person feature decomposition (more specifically, the availability of the feature [–2]) and underspecification. These two markers are then inserted wherever possible, discharging the matching features in T. The next specific marker is /t/. It is then inserted in all contexts in which the specification [–1] is available. Since the more specific markers do not bear the feature [–1], their insertion cannot possibly have led to a discharge of [–1]. However, the case is different with the impoverishment rules in (9), which both delete [–1] (and [+1]) in [–2] contexts, except for the SG.PRES domain. Consequently, /t/ is inserted in all 2. and 3. person contexts,

5 The effects of impoverishment rules can be imitated by rules of referral as they have been employed in recent Word-and-Paradigm approaches (see Stump 2001 and references cited there), and by optimization procedures (see Prince and Smolensky 2004) that give rise to optimal unfaithful feature specifications (with non-fatal MAX violations), as they have beeen argued for in recent versions of Minimalist Morphology (see, e.g., Wunderlich 2004). However, of these three means to derive system-wide patterns of syncretism, impoverishment seems to be the most restrictive one since it is standardly taken to be only able to bring about feature *deletion*. In contrast, rules of referral can effect feature deletion, feature insertion, feature change, and much more; and optimization procedures based on ranked and violable constraints can bring about feature deletion (violations of MAX constraints), feature insertion (violations of DEP constraints), and feature change (violations of IDENT constraints).

6 The / /-notation is supposed to indicate that the markers are abstract items that may undergo certain (morpho-) phonological changes.

except for those 3. person contexts that are subject to pre-morphological impoverishment. As a result, the /t/ in *Er glaub-t* ('He believes') and the /t/ in *Du glaub-s-t* ('You.SG. believe') emerge as one and the same marker. Finally, the least specific (in fact, radically underspecified) vocabulary item /(e)/ acts as a default marker that can in principle be inserted anywhere. I assume that such radically underspecified markers are inserted iff there is no other marker in the functional morpheme.[7] Furthermore, the vocabulary item /(e)/ is somewhat more abstract than the other markers in (10) in the sense that it requires a minimal indication of deviation from the present tense stem. Hence, /(e)/ → Ø whenever there is stem alternation in past tense contexts (strong verbs); and /(e)/ → *e* in present tense contexts when there is no other marker. The paradigm of verb inflection in German that arises in this way is shown in (11).

(11) *Vocabulary insertion into impoverished Ts in German*

T	[–past] [–strong]	 [+strong]
[+1,–2,–pl]	/e/	/e/
[–1,+2,–pl]	/s/-/t/	/s/-/t/
[–1,–2,–pl]	/t/	/t/
[~~+1~~,–2,+pl]	/n/	/n/
[–1,+2,+pl]	/t/	/t/
[~~–1~~,–2,+pl]	/n/	/n/

T	[+past] [–strong]	 [+strong]
[~~+1~~,–2,–pl]	/te/	Ø
[–1,+2,–pl]	/te/-/s/-/t/	/s/-/t/
[~~–1~~,–2,–pl]	/te/	Ø
[~~+1~~,–2,+pl]	/te/-/n/	/n/
[–1,+2,+pl]	/te/-/t/	/t/
[~~–1~~,–2,+pl]	/te/-/n/	/n/

This may suffice as an exposition of the gist of German verb inflection under the present analysis. It is worth emphasizing that the two impoverishment rules in (9) play a fundamental role in this approach. They account for the fact that there are system-wide patterns of syncretism in (7); and they are crucial if a maximally simple inventory of inflection markers in (10) is to be posited, with only one meaning (i.e., feature specification) for any given form (i.e., phonological information). What is more, the approach in terms of impoverishment is corroborated by diachronic evidence. Identity of inflection markers for 1.PL.PAST and 3.PL.PAST (as well as for 1./3.SG.PAST forms; see below) is also a property of the Middle High German system of verb inflection; however, identity of 1.PL.PRES and 3.PL.PRES is not yet. This is accounted for if ϕ-feature impoverishment was confined to past contexts in Middle High German, and has then been extended to present plural contexts on the way to Modern German. Interestingly, the marker for 3.PL.PRES contexts was /n/-/t/ in Middle High German, and this is exactly what we expect it to be under the present analysis: If there is no impoverishment rule deleting [–1] in 3.PL contexts, /t/ is predicted to show up. Thus, we find *hëlfen* (1.PL.PRES, 'we help') alongside *hëlfent* (3.PL.PRES, 'they help').

7 This also precludes unwanted iteration of /(e)/-insertion, as it would otherwise be wrongly predicted to be possible under (6); also compare Wunderlich's (1996) Monotonicity Principle.

3. A failed attempt

Returning to the issue of pro-drop, I would like to contend that it is the existence of impoverishment rules of the type in (9) that makes argumental pro-drop impossible in German and other languages without licensing of subject *pro* (i.e., such ϕ-feature impoverishment is responsible for the lack of sufficient morphological richness). However, if nothing else is said, this reasoning leads to a dilemma.

As has been observed by Bobaljik (2002a), properties of the morphological inventory cannot be held responsible for the operation of V-to-T movement in syntax if inflectional morphology is post-syntactic: The hypothesis according to which V-to-T movement takes place if a language has a morphological system of verbal inflection that is sufficiently rich (see Roberts 1993, Vikner 1997, Holmberg and Platzack 1995, and Rohrbacher 1999 for various suggestions that differ in detail but converge on the general idea) must therefore be given up in a Distributed Morphology approach. If there is any synchronically relevant correlation at all (see Alexiadou and Fanselow 2000), it must go in the other direction: Rich verbal morphology can be a *reflex* of movement, but not the *reason* for it. Bobaljik (2002a) argues that this is the correct result for V-to-T movement.

Evidently, the same situation occurs with licensing of *pro*: Assuming post-syntactic morphology, *pro*-licensing cannot be determined by morphological properties (like size of the morphological inventory, or, as I want to argue here, presence or absence of ϕ-feature neutralizing impoverishment operations) because these properties are not yet visible at the point of the derivation where they would be needed. Consequently, given a post-syntactic morphology, there are only two possibilites as far as the interaction with pro-drop is concerned: The first possibility is that pro-drop is a syntactic phenomenon which is independent of the properties of morphological inventories; see, e.g., Grimshaw and Samek-Lodovici (1998).[8] The second possibility is that pro-drop is a post-syntactic (PF) phenomenon (see, e.g., Adger 2003, Holmberg 2004). Both options strike me as worth pursuing in principle. Still, in what follows I presuppose that a case can be made for the standard view that pro-drop is determined by the morphological properties of verb inflection systems, and that it is a syntactic phenomenon involving the licensing of an argumental empty category *pro* (see Rizzi 1986 and Jaeggli and Safir 1989, among many others). How, then, can the dilemma sketched above be solved?

I suggest that it can be solved if we give up the assumption that inflectional morphology is post-syntactic, and make the opposite assumption that inflectional morphology is pre-syntactic. If so, impoverishment will be pre-syntactic too, and the relevant morphological information (i.e., whether or not T has been affected by ϕ-feature impoverishment) is available when the syntactic decision on *pro*-licensing must be made. It then remains to be shown that a pre-syntactic approach to inflectional morphol-

---

8 More generally, this would seem to be the standard approach in optimality-theoretic syntax; see also Vikner (2001a) on V-to-T movement. Alternatively, optimality-theoretic analyses may treat morphological marking as a reflex of syntax (see, e.g., Legendre et al. 1998 on the general issue, and Müller 2002 on the correlation of morphological richness and free word order). However, the idea of morphology-driven syntax seems to be much harder to reconcile with basic optimality-theoretic assumptions (but see Vikner 2001b for a recent attempt).

ogy that incorporates essential features of Distributed Morphology (underspecification, fission, and, especially, impoverishment) is both tenable and independently supported. I address this issue in the following section.

4. Pre-syntactic morphology

Given optimal design considerations (see Chomsky 2001; Chomsky 2005), a system of inflectional morphology that relies on post-syntactic vocabulary insertion looks like an imperfection. A first potential conceptual problem is that inflection markers are split into two separate units (f-morpheme and vocabulary item). A second one is that the insertion operation as such is clearly very different in nature and scope from the few elementary operations that are adopted in the minimalist program (Merge and Agree, perhaps also Move); it seems that the null hypothesis should be that the elementary operations of syntax are also active in morphology, and specific morphological operations like vocabulary insertion can be dispensed with. Finally, a third conceptual problem with late insertion is that it cannot possibly satisfy two highly general and independently motivated constraints (see Chomsky 1995; Chomsky 2000; Chomsky 2001) at the same time: the Inclusiveness Condition, according to which new elements (like features) cannot be introduced in the course of the derivation, and the Legibility Condition, from which one can derive that morpho-syntactic features can be present in some component of grammar only if they are interpretable in this component. This dilemma is created by features that are relevant in morphology but irrelevant (i.e., uninterpretable) in syntax, viz., inflection class features (like [±strong] in the present analysis of German verb inflection). As argued in detail in Alexiadou and Müller (2005), the Inclusiveness Condition is violated if class features are absent in syntax and added afterwards in a post-syntactic approach to inflection; and the Legibility Condition is violated if syntactically uninterpretable class features are present in syntax but only used afterwards, in the morphological component.

In view of this state of affairs, I would like to conclude that a morphological component of grammar that meets minimalist requirements should be pre-syntactic rather than post-syntactic. However, there does not seem to be a good reason to give up successful concepts like underspecification, fission, and impoverishment under this changed perspective. With respect to underspecifcation, the case is simple because the issue is orthogonal to the timing of inflection (pre-, post-, or intra-syntactic); accordingly, there are pre-syntactic approaches to morphology that embrace this concept (see, e.g., Wunderlich 1996). With respect to fission and impoverishment, the issue turns out to be slightly more involved. The first outlines of a pre-syntactic morphology that satisfies minimalist goals have been sketched in Alexiadou and Müller (2005). I will use this approach as a point of departure, and then integrate the concepts of fission and impoverishment into it.

The main assumption is that (inflectional) morphology takes place in a generative component of grammar that follows the lexicon but precedes syntax. Assuming the minimalist program, a natural assumption then is that inflectional morphology is located in the *numeration* (cf., e.g., Chomsky 1995; Chomsky 2000 for this notion).[9] Pre-syntactic inflection in the numeration is assumed to be driven by (syntactically) uninterpretable material, viz., inflection class features. As regards verb inflection, suppose that inflection class features, which are inherent to a stem, are borne by V (note that these features may in principle be trivial if there is only one inflection class in the language; see Aronoff 1994). In contrast, non-inherent features relevant for verb inflection in German (tense features, mood features, and, in particular, ϕ-features) are freely instantiated on T as T enters the numeration. Together, these sets of fully specified features provide the context. An uninterpretable class feature on V then acts as a probe; it triggers an Agree operation with a matching inflection marker (i.e., a compatible inflection marker, which does not have to be equipped with a class feature itself), which leads to Merge of V and the inflection marker (selection requirements ensure that it is V rather than T that is affected by Merge of the inflection marker). All of the inflection marker's features (including underspecified ϕ-features) are inherent. The effects of the Subset Principle follow if we make the (standard) assumption that Agree requires feature matching (see the compatibility requirement in (3i)), and the more specific assumption that Agree also requires a maximization of matching effects (see Chomsky 2001: 15). We may understand the latter condition in such a way that as many high-ranked features are affected by an Agree operation as possible, where quality outranks quantity (see the specificity requirement in (3ii), (4)). Thus, the most specific compatible inflection marker is used in pre-syntactic Agree-driven Merge operations that bring about verbal inflection.[10] We can then further assume that syntactically uninterpretable features of V (inflection class), and all morpho-syntactic features of the inflection marker (typically underspecified – hence, syntactically uninterpretable – information), are deleted in morphology; syntactically interpretable features of V and all features of T remain unaffected. Finally, the items of the numeration (inflected V, v, T, ...) enter syntax, bearing only fully specified and syntactically interpretable morpho-syntactic features.

At this point, the question arises how fission and impoverishment can be integrated into the approach. As concerns the former concept, I would like to suggest that it is the

9 Chomsky (2000, 2001) assumes that numerations are split up into lexical *subarrays*; these contain the lexical material that shows up in a phase (vP or CP). As will become clear shortly, morphological operations cannot be confined to the subarray under present assumptions, unless we assume that T and V may belong to the same subarray (hence, the same phase; see Richards 2004 for such a proposal). The reason is that T and V will both emerge as relevant for verb inflection. Consequently, the local domain for pre-syntactic morphology must either be the complete numeration, or, more likely, a subpart of it that corresponds to Grimshaw's (2000) extended projection of V (including T). The latter approach might eventually emerge as more elegant because it automatically ensures that a verb is inflected for the features of the T category that is its clause-mate in syntax (and not some other T node in a higher or lower clause). However, since nothing depends on this issue in the present context, I will continue to talk about numerations (rather than modified lexical subarrays) as the domains in which pre-syntactic morphology takes place.

10 Where do the inflection markers that are required by class feature probes come from? There are at least two options. One option is that they may simply be taken out of the lexicon when the need arises. This implies that operations in the numeration can still have access to the lexicon (and carry out search in the lexicon, selecting the most specific compatible item). Alternatively, the complete inventory of inflection markers of a given type (e.g., realization for V-T in German) may enter the numeration as a single set, and the inflection operations picks out one item of this set. This implies that not all material that enters a numeration will have to be used in the syntactic derivation.

defining property of fission that a class feature probe that has triggered an Agree operation does not delete immediately, but may trigger further Agree operations, and only deletes when no further Agree operation is possible anymore.[11] That leaves impoverishment. Impoverishment can be viewed as a pre-syntactic operation, provided that the features that are affected are invisible for morphology, but not for subsequent syntactic operations (e.g., a [–1] feature that is affected by impoverishment in German must still be visible in syntax, where 1. and 3. person *can* be distinguished in all numbers and tenses). On this view, impoverishment of T applies in the morphological component after T's non-inherent features have been added, but before Agree with an inflection marker is carried out. Thus, suppose that impoverishment *marks* features as morphologically unaccessible, but it does not actually *delete* them, and they remain accessible in syntax.[12] As an example of how this system of pre-syntactic verb inflection in German works, consider the following derivation of *glaubten* ('believe.3.PL.PAST') in the numeration:[13]

(12) a. Selection of *glaub*: {V, [–strong]} from the lexicon.
   b. Selection of T from the lexicon; addition of fully specified ϕ- and tense features: T: {[–1,–2,+pl,+past]}
   c. Application of impoverishment rule (9b) to T:
      T: {[–1,–2,+pl,+past]} → T: {[~~–1~~,–2,+pl,+past]}
   d. Agree-driven Merge of (matching, most specific) /te/ with V:
      glaub: {V, [–strong]}, T: {[~~–1~~,–2,+pl,+past]} + /te/: {[+past,–strong]} →
      glaub-te: {V, [–strong]}, T: {[~~–1~~,–2,+pl,+past]}
   e. Agree-driven Merge of (matching, next specific) /n/ with V:
      glaub-te: {V, [–strong]}, T: {[~~–1~~,–2,+pl,+past]} + /n/: {[–2,+pl]} →
      glaub-te-n: {V}, T: {[~~–1~~,–2,+pl,+past]}
   f. T and the inflected V enter syntax, with all remaining morpho-syntactic features (whether affected by impoverishment, affected by Agree, or not affected at all) accessible to syntactic operations.

To sum up so far, we have two main results. First, ϕ-feature impoverishment plays an important role in verb inflection (section 2). And second, ϕ-feature impoverishment can and, given basic tenets of the minimalist program, arguably should be conceived of as a pre-syntactic operation (section 4). With these results in place, we are now in a position to give a principled answer to the question why some languages exhibit subject argument pro-drop, and others do not.

---

11 In syntax, a similar approach suggests itself for multiple *wh*-movement in Bulgarian (assuming that such multiple movement is a homogenous phenomenon, but cf. Bošković 2002): The *wh*-probe feature on C does not delete after checking the first *wh*-phrase; it deletes when there is no further *wh*-phrase left.

12 Note that this is parallel to Chomsky's (1995) distinction between deletion and erasure.

13 Impoverished features are struck through; discharged features fission contexts are underlined.

5. Proposal

As indicated above, I suggest that whether or not subject argument *pro* is licensed by T depends on whether or not T is affected by a certain type of impoverishment operation in the numeration.[14] More specifically, I would like to advance the following generalization:[15]

(13) *Pro Generalization*:
An argumental *pro* DP cannot undergo Agree with a functional head α if α has been subjected (perhaps vacuously) to ϕ-feature neutralizing impoverishment in the numeration.

The ϕ-feature that is relevant in the present context is the person feature; one might speculate that it is cross-linguistically only this feature which is relevant for determining pro-drop (see Rohrbacher 1999). We can say that an impoverishment rule is "ϕ-feature neutralizing" in the sense of (13) when a distinction between two ϕ-categorizations – e.g., 1. person ([+1,–2]) vs. 3. person ([–1,–2]) cannot be made anymore. As a consequence of (13), only a system-defining syncretism (i.e., one that is derived by impoverishment) can preclude the licensing of *pro* by T. In contrast, syncretisms that are solely due to underspecification of particular inflection markers, as well as accidental homonomies, cannot block *pro* licensing.

With respect to German, it is now clear why the language does not exhibit subject argument *pro* (see (1d)) even though at first glance it seems to have a fairly rich system of verb inflection (see (7)): The reason is that German has the two impoverishment rules in (9), which neutralize ϕ-feature distinctions (and only one of the rules would have been sufficient to preclude pro-drop). Given that each T head is affected by these two rules in the numeration (perhaps vacuously so, e.g., if T is specified as [+2], or as [–pl, –past]), there is no T which esacpes the weakening effect of impoverishment, and subject argument *pro* is correctly predicted to be impossible throughout in this language.

In the remainder of this paper, I will briefly discuss three languages that are relevant for the main hypothesis, for different reasons. First, Icelandic is renowned among the Germanic languages for its rich inflectional morphology; still, it does not permit subject argument *pro*. Second, Modern Irish presents the reverse situation: The system of verb inflection is fairly poor, but *pro* is licensed. Finally, I discuss Russian, which raises interesting questions concerning both pro-drop and impoverishment.

---

14 Note that this information can be checked locally for every single T head in the numeration, which avoids a further problem typically faced by paradigm-based approaches to morphological richness and pro-drop, viz.: How can a global property of the language (richness of verb inflection) be encoded as a local property of the actual head that licenses *pro*? Note also that this reasoning implies that whereas the effect of a pre-syntactic impoverishment operation is invisible in syntax, the fact that the operation has applied is not: a T item to which (ϕ-feature neutralizing) impoverishment has applied in the numeration is a special T item in syntax.

15 This generalization may ultimately follow from more basic principles, but I will not pursue this matter here.

5.1 Icelandic

Subject argument pro-drop is not possible in Icelandic (except for questionable instances of pro-drop with meteorological predicates); see (14) (cf. Platzack 1987, Holmberg and Platzack 1995, and Rohrbacher 1999).

(14) a. *Hann dansar.* b. *pro *dansar.*
he dance-3.SG dance-3.SG

We are thus led to expect that Icelandic verb inflection is not morphologically rich. However, a superficial look at the actual paradigms does not confirm this expectation. Consider, e.g., the paradigms in (15) (see Kress 1982), which illustrate three arbitrarily selected inflection classes (two weak, one strong); all conjugations employ a rich inventory of inflection markers.

(15) a. *Weak conjugation, class 1* *krefja* ('demand') b. *Weak conjugation, class 4* *dansa* ('dance') c. *Strong conjugation, class 3* *sleppa* ('slip')

	Present	Past	Present	Past	Present	Past
1.SG	kref	krafði	dansa	dansaði	slepp	slapp
2.SG	krefur	krafðir	dansar	dansaðir	sleppur	slappst
3.SG	krefur	krafði	dansar	dansaði	sleppur	slapp
1.PL	krefjum	kröfðum	dönsum	dönsuðum	sleppum	sluppum
2.PL	krefjið	kröfðuð	dansið	dönsuðuð	sleppið	sluppuð
3.PL	krefja	kröfðu	dansa	dönsuðu	sleppa	sluppu

However, under the perspective adopted here, the decisive question is not whether Icelandic verb inflection uses a rich set of morphological markers, but rather whether there are system-wide patterns of syncretism that can be traced back to pre-syntactic impoverishment operations. And indeed, there are: On the one hand, the inflection markers for 1.SG.PAST and 3.SG.PAST contexts are identical in all inflection classes; i.e., the situation is exactly as in German in this respect. Accordingly, we can postulate the impoverishment rule in (16a), which deletes [±1] in SG.PAST contexts and thereby renders 1. and 3. person indistinguishable in these environments (see Frampton 2002). On the other hand, it turns out that 2.SG.PRES and 3.SG.PRES contexts also have identical morphological realizations in all conjugations. This, too, can be captured straightforwardly by postulating a simple impoverishment rule because 2. person and 3. person form a natural class (captured by [–1]) under the decomposition of person features in (8). Thus, (16b) deletes [±2] in SG.PRES contexts; as a result, there can be no morphological differentiation between 2. and 3. person in this environment.[16]

(16) *Impoverishment rules for Icelandic verb inflection*:
a. $[\pm 1] \rightarrow \emptyset / [-2, -\text{pl}, +\text{past}]\_\_$

[16] The same impoverishment rule may also take effect in Brazilian Portuguese (with identity of markers for 2. and 3. person) vs. European Portuguese (with two different markers for 2. and 3. person), and account for the rapid decline of pro-drop in the former. See Rohrbacher (1999: 246-250) for discussion.

b. [±2] → Ø/[–1,–pl,–past]__

As a consequence of these impoverishment rules affecting all T heads in the numeration, subject argument *pro* cannot be licensed in Icelandic, despite its seemingly "rich" verb inflection. Note that this conclusion is independent of the actual feature specifications for inflection markers that underlie the paradigms in (15).[17]

More generally, we can note that all Germanic languages, beginning with the earliest records, exhibit what looks like a system-wide syncretism in 1. and 3.SG.PAST contexts. Consequently, the prediction would seem to be that they all do not permit subject argument pro-drop. Whereas this conclusion is fairly uncontroversial with regard to the modern Germanic languages, there is some disagreement as to whether or not subject argument pro-drop is possible in the older Germanic languages; in particular, the question arises for Gothic, Old Norse, Old High German, and Old English. In these languages, there are cases where subject argument pronouns can be left without phonological realization (see Abraham 1991, Gelderen 2000, and Axel 2004); however, the distribution of these instances of argument omission is quite different from what is otherwise known for subject argument pro-drop, particularly so in Old High German and in Old English. I will have to leave this issue unresolved here, the simple options of analysis at this point being to assume (i) that *pro* is not licensed in one or more of the languages under consideration, and/or (ii) that the general nature of the 1./3.SG.PAST syncretism has not yet become manifest in a system-defining impoverishment rule in one or more of the relevant languages. Still, in this context it is instructive to note that the domain in which system-wide patterns of syncretism can be found in verb inflection grows steadily from the earliest records in Gothic and Old High German (1./3.SG.PAST realizations are identical) through Middle High German (1./3.SG.PAST and 1./3.PL.PAST realizations are identical) to Modern German (1./3.SG.PAST, 1./3.PL.PAST, and 1./3.PL.PRES realizations are identical); in the present approach, this corresponds to a proliferation of impoverishment rules. Concurrently, subject argument omission becomes more and more restricted.

5.2 Modern Irish

Modern Irish exhibits subject argument pro-drop; cf. (17) (from McCloskey and Hale 1984: 488).

(17)	*Dá*	*gcuirfeá*	pro	*isteach*	*ar*	*an*	*phost sin*	*gheobhfá*	pro	*é.*
	if	put-2.SG.COND		in	on	that	job	get-2.SG.COND		it

'If you applied for that job, you would get it.'

[17] I do not attempt to provide a full-fledged analysis of Icelandic verb inflection here; suffice it to say that the concepts of underspecification and fission strongly suggest themselves for this system, too. (Consider, e.g., the distribution of theme vowels in inflectional endings – an alternation like /u/-/r/, /a/-/r/, /i/-/r/ in 2. person contexts clearly argues for a subanalysis of the markers, and thus lends itself to an approach in terms of fission.)

However, the system of verb inflection does not qualify as morphologically rich under standard assumptions, and this poses a problem for approaches in which the number of distinct inflection markers in paradigms determines whether or not subject argument pro-drop is possible (see McCloskey and Hale 1984 and Jaeggli and Safir 1989). Consider the conditional paradigm and the present indicative paradigm of the verb *cuir* ('put') in (18), which contain so-called analytic and synthetic forms; only the latter permit subject pro-drop. The analytic forms are *chuirf-eadh* and *chuir-eann*, respectively. The conditional paradigm has three different synthetic forms, the indicative paradigm only one. Thus, both German and Icelandic have a larger inventory of verb inflection markers.

(18) a. *Conditional conjugation*
cuir ('put')

1.SG	chuirf-inn
2.SG	chuirf-ea
3.SG.MASC	chuirf-eadh
3.SG.FEM	chuirf-eadh
1.PL	chuirf-imis
2.PL	chuirf-eadh
3.PL	chuirf-eadh

b. *Indicative conjugation*
cuir ('put')

1.SG	chuir-im
2.SG	chuir-eann
3.SG.MASC	chuir-eann
3.SG.FEM	chuir-eann
1.PL	chuir-eann
2.PL	chuir-eann
3.PL	chuir-eann

Modern Irish verb inflection raises several problems for morphological analysis. For instance, it is not quite clear whether the analytic forms in (18) should be treated on a par with the synthetic forms in morphology, as involving *radically* underspecified default markers (on this general view, the markers cannot be *partially* underspecified with respect to ϕ-features because the occurrences of the analytic forms do not form a natural class); whether we are dealing with genuine paradigmatic gaps here that are filled by external forms; or whether some kind of blocking is involved. However, even assuming that the first analysis is on the right track, i.e., that the syncretism in (18) is due to radical underspecification in both cases, it seems clear that there are no general, system-defining impoverishment rules involved that would give rise to trans-paradigmatic patterns. Consequently, the availability of subject argument pro-drop with the synthetic forms is expected under present assumptions, even though "Irish is not a language which is in any general sense rich in its system of person-number marking morphology for verbs" (McCloskey and Hale 1984: 492).[18]

[18] More must eventually be said about the absence of pro-drop with the analytic forms; but any such analysis will have to clarify the morphological questions just mentioned in the main text first.

5.3 Russian

Consider the present tense paradigm of Russian verb inflection in (19).

(19) *Present tense conjugation:*
čitat' ('read')

1.SG	čita-ju	1.SG	čita-em
2.SG	čita-eš'	2.SG	čita-ete
3.SG	čita-et	3.SG	čita-jut

There is no syncretism here; hence, there can be no impoverishment rule underlying the Russian system of verb inflection in the present tense. Based on this evidence, we may expect subject argument pro-drop to occur in Russian. And indeed, relevant examples are typically well formed; see, e.g., (20).

(20) pro *ponimaju* [$_{\text{CP}}$ *čto* [$_{\text{TP}}$ pro *imeeš'* *v vidu idti v koncert*]].
understand-1.SG.PRES that have-2.SG.PRES in mind go to concert

Accordingly, it has often been suggested that Russian permits subject argument *pro* licensing (see, e.g., Růžička 1986, Müller 1988, Demjjanow and Strigin 2000, Perlmutter and Moore 2002). However, there is also a different view that holds that subject argument *pro* is not possible in Russian (see, e.g., Franks 1995: ch. 7, Avrutin and Rohrbacher 1997, and Junghanns 2005); on this latter view, instances of subject argument omission as in (20) involve contextually licensed ellipsis. Among the main reasons for not assuming Russian to have regular subject *pro* is the fact that "pro-drop in Russian is subject to discourse conditions that make it much less common than pro-drop in Italian or Spanish" (see Perlmutter and Moore 2002: 632), or, for that matter, in other Slavic languages like Czech or Polish; in particular, Russian differs from some other pro-drop languages in that overt subject pronouns can be unmarked in non-emphatic contexts. Furthermore, as noted by Růžička (1986), there is an asymmetry between 1./2. and 3. person pro-drop, with the latter option emerging as more marked (a similar situation obtains in Hebrew, though). However, I do not consider these (and other) arguments to be convincing counter-evidence to the hypothesis that Russian has subject argument *pro* licensing by T; and I believe there is at least one strong argument in support of a *pro* analysis of subject argument omission in finite clauses in Russian (see Müller 1988: 98–100). It turns out that in some cases, using *pro* is in fact the only possibility to achieve a certain reading of a sentence; the counterpart with a phonologically realized pronoun necessarily means something else. The generic interpretation of subject argument pronouns as 'one' is a case in point. The examples in (21) can all have a generic interpretation of the subject pronoun argument; this reading is not available when the pronoun is overtly realized (see Suñer 1983 for the same effect in Spanish). It seems clear that a PF-approach in terms of contextually licensed ellipsis cannot possibly be held responsible for the generic interpretation of the 2.SG. and 3.PL subject arguments in (21).

(21) a. *Ob ėtom* pro *mnogo govorjat.*
about this – much talk-3.PL
'There is much talk about this.'

b. pro *prosjat ne kurit'.*
– ask-3.PL not to smoke
'It is requested that there is no smoking.'

c. *Ėtu knigu* pro *pročityvaeš' za dva časa.*
this book – read-2.SG in two hours
'This book can be read in two hours.'

So far, I have only considered present tense contexts. Things are slightly more difficult in past tense environments. At first sight, it looks as though there is radical person impoverishment in the past tense; as shown in (22), past tense forms are marked only for gender (which is not marked in the present conjugation) and number.

(22) *Past tense conjugation*:
čitat' ('read')

	MASC	FEM	NEUT
1.SG	čita-l	čita-l-a	čita-l-o
2.SG	čita-l	čita-l-a	čita-l-o
3.SG	čita-l	čita-l-a	čita-l-o

	MASC	FEM	NEUT
1.PL	čita-l-i	čita-l-i	čita-l-i
2.PL	čita-l-i	čita-l-i	čita-l-i
3.PL	čita-l-i	čita-l-i	čita-l-i

Assuming (as I have done so far) that the presence of one impoverishment rule affecting T (perhaps vacuously) suffices to block *pro* licensing by T in general, an impoverishment approach to (22) would be incompatible with the claim that Russian has subject argument pro-drop. What is more, pro-drop can also take place in past tense contexts, even though the forms are not specified for person features at all (see Müller and Rohrbacher 1989, Růžička 1986); cf. (23).[19]

(23) *Anna postupila verno* [$_{CP}$ *čto* pro *rešila* [$_{CP}$ *stat' vračom*]].
Anna acted correctly that decided-3.SG.PAST to become doctor

As a key to a solution of this problem, a diachronic perspective can be taken: The inflected past tense forms in (23) are historically *l*-participles (with the originally accompanying auxiliaries dropped in modern Russian); as such, they never had any person features in the first place (those features had been located on the auxiliaries). Hence, it seems that the most plausible analysis of these forms is not that an impoverishment rule deletes all person features in past tense contexts, but rather that the inflection markers for past tense are not specified for person features for essentially historical reasons: The markers did not bear person features to begin with; they have been re-analyzed as finite markers in parallel with the disappearance of the associated auxiliaries, and they simply have not acquired person features in their new function.

19 However, the issue of markedness may arise here again. For instance, in Dobrova's (2002) discussion of acquisition of pronouns by Russian children, it is presupposed that there is an asymmetry between past and present tense as concerns the occurrence of pro-drop in input data.

This then explains why subject pro-drop is possible in Russian, and in all tenses.[20] Of course, a lot more would eventually have to be said about pro-drop in Russian, but I will leave it at that, and draw a conclusion.[21]

6. Conclusion

The main goal of this paper has been to sketch the outlines of a theory of how morphological properties of verb inflection determine subject argument pro-drop. I have suggested that the notion of morphological richness that is relevant in this context does not rely on counting forms of reference grammar paradigms, but rather involves abstract restrictions on morphological systems that have been independently suggested in Distributed Morphology to capture system-wide patterns of syncretism; viz., impoverishment rules – more specifically, those impoverishment rules that lead to a neutralization of ϕ-feature distinctions. However, to ensure that morphological impoverishment can have an effect on syntax, it is imperative for this operation to apply pre-syntactically, rather than post-syntactically (as is standardly assumed in Distributed Morphology). Accordingly, I have developed a pre-syntactic approach to inflection that incorporates main features of Distributed Morphology (underspecification, fission, impoverishment), and that I have argued to be independently motivated, given minimalist assumptions. The empirical focus of the study has been on German first; I have then extended the analysis to Icelandic, Modern Irish, and Russian.

To end this paper, I would like to emphasize that many of the specific results here should be viewed as tentative. The underlying assumption of this paper is that recent advances in theoretical morphology can be put to good use in theoretical syntax, and the main claim is that the abstract notion of impoverishment may offer a better means of measuring morphological richness (as relevant for pro-drop) than is available in more traditional conceptions of morphology. I have presented a specific version of this general approach that makes a number of verifiable predictions; but it seems quite likely that further analysis along these lines will show that significant modifications of the generalization in (13) are necessary.

20 Note that the analysis here is similar to that which can be given for pro-drop languages like Japanese, Korean, and Chinese in the present approach: Assuming that instances of subject pronoun omission in these languages can indeed involve an empty category *pro* that undergoes Agree with T (rather than discourse-governed ellipsis), the generalization in (13) can be met because if there are no ϕ-features in the first place, there is no motivation for invoking a designated impoverishment rule that deletes ϕ-features – and it is only the presence of such impoverishment rules that may block *pro* licensing under present assumptions.

21 Except for one additional remark: Suppose – counterfactually, assuming the validity of the argument presented above – that Franks (1995) is right in concluding that Russian does not exhibit genuine pro-drop (in any tense), whereas Czech, Polish, and other Slavic languages do. Would it be possible to account for such a situation under present assumptions? The answer is yes: We could stipulate that there is ϕ-feature neutralizing impoverishment in the past tense in Russian after all; then pro-drop is predicted to be impossible, given (13). The existence of pro-drop in Slavic languages like Czech and Polish could then be traced back to the fact that these languages have retained auxiliary clitics (for 1. and 2. person) with their *l*-participle tenses. Since there would then be no comparable ϕ-feature impoverishment operation here, T could undergo Agree with *pro*.

References

Abraham, W. 1991. Null subjects: from Gothic, Old High German and Middle High German to Modern German. *Groninger Arbeiten zur Germanistischen Linguistik* 34, 1–28.

Adger, D. 2003. *Core Syntax*. Oxford, New York: Oxford University Press.

Alexiadou, A. and E. Anagnostopoulou. 1998. Parametrizing Agr: word order, V-movement, and EPP-checking, *Natural Language and Linguistic Theory* 16, 491–539.

Alexiadou, A. and G. Müller. 2005. Class features as probes. Ms., Universität Stuttgart and Universität Leipzig. To appear in: A. Bachrach and A. Nevins (eds.), *Paradigms*. Oxford University Press.

Alexiadou, A. and G. Fanselow. 2000: On the correlation between morphology and syntax: the case of V-to-I. In: W. Abraham and J.-W. Zwart (eds.), *Proceedings from the 15th Comparative Germanic Syntax Workshop*, 219–242. Amsterdam: John Benjamins.

Aronoff, M. 1994. *Morphology by Itself*. Cambridge, Mass.: MIT Press.

Avrutin, S. and B. Rohrbacher. 1997. Null subjects in Russian inverted constructions. In: *Proceedings of the Fourth Annual Workshop on Formal Approaches to Slavic Linguistics*, 32–53. Ann Arbor: Michigan Slavic Publications.

Axel, K. 2004. Null subjects and verb placement in Old High German. Ms., Universität Tübingen.

Bierwisch, M. 1961. Zur Morphologie des deutschen Verbalsystems. PhD thesis, Universität Leipzig.

Bierwisch, M. 1967. Syntactic features in morphology: general problems of so-called pronominal inflection in German. In: *To Honour Roman Jakobson*, 239–270. The Hague/Paris: Mouton.

Bobaljik, J. 2002a. Realizing Germanic inflection: why morphology does not drive syntax. *Journal of Comparative Germanic Linguistics* 6, 129–167.

Bobaljik, J. 2002b. Syncretism without paradigms: remarks on Williams 1981, 1994. In: G. Booij and J. van Marle (eds.), *Yearbook of Morphology 2001*, 53–85. Dordrecht: Kluwer.

Bonet, E. 1991. *Morphology after Syntax*. PhD thesis, MIT, Cambridge, Mass.

Borer, H. 1986. I-Subjects. *Linguistic Inquiry* 17, 375–416.

Bošković, Ž. 2002. On multiple wh-fronting. *Linguistic Inquiry* 33, 351–383.

Chomsky, N. 1982. *Some Concepts and Consequences of the Theory of Government and Binding*. Cambridge, Mass.: MIT Press.

Chomsky, N. 1995. *The Minimalist Program*. Cambridge, Mass.: MIT Press.

Chomsky, N. 2000. Minimalist inquiries: the framework. In: R. Martin, D. Michaels, and J. Uriagereka (eds.), *Step by Step*, 89–155. Cambridge, Mass.: MIT Press.

Chomsky, N. 2001. Derivation by phase. In: M. Kenstowicz (ed.), *Ken Hale. A Life in Language*, 1–52. Cambridge, Mass.: MIT Press.

Chomsky, N. 2005. On phases. Ms., MIT, Cambridge, Mass.

Demjjanow, A. and A. Strigin. 2000. Case assignment to conceptual structures: the Russian instrumental. In: M. Kracht and A. Strigin (eds.), *Papers on the Interpretation of Case*. Vol. 10 of *Linguistics in Potsdam*, 75–107. Universität Potsdam, Institut für Linguistik, Potsdam.

Dobrova, G. 2002. "Ontogenetic universals" and cross-linguistic variations in children's acquisition of personal pronouns and kinship terms. Ms., University of Russia.

Eisenberg, P. 2000. *Grundriß der deutschen Grammatik. Band 1: Das Wort*. Stuttgart: Metzler.

Fanselow, G. and S. Frisch. 2005. Effects of processing difficulty on judgments of acceptability. Ms., Universität Potsdam.

Frampton, J. 2002. Syncretism, impoverishment, and the structure of person features. In: M. Andronis, E. Debenport, A. Pycha, and K. Yoshimura (eds.), *Papers from the Chicago Linguistics Society Meeting*. Vol. 38, 207–222. Chicago: CLS.

Franks, S. 1995. *Parameters of Slavic Morphosyntax*. New York and Oxford: Oxford University Press.

Gelderen, E. van. 2000. The role of person and position in Old English. In: O. Fischer, A. Rosenbach, and D. Stein (eds.), *Pathways of Change: Grammaticalization in English*, 187–206. Amsterdam: John Benjamins.

Grewendorf, G. 1989. *Ergativity in German*. Dordrecht: Foris.

Grewendorf, G. 1990. Small pro in German. In: G. Grewendorf and W. Sternefeld (eds.), *Scrambling and Barriers*, 295–315. Amsterdam: John Benjamins.

Grimshaw, J. 2000. Extended projection and locality. In: P. Coopmans, M. Everaert, and J. Grimshaw (eds.), *Lexical Specification and Insertion*, 115–133. Amsterdam: John Benjamins.

Grimshaw, J. and V. Samek-Lodovici. 1998. Optimal subjects and subject universals. In: P. Barbosa, D. Fox, P. Hagstrom, M. McGinnis, and D. Pesetsky, eds., *Is the Best Good Enough?* 193–219. Cambridge, Mass.: MIT Press and MITWPL.

Haider, H. 1988: Pro Pro-Drop Drop. *Wiener Linguistische Gazette* 42/43, 57–79.

Halle, M. 1997: Distributed Morphology: impoverishment and fission. In: B. Bruening, Y. Kang, and M. McGinnis (eds.), *Papers at the Interface*. Vol. 30, MITWPL, 425–449.

Halle, M. and A. Marantz. 1993: Distributed Morphology and the pieces of inflection. In: K. Hale and S.J. Keyser (eds.), *The View from Building 20*, 111–176. Cambridge, Mass.: MIT Press.

Halle, M. and A. Marantz. 1994. Some key features of Distributed Morphology. In: A. Carnie, H. Harley, and T. Bures (eds.), *Papers on Phonology and Morphology*. Vol. 21 of *MIT Working Papers in Linguistics*, MITWPL, Cambridge, Mass., 275–288.

Harley, H. and R. Noyer. 2003. Distributed Morphology. In: L. Cheng and R. Sybesma (eds.), *The Second GLOT International State-of-the-Article Book* 463–496. Berlin: Mouton de Gruyter.

Holmberg, A. 2004. Null subjects and uninterpretable features: evidence from Finnish. Ms., University of Durham.

Holmberg, A. and C. Platzack. 1995. *The Role of Inflection in the Syntax of the Scandinavian Languages*. Oxford: Oxford University Press.

Jaeggli, O. and K. Safir. 1989. The Null Subject Parameter and parametric theory. In: O. Jaeggli and K. Safir (eds.), *The Null Subject Parameter*, 1–44. Dordrecht: Kluwer.

Jakobson, R. 1962. Beitrag zur allgemeinen Kasuslehre. Gesamtbedeutungen der russischen Kasus. In: *Selected Writings*. Vol. 2, 23–71. The Hague and Paris: Mouton.

Junghanns, U. 2005. Bedingungen für die Argumentrealisierung in slavischen Sprachen. Ms., Universität Leipzig.

Kress, B. 1982. *Isländische Grammatik*. 1. edn. Leipzig: VEB Verlag Enzyklopädie.

Legendre, G., P. Smolensky, and C. Wilson. 1998. When is less more? Faithfulness and minimal links in wh-chains. In: P. Barbosa, D. Fox, P. Hagstrom, M. McGinnis, and D. Pesetsky (eds.), *Is the Best Good Enough?* 249–289. Cambridge, Mass.: MIT Press and MITWPL.

Lumsden, J. 1992. Underspecification in grammatical and natural gender, *Linguistic Inquiry* 23, 469–486.

McCloskey, J. and K. Hale. 1984. On the syntax of person-number inflection in Modern Irish. *Natural Language and Linguistic Theory* 1, 487–533.

Müller, G. 1988. Zur Analyse subjektloser Konstruktionen in der Rektions-Bindungs-Theorie. That-trace-Effekte und pro-Lizensierung im Russischen. Ms., Universität Frankfurt/Main.

Müller, G. 2002. Free word order, morphological case, and sympathy theory. In: G. Fanselow and C. Féry (eds.), *Resolving Conflicts in Grammars: Optimality Theory in Syntax, Morphology, and Phonology* (special issue of *Linguistische Berichte*), 9–48. Hamburg: Buske.

Müller, G. 2005. Syncretism and iconicity in Icelandic noun declensions: a Distributed Morphology approach. In: G. Booij and J. van Marle (eds.), *Yearbook of Morphology 2004*, 229–271. Dordrecht: Springer.

Müller, G. 2006. Subanalyse verbaler Flexionsmarker. In: E. Breindl, L. Gunkel, and B. Strecker (eds.), *Grammatische Untersuchungen*. Tübingen: Narr.

Müller, G. and B. Rohrbacher. 1989. Eine Geschichte ohne Subjekt. Zur Entwicklung der pro-Theorie. *Linguistische Berichte* 119, 3–52.

Noyer, R. 1992. *Features, Positions, and Affixes in Autonomous Morphological Structure*. PhD thesis, MIT, Cambridge, Mass.

Perlmutter, D. 1971. *Deep and Surface Structure Constraints in Syntax*. New York: Holt, Rinehart and Winston.

Perlmutter, D. and J. Moore. 2002. Language-internal explanation: the distribution of Russian impersonals. *Language* 78, 619–650.

Platzack, C. 1987. The Scandinavian languages and the Null Subject Parameter. *Natural Language and Linguistic Theory* 5, 377–402.

Prince, A. and P. Smolensky. 2004. *Optimality Theory. Constraint Interaction in Generative Grammar*. Oxford: Blackwell. (Original version from 1993).

Richards, M. 2004. *Object Shift and Scrambling in North and West Germanic: A Case Study in Symmetrical Syntax*. PhD thesis, University of Cambridge, Cambridge, UK.

Rizzi, L. 1986. Null objects in Italian and the theory of 'pro'. *Linguistic Inquiry* 17, 501–557.

Roberts, I. 1993: *Verbs and Diachronic Syntax*. Dordrecht: Kluwer.

Rohrbacher, B. 1999. *Morphology-Driven Syntax: A Theory of V to I Raising and Pro-Drop*. Amsterdam: John Benjamins.

Růžička, R. 1986: Funkcionirovanie i klassifikacija pustyx kategorij v russkom literaturnom jazyke. *Zeitschrift für Slavistik* 31, 388–392.

Stump, G. 2001: *Inflectional Morphology*. Cambridge: Cambridge University Press.

Suñer, M. 1983. Pro_{arb}. *Linguistic Inquiry* 14, 188–191.

Vikner, S. 1997. V-to-I movement and inflection for person in all tenses. In: L. Haegeman (ed.), *The New Comparative Syntax*, 189–213. London: Longman.

Vikner, S. 2001a. V-to-I movement and *do*-insertion in Optimality Theory. In: G. Legendre, J. Grimshaw, and S. Vikner (eds.), *Optimality-Theoretic Syntax*, 424–464. Cambridge, Mass.: MIT Press.

Vikner, S. 2001b. *Verb movement variation in Germanic and Optimality Theory.* Habilitation thesis, Universität Tübingen.

Wiese, B. 1994. Die Personal- und Numerusendungen der deutschen Verbformen. In: K.-M. Köpcke (ed.), *Funktionale Untersuchungen zur deutschen Nominal- und Verbalmorphologie*, 161–191. Tübingen: Niemeyer.

Wiese, B. 1999: Unterspezifizierte Paradigmen. Form und Funktion in der pronominalen Deklination, *Linguistik Online* 4. (www.linguistik-online.de/3_99).

Williams, E. 1994. Remarks on lexical knowledge. *Lingua* 92, 7–34.

Wunderlich, D. 1996. Minimalist Morphology: the role of paradigms. In: G. Booij and J. van Marle (eds.), *Yearbook of Morphology 1995*, 93–114. Dordrecht: Kluwer.

Wunderlich, D. 2004. Is there any need for the concept of directional syncretism? In: G. Müller, L. Gunkel and G. Zifonun (eds.), *Explorations in Nominal Inflection*, 373–395. Berlin: Mouton de Gruyter.

Part II: Structure

A Note on Targets of A'-Movement in the Left Periphery of German Sentences*

Josef Bayer

1. The left edge of the German clause

In his 2002 monograph *Minimalistische Syntax*, Günther Grewendorf proposes a C-domain for German which is positionally richer than traditionally assumed in generative syntax. In following leading ideas by Rizzi (1997), Grewendorf assumes a structure for German in which the C-domain may be split up into different functional positions among which there are two potential landing sites for *wh*-phrases. There is a lower one which is the specifier of the so-called FINITE PHRASE (SpecFinP), and a higher one which is the specifier of the so-called FOCUS PHRASE (SpecFocP). These two will figure in the following discussion.

According to Grewendorf, movement of a *wh*-phrase to SpecFinP is like any other movement to the familiar "Vorfeld" of the German clause. It is triggered by an EPP-requirement. The *wh*-phrase acquires its operator scope only after another step from SpecFinP to SpecFocP. Thus, the actual variable which is bound by the *wh*-phrase is in SpecFinP (or is transmitted from there to the foot of the chain). (1) shows the relevant steps of *wh*-movement in a simplex interrogative clause.

(1)

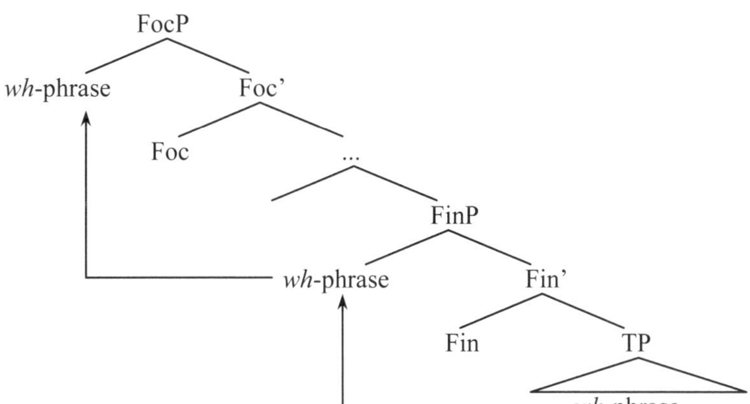

* I am indebted to Werner Frey for discussions about some of the issues of the present contribution. Many thanks also to Eric Fuß and Patrick Brandt for their suggestions which led to various improvements. The responsibility for any errors and misconceptions is my own. The present work has been supported by the *Deutsche Forschungsgemeinschaft* via the Konstanz Sonderforschungsbereich 471 *Variation and Evolution in the Lexicon*.

2. Weak Crossover

A major motivation for this distinction of landing sites comes from data on WEAK CROSSOVER (WCO), i.e. an ungrammaticality that arises due to a variable which is coindexed with a pronoun to its left such that the pronoun does not c-command the variable (as is the case in STRONG CROSSOVER). It has been observed by Frey (1990: ch.6) that German displays clear WCO-effects only in long *wh*-movements. (For relevant discussion of (W)CO in German see also Grewendorf 1988 and Haider 1993). *Wh*-movement in the simplex clause, short *wh*-movement, does not lead to a comparable loss of grammaticality.[1] The following examples are from Grewendorf (2002).

(2) *Wen$_1$ hat sein$_1$ bester Freund* t$_1$ *an die Polizei verraten?*
who has his best friend to the police betrayed
'*Who did his best friend betray to the police?' (grammatical in German)

(3) **Welchen Studenten$_1$ hat sein$_1$ Vater gesagt, dass Maria* t$_1$ *gesehen hat?*
which student has his father said that Maria seen has
'*Which student did his father say that Maria has seen?'
(ungrammatical in German)

Following Lasnik and Stowell's (1991) analysis of the absence of a WCO-effect in English relative clauses as in *John$_1$, who$_1$ his$_1$ father likes* t$_1$, he suggests that the WCO-effect appears only if the A'-binder has been moved to a genuine operator position. The distinction between SpecFinP and SpecFocP as potential landing sites for *wh*-phrases enables him to argue that only SpecFocP but not SpecFinP is a genuine operator position. The absence of the WCO-effect in (2) follows because in crossing over the possessive pronoun *sein, wen* has been moved to the non-offending position SpecFinP and only later to SpecFocP. The presence of the WCO-effect in (3) follows because in crossing over the possessive pronoun *sein, wen* has now been moved from the potentially offending position SpecFocP to the matrix clause's SpecFocP. One background assumption here is that the constraint against improper movement prevents movement from SpecFocP to SpecFinP. Grewendorf's suggestion is that in this situation the FinP of the matrix clause is not projected at all. A more natural solution comes to mind rather easily: Since the *wh*-phrase does not terminate in the first position which is reached after A'-movement, one could equally well expect that FinP will not be projected in the embedded clause. Thus, the *wh*-phrase would move to SpecFocP directly. According to Grewendorf's assumptions this solution is not viable, however, because we would lose the explanation for the absence of the WCO-effect in the simplex clause. The new account of *wh*-movement in German is, thus, characterized by the two generalizations in (4).

(4) a. The target of short *wh*-movement is the non-operator position SpecFinP.
b. The target of long *wh*-movement is the operator position SpecFocP.

---

1 Although the constructions appear to be semi-grammatical or marked for certain speakers, a judgment which I do no share myself.

Another important aspect of the analysis is that EPP-checking is assumed only for SpecFinP, i.e. for the position whose adjacent head hosts the finite verb. Once *wh*-movement has passed the complementizer, there is no EPP-checking, and the assumption is that in this case, *wh*-movement leads directly to SpecFocP. The contrast which is adduced in support of this is the one shown in (5), the judgments being Grewendorf's.[2]

(5) a. ? *Wen*$_1$ *glaubt* *Maria* [$_{FinP}$ t_1 *hat sein*$_1$ *bester Freund* t_1 *an die Polizei*
who believes Maria has his best friend to the police
verraten]?
betrayed

b. ?? *Wen*$_1$ *glaubt* *Maria* [$_{FocP}$ t_1 *dass sein*$_1$ *bester Freund* t_1 *an die Polizei*
who believes Maria that his best friend to the police
verraten hat]?
betrayed has

While the distinction between a low A'-position such as SpecFinP and a high A'-position such as SpecFocP strikes me as well-motivated in general, I do not share the judgments in (5). In my view, both questions are equally grammatical, i.e. none of them involves a WCO-violation. The difference seems to rest in the increased complexity of (5b) as compared to (5a).[3] On the other hand, there is evidence that movement across the overt complementizer does target a higher A'-position such as SpecFocP.[4] If I am right with this evaluation, we have to look for another explanation of the partial absence of the WCO-effect in German. In the next section, I will sketch an alternative which can be reconciled with Grewendorf's general and in my view correct distinction as described in (4).

3. A-Scrambling

Grewendorf and Sabel (1999) suggest with many others that scrambling in German is some kind of A'-movement, and that as a consequence *wh*-movement cannot take the scrambling position as the position of the variable to be bound by the *wh*-operator.[5] Thus, the absence of WCO effects with short *wh*-movement has to be relegated to the left peripheral positions discussed in the previous section. Various linguists have argued, however, that at least a subset of scrambling operations in German shows properties of

---

2 The examples are Grewendorf's (207a,b) which I enrich here with labelled brackets for the sake of clarity.

3 This would be especially so if one were to follow Reis' (1995) arguments in favor of analyzing the V1-construction *glaubt Maria* in (5a) as a parenthesis. If Reis is correct, only (5b) shows genuine long movement.

4 See, for instance, Bayer (2001; 2005a,b) and Frey (2005a). Frey suggested the term *Kontrast Phrase* (KontrP) instead of FocP, which could help avoiding the association that material is moved to SpecFocP for focus checking. I assume that in German, information focus must be checked in the middle field to the left of the verb as in most languages with a head-final VP. The transfer to SpecFocP or SpecKontrP must then have an independent reason.

5 The reason is a violation of the constraint against improper movement, as discussed in detail by Müller and Sternefeld (1993).

A-movement.[6] From this perspective, the absence of a WCO-effect as seen in (2) can be explained quite naturally: A-movement of the object may cross the subject without interfering with variable binding. Under the plausible assumption that *wh*-scrambling is not performed for the purpose of *wh*-checking, there is no reason for the scrambling trace to count as a variable, which of course needs to be A-free, i.e. must not be coindexed with a pronoun in some A-position. This is shown in (6), a reanalysis of (2) in which the step from t_1 to t'_1 should be conceived of as an instance of A-scrambling.

(6) *Wen*$_1$ *hat* t'_1 *sein*$_1$ *bester Freund* t_1 *an die Polizei verraten?*

The correlation between the scrambling option and the WCO-circumvention which is found in German but not in English should be a strong conceptual motivation for retaining such an approach. The fact that long *wh*-movement may induce a WCO-violation as seen in (3) can be accounted for by noticing that German does not allow scrambling (or at least A-scrambling) out of CP. In (3), A-scrambling of the *wh*-object would not rescue the derivation because the WCO-violation would invariably occur in the matrix clause.[7] The predictions are exactly the same as those made by Grewendorf's system. This is not so in (5b), however. Grewendorf predicts a WCO-effect due to direct movement to SpecFocP, whereas the A-scrambling approach predicts that the structure is rescued by first scrambling the *wh*-operator, in the present case *wen*, across the potentially offending pronoun, in the present case *sein*. Since to my ear (5b) is perfectly grammatical, I conclude that the A-scrambling approach should be accepted.

4. Do landing sites determine operator status?

Provided that there are at least two different landing sites for A'-movement, the next question to ask is whether these landing sites determine the operator status of the phrase to be moved. Recall the assumption in Grewendorf (2002) that operator status is yielded by movement to SpecFocP but not by movement to SpecFinP. Material in the latter can only agree for an EPP- or V2-feature. On the other hand, movement beyond the complementizer should invariably invoke SpecFocP. The expectation is nourished that movement to SpecFocP assigns operator status to the affected material.[8] I would like to demonstrate now that neither of these expectations is fulfilled.

---

6 For example Fanselow (1990), Bayer (1996), Hinterhölzl and Pili (2003) among many others. Bayer and Kornfilt (1994) present a number of arguments against scrambling as A'-movement. There is also a more marked type of A'-scrambling which has been referred to by notions like "focus scrambling" and "I-topicalization". This second type does not concern us here.

7 Alternatively, one could argue that in this situation scrambling does not occur as it would be a superfluous step. However, in this case some look-ahead device is required by which it is determined whether or not a crossover violation will occur in an upper cycle. Given that *wh*-movement is in terms of linear distance "unbounded" (cf. *Welchen Studenten*$_1$ *hat sein*$_1$ *Vater gesagt, dass Waldemar meint, dass Ernestine glaubt, dass ... Maria* t_1 *gesehen hat?*), a solution which involves such a powerful device seems to be illusory. Thanks to Eric Fuß for drawing my attention to this problem. The issue is a really interesting one. For initial discussion cf. Bayer (2005a, §9).

8 We need to be careful with this interpretation because Grewendorf (2002) is not explicit about it.

Assume a V2-complement as in (7) in which the subject has been raised to SpecFinP for EPP- or V2-checking.

(7) *Wer hat behauptet* [$_{\text{FinP}}$ *Heinz-Rüdiger sei in die Toskana ausgewandert*]?
who has asserted Heinz-Rüdiger be$_{\text{subj}}$ in the Tuscany emigrated

If the subject *Heinz-Rüdiger* in (7) is replaced by the *wh*-word *wer* ('who'), we should expect association with the matrix *wh*-operator. The reason is that the lower *wh*-word has not taken an operator or scope position yet. Thus, replacement of the name by a *wh*-word should lead to a well-formed multiple question. However, (8) is clearly deviant.

(8) **Wer hat behauptet* [$_{\text{FinP}}$ *wer sei in die Toskana ausgewandert*]?
who has asserted who be$_{\text{subj}}$ in the Tuscany emigrated

There are two conclusions one can draw from this: (i) The lower *wer* has been moved one step up into SpecFocP and has therefore undergone scope freezing, which leads to a semantic selectional conflict with the matrix predicate. Alternatively, (ii) *wer* has taken scope in SpecFinP and has undergone scope freezing right there, which leads to the same selectional conflict. I believe that the first conclusion is not very plausible. Projectional economy would bar the projection of the FocP-layer if there is no reason for the *wh*-word to take scope in the lower clause. Apart from this, SpecFinP must be a scope position for other types of operators such as quantified DPs. This is at least the most parsimonious assumption.[9] So the conclusion would be that even SpecFinP is a possible scope position, and that a *wh*-phrase which moves there for reasons of EPP-checking would in this position automatically display *wh*-scope.[10]

One can equally show that movement to SpecFocP does not automatically assign operator status to the affected material. Relative clauses are a case in point. Given the assumption that long A'-movement from a CP (with the overt complementizer *dass*) has to target SpecFocP one could be led to the expectation that long relativization exhibits a WCO-effect. This expectation is not borne out, however. (9) is, with the intended interpretation, a grammatical sentence, showing at best "weakest" crossover.

(9) *Heinz-Rüdiger ist* [*ein Doktorand*]$_1$ [*den*$_1$ [*seine*$_1$ *Doktormutter vorschlägt*
Heinz-Rüdiger is a PhD-student who his adviser proposes
[*dass wir bald* t$_1$ *zu einem Kolloquium einladen sollten*]]].
that we soon to a colloquium invite should
'Heinz-Rüdiger is a PhD-student who his adviser proposes that we should soon invite for a colloquium.'

(9) shows that if the relative pronoun is a null constant as Lasnik and Stowell (1991) have argued, it remains such a null constant whichever A'-position it moves to. In short,

---

9 I do not want to argue against an extra step of "short QR" which could be motivated for purely semantic reasons. I see, however, no empirical justification for identifying such an operation with the syntactically motivated landing sites which the split-C system has in stock.

10 One worry with this conclusion could be that one loses the explanation of long *wh*-movement from V2-complements. This worry is likely to disappear if one follows Reis (1995) in assuming that the bridge construction in these cases can be treated as a special kind of parenthesis (see also note 3).

the determinant of ± operator status is the phrase which is moved rather than the landing site.

5. On the relevance of different A'-positions

In this section, I would like to show that my previous concerns do not in any way discredit the assumption of different A'-landing sites for German. Lenerz (1994:162) has drawn the attention to examples such as (10).

(10) *Ihr Geld ist ja nicht weg, meine Damen und Herren.*
your money is PRT not away my ladies and gentlemen.
Es haben jetzt nur andere.
it have now only others
'Your money hasn't gone, ladies and gentlemen. It is only in the possession of others.'

The positioning of an object-*es* in SpecFinP is remarkable because it shows that the ban on *es*-preposing to the "Vorfeld" that had first been pointed out by Travis (1984) is not absolute. The critical sentence can be changed into a so-called extraction from V2 without any loss of grammaticality.

(11) *Es glaube ich haben jetzt nur andere.*
it believe I have now only others

However, as Frey (2005a:153) has pointed out, extraction of *es* from a *dass*-CP leads to severe ungrammaticality. Changing (11) into a case of genuine long extraction as in (12) would be inconceivable.[11]

(12) **Es glaube ich dass jetzt nur andere haben.*
it believe I that now only others have

The same is true for other non-focusable or non-contrastable material. In the first class, there would be the indefinite subject pronoun *man* ('one', similar to French *on*). In the second class, there would be speaker oriented adverbs like *leider* ('unfortunately'), which Frey has shown cannot be contrasted. Consider Frey's (2005a) examples in (13a) and (13c) to which I add the grammatical example in (13b).

(13) a. *Leider*$_1$ *hat* t_1 *keiner dem alten Mann geholfen.*
unfortunately has nobody the old man helped
'Unfortunately, nobody helped the old man.'
b. *Leider*$_1$ *sagte Karl hat* t_1 *dem alten Mann keiner geholfen.*
unfortunately said Karl has the old man nobody helped
c. **Leider*$_1$ *sagte Karl dass* t_1 *keiner dem alten Mann geholfen hat.*
unfortunately said Karl that nobody the old man helped has

[11] For extensive discussion of German object-*es* in the pre-field see Frey (this volume). Even if (11) is slightly degraded in comparison with (10) as Eric Fuß (p.c.) feels, (12) is entirely ungrammatical. Judgments may deviate in degree but they do not deviate in direction.

Again, we see that in simplex clauses, movement to SpecFinP comes without any requirements as to operator status, focus status, contrastiveness etc. The same is true for apparent movement from a V2-clause (cf. 13b), which in my view strongly confirms Reis' (1995) claim that such cases have to be analyzed as simplex clauses with a parenthetical insertion (cf. notes 3 and 9). Movement beyond the complementizer *dass*, however, must target a position with more constraints than SpecFinP. (13c) can only be interpreted with *leider* being associated with the matrix clause, not with the filler-gap relation which is displayed in the current representation. This position is what Grewendorf (2002) calls SpecFocP and Frey (2005a,b) SpecKontrP. If movement to the left of *dass* is movement to such a distinct position, the ultimate landing sites must share the properties of SpecFocP / SpecKontrP. Otherwise we would expect effects of improper movement.

I share with Grewendorf's analysis the view that movement to the specifier of *dass* invokes something like SpecFocP. In a recent account of the *that*-trace phenomenon, Bayer (2005a,b), I argue that in Bavarian, a dialect which disrespects the DOUBLY-FILLED-COMP-FILTER (DFCF), movement to the specifier of *dass* can only affect *wh*-phrases which originate in the comment domain of the clause. In the presence of an individual-level predicate, the subject originates in a higher position. Assuming that in this case the feature for checking contrastivity is not available, the expectation is that a *dass*-trace effect appears. According to my intuitions, this expectation is borne out.

(14) a. *I woass* [*wer*$_1$ [*dass neili* t$_1$ *b'suffa woa*]]. STAGE LEVEL
I know who that recently drunk was
'I know who was recently drunk.'

b. ?* *I woass* [*wer*$_1$ [*dass* t$_1$ *schlau is*]]. INDIVIDUAL LEVEL
I know who that clever is
'I know who is clever.'

In (14a) the *wh*-phrase originates in the comment domain following an adverbial stage topic (*neili*). This entitles the subject to bear an appropriate feature for checking focus/contrast in a position to the left of *dass*. In (14b), the *wh*-phrase arguably originates in the topic rather than in the comment domain because no stage topic would be compatible with the individual level predicate.[12] As a consequence, the *wh*-phrase does not have the right feature for checking focus/contrast to the left of *dass*.

Since *wh*-movement in Standard German, a language which does respect the DFCF, does not show any such contrast,[13] the conclusion should be that in Standard German, short *wh*-movement is generally movement to SpecFinP. Grewendorf's generalization in

---

12 Speakers of English can verify this by considering the semantic ill-formedness of sentences like **There is someone intelligent* (which, of course, requires ignoring the homophonous but structurally different expression with the meaning 'There is someone who is intelligent').

13 No difference in grammaticality can be observed between (i) and (ii):

(i) *Ich weiß wer neulich betrunken war.* STAGE LEVEL
I know who recently drunk was

(ii) *Ich weiß, wer intelligent ist.* INDIVIDUAL LEVEL
I know who intelligent is

(4a) will then extend to cases in which *wh*-movement is not triggered by the V2-requirement. To the extent that this step is feasible, the difference between DFCF-languages like modern German and non-DFCF-languages like Bavarian consists of more than just a low-level parameter by which the neutral complementizer *dass* can either surface or remain silent. Indirect further support for this result will be presented directly.

6. Emphatic topicalization

Bavarian EMPHATIC TOPICALIZATION as described by Lutz (1997; 2004) and Bayer (2001) is a process by which a contrastable XP moves into the specifier of a complementizer, activates a +emphasis feature which is a feature of the root clause and, as a consequence, forces the entire CP to undergo pied-piping to the specifier of the root clause.[14] The process is shown with an example from Bayer (2001) in (15).

(15) a. *I häid ned g'moant [dass-e an Fünfer kriag].* E-TOPICALIZATION ⇨
I had$_{subj}$ not thought that-I a five get
'I didn't think I would get (the rather bad) grade 5.'
b. **I häid ned g'moant [an Fünfer*$_1$ *[dass-e* t$_1$ *kriag]].* PIED-PIPING ⇨
c. *[An Fünfer*$_1$ *[dass-e* t$_1$ *kriag]]*$_2$ *häid-e ned g'moant* t$_2$.

Non-contrastive XPs as discussed in (10) through (12) are excluded from emphatic topicalization: **Man*$_1$ *dass* t$_1$ *nicht rauchen soll* ...('that one should not smoke'), **Leider*$_1$ *dass* t$_1$ *keiner dem alten Mann geholfen hat* ...('that unfortunately nobody has helped the old man') are sharply ungrammatical. The most remarkable property of emphatic topicalization, however, is that once it has applied it triggers CP pied-piping as seen in (15c). This property is not shared by *wh*-complements. (14a) is grammatical as it is.[15] Thus emphatic topicalization must involve another feature than *wh*-movement. The split C-domain which Grewendorf (2002) proposed for German invites the conclusion that Bavarian involves a third layer above FocP/KontrP, namely an EMPHASIS PHRASE (EmpP). In Bayer (2001) I followed a more conservative approach by which the C-position itself can be associated with features for *wh* and emp. The expectation is then that either a *wh*-phrase appears in SpecCP or an emphatic topic but not both. Run-of-the-mill examples from spoken language seem to confirm this expectation.[16] I cannot exclude the possibility, however, that *wh*-phrase and emphatic topic may appear simultaneously. The sentence in (16) strikes me as exceptional but could well be acceptable for certain speakers.

---

14 What the "specifier of the root clause" could be in a split-CP system will be made more precise directly.

15 Preposing the *wh*-CP may apply but does not need to.

16 See Bayer (2001, note 16).

(16) [*An Heinz-Rüdiger*$_2$ [*wer*$_1$ [*dass* [t$_1$ *vo uns* t$_2$ *kennt*]]]]$_3$ *kannt-a-da*
the Heinz-Rüdiger who that of us knows could$_{subj}$-I-you
ned song t$_3$.
not tell
'As for Heinz-Rüdiger, who of us would know him I could not tell you.'

Given the evidence for two distinct landing sites for A'-movement in the left periphery of the German clause, it would not be very surprising to see dialects which allow also for a tripartite structure as in (16). Detailed investigations of speakers' intuitions would be necessary, however, to ground such a conclusion on a serious data base. Whichever answer can be given, the landing site for the pied-piped CP must match that emphatic interpretation that is yielded by e-topicalization. Thus, the pied-piped CP cannot land in SpecFinP but either in SpecKontrP or in an even higher position which is reserved for emphatic material.

7. Conclusion

With his adaptation of the split-CP hypothesis for German, Grewendorf (2002) has initiated a rethinking of traditional assumptions about the left periphery of the German sentence which showed a fairly stable continuum from the topological fields model to generative work in GB-theory. In the previous sections, I gave additional evidence in favor of the split-CP hypothesis while on the other hand I tried to overcome certain assumptions concerning operator status and weak crossover which were considered to be constitutive for the approach.

References

Bayer, J. 1996. *Directionality and Logical Form*. Dordrecht: Kluwer.

Bayer, J. 2001. Asymmetry in emphatic topicalization. In: C. Féry and W. Sternefeld (eds.), *Audiatur Vox Sapientiae*, 15-47. Berlin: Akademie Verlag.

Bayer, J. 2005a. Was beschränkt die Extraktion? Subjekt – Objekt vs. Topic – Fokus. In: F.-J. d'Avis (ed.), *Deutsche Syntax: Empirie und Theorie. Göteborger Germanistische Forschungen* 46, 233-257.

Bayer, J. 2005b. *That*-trace without reference to the subject. Invited lecture. *20th Comparative Germanic Syntax Workshop*, Tilburg, 10–11 June 2005.

Bayer, J. and J. Kornfilt. 1994. Against scrambling as an instance of move-alpha. In: H. van Riemsdijk and N. Corver (eds.), *Studies on Scrambling*, 17-60. Berlin: de Gruyter.

Fanselow, G. 1990. Scrambling as NP-movement. In: G. Grewendorf and W. Sternefeld (eds.), *Scrambling and Barriers*, 113-140. Amsterdam: John Benjamins.

Frey, W. 1990. Syntaktische Bedingungen für die Interpretation. *Arbeitspapiere des Instituts für maschinelle Sprachverarbeitung* 01-90. Universität Stuttgart.

Frey, W. 2005a. Zur Syntax der linken Peripherie im Deutschen. In: F.-J. d'Avis (ed.), *Deutsche Syntax: Empirie und Theorie. Göteborger Germanistische Forschungen* 46, 147-171.

Frey, W. 2005b. Zur Syntax und Pragmatik der linken Peripherie im Deutschen. Invited lecture. *Sonderforschungsbereich 471*. Konstanz, 08.12.2005.

Frey, W. (this volume). How to get an object-*es* into the German prefield.

Grewendorf, G. 1988. *Aspekte der deutschen Syntax*. Tübingen: Gunter Narr Verlag.

Grewendorf, G. 2002. *Minimalistische Syntax*. Tübingen: Francke Verlag.

Grewendorf, G. and J. Sabel. 1999. Scrambling in German and Japanese: adjunction versus multiple specifiers. *Natural Language and Linguistic Theory* 17, 1-65.

Haider, H. 1993. *Deutsche Syntax generativ*. Tübingen: Gunter Narr Verlag.

Hinterhölzl, R. and D. Pili. 2003. Argument shift phenomena across language types. Ms., Humboldt Universität zu Berlin and Università di Siena.

Lasnik, H. and T. Stowell. 1991. Weakest crossover. *Linguistic Inquiry* 22, 687-720.

Lenerz, J. 1994. Pronomenprobleme. In: B. Haftka (ed.), *Was determiniert Wortstellungsvariation?* 161-173. Opladen: Westdeutscher Verlag.

Lutz, U. 1997. Parasitic Gaps und Vorfeldstruktur. In: F.-J. d'Avis and U. Lutz (eds.), *Zur Satzstruktur des Deutschen. Arbeitspapiere des Sonderforschungsbereichs 340*, 55-80. Tübingen.

Lutz, U. 2004. ET, parasitic gaps, and German clause structure. In: H. Lohnstein and S. Trissler (eds.), *Syntax and Semantics of the Left Periphery*, 265-312. Berlin: de Gruyter.

Müller, G. and W. Sternefeld. 1993. Improper movement and unambiguous binding. *Linguistic Inquiry* 24, 461-507.

Reis, M. 1995. Extractions from verb-second clauses in German? In: U. Lutz and J. Pafel (eds.), *On Extraction and Extraposition in German*, 45-88. Amsterdam: John Benjamins.

Rizzi, L. 1997. The fine structure of the left periphery. In: L. Haegeman, (ed.), *Elements of Grammar*, 281-337. Dordrecht: Kluwer.

Travis, L. .1984. *Parameters and Effects of Word Order Variation*. Ph.D. dissertation, MIT, Cambridge, Mass.

Extending Doubling to Non-Local Domains: Complete vs. Partial Copying + Deletion and Related Reconstruction Issues

Adriana Belletti

Ever since the original work of Cinque (1977), data like the following have been presented to illustrate the island sensitivity of Clitic Left Dislocation (CLLD):

(1) a. ?(?) *Gianni, temo la possibilità che lo arrestino.*
Gianni, I.am.afraid.of the possibility that him(CL) they.will.arrest
b. ?? *A Gianni, sono certo del fatto che gli parleranno.*
To Gianni, I.am sure of.the fact that to.him(CL) they.will.talk

The relatively better status of (1a) can be related to the interference with the so-called Hanging Topic (HT) construction, discussed in Cinque (1990), which shows no island sensitivity at all. (2) illustrates the HT construction.

(2) a. *Quanto a Gianni, temo la possibilità che lo arrestino.*
As for Gianni, I.am.afraid.of the possibility that him(CL) they.will.arrest
b. *Quanto a Gianni, sono certo del fatto che gli parleranno.*
As for Gianni, I.am sure of.the fact that to.him(CL) they.will.talk

The traditional approach to these facts assumes that CLLD involves movement, while HT does not.[1] Other approaches, like the one recently developed by Frascarelli (2003), assume base generation of CLLD (much as HT) combined with the hypothesis that sensitivity to islands should come from the fact that the dislocated phrase of CLLD undergoes further movement from the left peripheral topic position where it is directly merged into a higher topic position (cf. also Iatridou 1990 for closely related ideas). Following Belletti (2005), I will assume an approach closer to the traditional one that I elaborate along the lines originally developed by Cecchetto (2000). According to this approach, CLLD involves a doubling movement derivation, starting out from an original big DP containing both the clitic and the dislocated phrase. Among the clearest advantages of this approach is the fact that it does not require any special machinery to ensure the establishment of the correct relations between the dislocated phrase and the clitic with respect to Case, thematic and discourse related interpretation, as discussed in

1 Cinque (1990). In HT, the relation between the Hanging Topic and the clitic is not created through movement but is assumed to be established via some kind of binding relation. Since the Hanging Topic can also be a simple DP and does not need to be introduced by an "as for" phrase, HT and CLLD are hardly distinguishable when the left peripheral phrase is a direct object.

Belletti (2005) in detail. Grewendorf (2002) has also developed a similar approach to Left Dislocation in German; he assumes Left Dislocation to involve movement as well as doubling, where the doubling pronominal element can be a "d-pronoun".[2]

As is expected under the movement analysis, CLLD gives rise to reconstruction effects. Consider the example in (3) from Cecchetto (2000) as a case of a principle C violation induced by reconstruction as well as the examples in (4) illustrating pronoun binding, which appears to be available due to reconstruction:

(3) *L'opera prima di [uno scrittore]*$_i$, pro$_{*i}$ *la scrive sempre volentieri.*
The first work of a writer, *pro* it(CL) writes always with.pleasure
(Cecchetto 2000)

(4) a. *I suoi*$_i$ *studenti, ogni*$_i$ *professore li*$_i$ *promuoverebbe.*
his students, every professor them would.pass
b. *Al suo*$_i$ *studente, ogni*$_i$ *professore sarà autorizzato a parlargli*$_i$.
to his students every professor will.be authorized to talk.to.them

Adopting the doubling approach to CLLD, these facts follow directly: the dislocated phrase originates within the big DP which is located under the scope of the quantifier in (4) and is c-commanded by the preverbal (pro) subject in (3). Assuming a minimalist approach to movement and reconstruction (Chomsky 1993 and subsequent work) in terms of copying and deletion, this means that the dislocated phrase is in fact under the scope of the quantifier in (4) and is c-commanded by the subject in (3): it is just pronounced in the dislocated position and its copy within the original big DP is deleted.

So far we have reviewed and rephrased relatively standard approaches to the analysis of CLLD in current minimalist terms; the core feature of the analysis is the doubling operation that we assume here as well.[3] Let us now consider some further questions against this background. It is often observed that in non-local domains, resumptive pronouns may "save" island violations. The following Italian data illustrate the point with some examples of restrictive relative clauses where the relativized element is inside a complex NP island. There is a very sharp contrast between the examples in (5), where a gap is present in the clause where the relative head originates, and those in (6), where a resumptive pronoun is present:

(5) a. **L'uomo che temo il pericolo che arresteranno...*
the man whom I.am.afraid.of the danger that they.will.arrest
b. **L'uomo a cui sono certo de fatto che parleranno...*
the man to whom I.am.sure.of the fact that they.will.talk

(6) a. ?(?)*L'uomo che temo il pericolo che lo arresteranno...*
the man whom I.am.afraid.of the danger that him(CL) they.will.arrest
b. ?(?)*L'uomo che sono certo de fatto che gli parleranno...*

[2] It is maintained here that HT does not involve movement of the topic phrase to the peripheral position. See the discussion below that also addresses the (un)availability of reconstruction.

[3] See again Cecchetto (2000), where the reconstruction problem is analyzed in subtler terms uncovering differences in acceptability judgements which he accounts for in terms of reconstruction to the clause internal intermediate topic position.

the man whom I.am sure of.the fact that to.him(CL) they.will.talk

The contrast between (5) and (6) is particularly significant in a language like standard Italian which does not systematically form relative clauses with the resumptive pronoun strategy. Relatives like those in (7) are excluded in standard Italian; they are only acceptable at a substandard level (indicated with the diacritic "%/*"):

(7) a. %/**L'uomo che lo arresteranno se continua così...*
the man that him(CL) they.will.arrest if he.goes.on like.that
b. %/**L'uomo che gli parleranno di sicuro...*
the man that to.him(CL) they.will.talk for sure

Thus, the status of (6) must be evaluated in comparison to (5) and (7): even if it is substandard since it involves use of a resumptive pronoun within the relative clause, (6) is significantly better than both (5) and (7), respectively, for the following reasons. (7) is marginal since it is substandard; (5) is strongly deviant since it violates an island. Use of the substandard resumptive pronoun strategy brings about an effect of amelioration with respect to the island violation (cf. (6)), while the effect of degradation is quite strong if there is no independent need for the use of the resumptive strategy (cf. (7)). "Saving" the structure from an island violation may be just such an independent need.[4]

It is tempting to analyze the sentences in (6) as involving a doubling derivation, with the head of the relative clause raised to the relevant relative head position and the clitic stranded within the clause. To the extent that CLLD structures have also been assimilated to (clitic) doubling as concerns essential aspects of their computation (cf. again Cecchetto 2000, Belletti 2005), this amounts to claiming that relative clauses that are formed via the resumptive pronoun strategy are related to CLLD in crucial respects.[5]

Notice now that we seem to come close to an observational dilemma: on the one hand, as noted at the outset, CLLD appears to be sensitive to islands (cf. (1)); on the other hand, adoption of a strategy which parallels CLLD in important respects, such as the use of resumptive pronouns in relatives, leads to the possibility of ameliorating, ultimately saving, the structures from island violations (cf. (6)). How can this be possible? We are faced here with the necessity of providing a better idealization of the relevant data.

The island sensitivity of CLLD of examples like (1) is felt to be relatively strong when compared to the perfect status of HT in (2) (and, more generally, to CLLD in non island contexts), while, as noted, (6) is felt to be relatively acceptable when compared to the clear island violations instantiated in the relative clauses in (5). However, taken in absolute terms, the status of (1) and (6) is essentially comparable: both are marginal as

---

4 This is a very general characterization, valid in a language like standard Italian. See Bianchi (2004) for a detailed discussion of different classes of relative clauses correlated with possible/obligatory presence vs. absence of resumptive pronouns in different languages.

5 A raising analysis of relative clauses (Bianchi 1999, Kayne 1994) is particularly appropriate to express this derivation. The only difference between the movement taking place in relatives and the one taking place in CLLD should concern the landing site in the left periphery of the relative head and of the dislocated phrase respectively (Rizzi 1997). In some cases these might coincide, as in appositive relatives. See also Cinque (1981). See also Boeckx (2003) for a general discussion of resumption as involving stranding.

they involve an island violation. We should furthermore add that a substandard flavor is added to (6), due to the presence of the resumptive pronoun in the relative clause.
We are now ready to address the following question: how can (6), despite its marginality and substandard flavor, be nevertheless felt as better than (5), with the plain island violation? This amounts to asking the following: why is there such a sharp difference between the case of a simple gap in the original Merge position of the relative head (cf. (5)) and the case where a resumptive pronoun is present in the same position, modulo cliticization (cf. (6))? Notice that since movement of the head of the relative is involved in both cases, as is implied by the doubling derivation of relatives involving resumption that we assume, the different status cannot be due to occurrence vs. non-occurrence of movement. According to our analysis, the only difference between the two cases is that one involves doubling, while the other does not. The one which does not gives rise to a significantly stronger island violation.

Assuming the copy and deletion approach to movement, the following difference between the two derivations is rather clear: in the resumptive relatives in (6), the original big constituent from which the doubling derivation starts out undergoes deletion of the lexical part which moves to the position of the relative head; the remaining part containing the pronoun does not move out of the clause. Thus, part of the original DP is never extracted out of the island configuration. Intrinsic to the doubling approach is the idea that deletion of the original constituent as a result of movement is necessarily partial, in that the pronoun remains present and is pronounced in the Merge position (modulo cliticization). I would like to propose that this is the crucial factor which helps reduce the locality violation. Since the part of the original big DP containing the pronoun does not move out of the island, an overt part of the big DP never quits the original position within the offending configuration. The pronoun undergoes movement within the island, to perform cliticization in the examples in (6),[6] but does not move out of it.

When no doubling derivation but movement of the whole constituent takes place, however, the resulting situation is very different. Deletion of the original copy of the relative head is "complete": a full gap is left behind in the base position within the island. In line with the proposal above, this has the consequence that there is no possibility of reducing the locality violation. In sum, I propose that complete vs. partial deletion of the original constituent is at the source of the sharp contrast between (5) and (6).

Since the assumed doubling computation assimilates CLLD and resumptive relatives in important respects,[7] the assumption must be that in CLLD as well, the clause out of which movement of the dislocated constituent occurs always contains part of the original big constituent, namely the clitic pronoun. However, as already noted, the outcome in CLLD is not felt to be as acceptable as in resumptive relatives. This is to be related to the fact that in this case, the status of the resulting structure does not compare to that of a more deviant structure like that of non-resumptive relatives, but rather to that of a

---

6 The question whether cliticization is directly performed by the pronoun or by a silent *pro* with the clitic directly generated within a clitic head is orthogonal to the discussion here (on cliticization see Sportiche 1998, Belletti 1999 and references cited therein).

7 Modulo the difference in landing site of the left dislocated constituent and the relative head which I have assumed not to play a role in the locality issue addressed here.

completely well-formed one like that involved in Hanging Topic. Hence the persistent impression of deviance in CLLD out of an island.

Summarizing the discussion so far, I propose that complete vs. partial deletion of the copy of the moved constituent is the crucial factor accounting for the amelioration effect caused by having a resumptive pronoun in relative clauses where the relativized element is inside an island. This strategy is assumed to involve movement of the relative head out of a big DP within the island and a regular doubling computation along the same lines as assumed for CLLD, modulo independent differences between the two processes.[8]

As pointed out in connection with (2), repeated here in (8), no marginality whatsoever is associated with Hanging Topic structures involving an island configuration such as, e.g., a complex noun phrase:

(8) a. *Quanto a Gianni, temo la possibilità che lo arrestino.*
As for Gianni, I.am.afraid.of the possibility that him(CL) they.will.arrest

b. *Quanto a Gianni, sono certo del fatto che gli parleranno.*
As for Gianni, I.am sure of.the fact that to.him(CL) they.will.talk

Hanging Topic thus minimally contrasts with CLLD structures as in (1). Why should this be so? As said above, it could be assumed that the contrast arises from the movement vs. non-movement nature of CLLD and HT, respectively. Although this is certainly a natural hypothesis to make, I would like to tentatively explore a slightly subtler approach now. If we abstract away from island sensitivity, CLLD and HT share a strong resemblance as concerns their syntactic make-up: in both types of structure, a clitic pronoun is present in the clause following the left peripheral phrase which predicates some property of it. It would be most welcome if the account of the two structures could capture this aspect of their resemblance. Within this spirit, let me assume that HT, much as we assumed for CLLD, involves a doubling computation; suppose, however, as is assumed in current accounts (notably, Cinque 1990), that the peripheral phrase is directly merged in its left peripheral position and that the doubling original big DP contains an empty DP[9] and the clitic. The doubling derivation should involve movement of the empty phrase to a left peripheral position within the predicate clause. Furthermore, some discourse principle should ensure that a relation be established between the HT and the empty constituent in the left periphery of the predicate clause to the effect that they are interpreted as referring to the same individual in the discourse context. The line of analysis just sketched preserves the idea that HT is not directly moved from inside the predicate clause, differently from what happens in CLLD, and at the same time interprets the presence of the clitic inside the predicate clause in a way identical to the one assumed in CLLD, modulo the non overt nature of the moved portion of the original big DP in the HT case. As for the lack of island sensitivity of HT, an analysis ac-

---

8 Such as the more or less standard level of the resulting sentences.

9 I will not go into a deeper discussion of the status of the empty DP at this point (i.e., whether it is an operator or a pronominal element).

counting for it could be phrased in the following terms. Suppose that movement of the empty phrase takes place within the island so that no extraction of material from the local domain is performed. The discourse relation between the empty phrase and the HT can be assumed to be established in a way that is not sensitive to the same locality constraint(s) as movement operations. The following schemas in (9) illustrate the essence of the proposed derivation for the examples in (2)/(8):

(9) a. *Quanto a Gianni*..........[*la possibilità* [*che* DP ..... *arrestino* [*lo* DP]]]

b. *Quanto a Gianni* ....[...*il fatto* [*che* DP ... *parleranno* [*gli* DP]]]

The movement relation is indicated by the line at the bottom of each schema, the interpretive relation between the empty phrase (DP in the schema) and the Hanging Topic is indicated by the upper line terminating with the arrows. As movement takes place inside the island while the relation between the Hanging Topic and the empty phrase occurs outside the island, there is an overall impression of island insensitivity in HT constructions. One major appeal of the outlined approach is that it does not assume special mechanisms to ensure the establishment of the required relation between the HT and the clitic in the predicate clause. The relation is indirectly established through the discourse relation between the HT and the empty DP. On the other hand, the origin of the clitic in the predicate clause is not different from that assumed in CLLD, involving doubling. This is a welcome result in that, as for the presence of the clitic, HT and CLLD should be assimilated to the greatest possible extent. The analysis sketched here assimilates the two structures by having both involve a doubling computation.

Let us make some final, although fairly inconclusive remarks concerning the reconstruction properties of the different structures analyzed here. We already noted in (3)-(4) above the often discussed observation that CLLD gives rise to reconstruction. The proposed analysis of HT gives rise to the expectation that HT as well should admit reconstruction to some extent, at least for those aspects of it relating to the movement nature of the assumed computation. Although clear judgments are hard to get,[10] some speakers (including myself) do not exclude the bound reading of the clitic pronoun in cases like those in (10), the HT version of the CLLD examples in (4), where reconstruction is clearly possible for all speakers:

---

10 Various independent factors are involved which should ultimately account for the extent to which reconstruction is available in different structures. I take the view that movement is just a necessary condition, but not a sufficient one. See Aoun et al. (2001) for relevant discussion. If movement is involved in a given derivation, reconstruction is to be expected; however, other factors, yet to be discovered, may make it available to various extents in different syntactic contexts, up to being totally unavailable. The latter seems to be the case in relative clauses (cf. footnote 12).

(10) a. ?(?) *Quanto ai suoi$_i$ studenti, ogni$_i$ professore li$_i$ promuoverebbe.*
As for his students, every professor them would.pass
b. ?(?) *Quanto al suo$_i$ studente, ogni$_i$ professore sarà autorizzato a parlargli$_i$.*
As for his student every professor will.be authorized to talk.to.him

We tentatively suggest that the extent to which reconstruction is felicitously available in HT might depend on the more or less detailed internal analysis that different speakers associate with the moved empty DP, to which the Hanging Topic is related through discourse.

Furthermore, to the extent that CLLD out of an island can be parsed as relatively acceptable,[11] speakers also admit reconstruction of the dislocated phrase inside the island, to an extent varying somewhat among different speakers. Consider (11) in this respect:

(11) ?(?) *I suoi$_i$ studenti, temo la possibilità che ogni$_i$ professore li$_i$ promuoverebbe.*
His students, I.am.afraid.of the possibility that every professor them would.pass

Availability of the bound variable reading in (11) is expected since (11) differs from (4) only as far as the locality of movement of the dislocated phrase is concerned.[12]

---

11 Although clearly marginal, as any island violation; see the discussion surrounding examples (1), (5) and (6).

12 For reasons which remain to be understood and which are probably orthogonal to the movement/non movement issue, reconstruction to a position inside the island is not readily available for all speakers in resumptive relatives. Sentences like in (i) are hard to parse and excluded by the speakers consulted:

(i) * *Quei suoi$_i$ studenti, che temo la possibilità che ogni$_i$ professore li$_i$ promuoverebbe.*
Those of.his students whom I.am.afraid.of the possibility that every professor them would.pass

The interesting issue arises as to how the distinction between CLLD and resumptive relatives should be characterized in this respect, a question which we leave open here. A possibly related issue concerns the fact that all speakers, including those who do not exclude (10), do not admit reconstruction in HT structures which also involve an island violation, cf. (ii):

(ii) * *Quanto ai suoi$_i$ studenti, temo la possibilità che ogni$_i$ professore li$_i$ promuoverebbe.*
As for his students, I.am.afraid.of the possibility that every professor them would.pass

The issue is clearly complex and requires further study. Its complexity suggests that the movement/non-movement distinction is only one aspect concerned in reconstruction; others are also at play, as hinted at in footnote 10.

References

Aoun, J., L. Choueiri, and N.Hornstein. 2001. Resumption, movement and derivational economy. *Linguistic Inquiry* 32, 371–403.

Belletti, A. 1999. Italian/Romance clitics: structure and derivation. In: H. van Riemsdijk (ed.), *Clitics in the Languages of Europe*, 543–579. Berlin: de Gruyter.

Belletti, A. 2005. Extended doubling and the VP periphery. *Probus* 17,1, 1–35.

Bianchi, V. 1999. *Consequences of Antisymmetry: Headed Relative Clauses*. Berlin: de Gruyter.

Bianchi, V. 2004. Resumptive relatives and LF chains. In: L. Rizzi (ed.), *The Structure of IP and CP. The Cartography of Syntactic Structures, Volume 2*, 76–114. New York: Oxford University Press.

Boeckx, C. 2003. *Islands and Chains: Resumption as Stranding*. Amsterdam: John Benjamins.

Cecchetto, C. 2000. Doubling structures and reconstruction. *Probus* 12, 1, 93–126.

Cinque, G. 1977. The movement nature of left dislocation. *Linguistic Inquiry* 8, 397–411.

Cinque, G. 1981. On the theory of relative clauses and markedness. *The Linguistic Review* 1, 247–294.

Cinque, G. 1990. *Types of A' Dependencies*. Cambridge, Mass.: MIT Press.

Chomsky, N. 1993. A minimalist program for linguistic theory. In: K. Hale and S.J. Keyser (eds.), *The View from Buidling 20*, 1–52. Cambridge, Mass.: MIT Press.

Frascarelli, M. 2003. 'Dislocazioni' e ripresa clitica: un'analisi comparata della topicalizzazione in italiano. Ms. University of Rome (talk given at the Interdepartmental Center for Cognitive Studies on Language (ciscl), University of Siena).

Grewendorf, G. 2002. Left dislocation as movement. In: S. Manck and J. Mittelstaedt (eds.), *Georgetown University Working Papers in Theoretical Linguistics*, 31–81.

Iatridou, S. 1990. Clitics and island effects. *Penn Working Papers in Linguistics* 2, 11–30.

Kayne, R. 1994. *The Antisymmetry of Syntax*. Cambridge, Mass.: MIT Press.

Rizzi, L. 1997. The fine structure of the left periphery. In: L. Haegeman (ed.), *Elements of Grammar*, 281–337. Dordrecht: Kluwer.

Sportiche, D. 1998. *Partitions and Atoms in Clause Structure*. London: Routledge.

On Pure Syntax (Uncontaminated by Information Structure)*

Gisbert Fanselow

0. Introduction

Word order correlates with distinctions of information structure, and Günther Grewendorf's work has contributed much to our understanding of what the pertinent regularities are in German and other languages such as Italian or Japanese, and it has also shaped our understanding of how these regularities are linked to grammar. Grewendorf (1980) constitutes one of the first concrete proposals of capturing the impact of informational distinctions on German word order. Grewendorf and Sabel (1994, 1999) developed one of the most detailed models of word order variation in the middle field. Grewendorf has also contributed substantially to recent developments concerning the left periphery of clauses, based on the view that categories like focus and topic are directly represented in the syntax. Rizzi (1997) proposed that there are Topic and Focus heads situated in the higher functional layers of the clauses, and that the specifiers of these heads host topic and focus phrases, respectively – if not already in the surface representation, then at least at LF. Rizzi's view has been elaborated for German by Frey (2004), Haftka (1995), Pili (2000), and, of course, Grewendorf (2005a,b).

Such approaches in which information structure is directly coded in the syntax constitute one of two extreme ends of a continuum of models of the syntax-information structure interaction. Correlations between positions and informational functions do not necessarily imply that information structure and word order are directly related by syntactic laws. Rather, informational concepts might be correlated with other properties of phrases (say, their length), which could then be the ones really affected by the forces determining word order, such as processability, see, e.g., Hawkins (1994) for such a view.[1]

In this paper, I will argue that information structure concepts indeed do not play an immediate role in syntax. While word order can reflect information structure categories, it does so because these categories are encoded phonologically in a certain form (or because of the semantic consequences of information structure distinctions) to which

* The research reported here has been supported by a grant of the Deutsche Forschungsgemeinschaft DFG to the SFB 632 "Information Structure" at the University of Potsdam (Project A1). For discussions, comments, and other help that shaped the present paper, I would like to thank Sigrid Beck, Eva Engels, Ingo Feldhausen, Caroline Féry, Werner Frey, Katharina Hartmann, Shin Ishihara, Manfred Krifka, Denisa Lenertová, Gereon Müller, Stefan Müller, Peter Staudacher, Thomas Weskott, and, of course, the anonymous reviewer. For technical support, thanks go to Pawel Logatschew, Julia Vogel, and Nic Werner.

1 The experimental findings of Arnold et al. (2000) show, however, that reference to both length *and* information structure is needed in an account of English word order variation.

syntax is sensitive. Syntax interacts with phonology (and semantics), but not with information structure. We will proceed as follows. The paper begins with two conceptual arguments. Section 2 summarizes the empirical observations of Fanselow and Lenertová (2006) showing that "focus" does not trigger syntactic movement. Section 3 presents some empirical arguments against the existence of a topic position in German clauses. Section 4 then gives the results of an acceptability rating experiment testing one of the empirical claims made by Frey (2004) in the context of a Topic position. A few remarks on possible extensions and a program for future research can be found in section 5.

1. Some properties of narrow syntax

Whether concepts of information structure figure in syntactic computations is, of course, an entirely empirical issue. Nevertheless, one should be aware that an integration of information structure notions into formal syntax leads beyond limits otherwise respected by syntax in at least two respects: syntactic features are context-invariant, and movement is in principle obligatory when it is possible.

Syntactic categories and syntactic relations are typically context-independent. They encode properties that an element has because of its lexical specification, or because it stands in a formal relation to other elements in the sentence. A word is a noun independent of the context of the utterance it occurs in, it c-commands another element because of the structural relations holding between the two items, and its prosodic properties (being accented/de-accented) can also be determined sentence-internally by comparing it to other elements within the same intonation or prosodic phrase. Chomsky (1995) introduced a further restriction on the nature of features that may figure in the syntax by postulating an "inclusiveness" condition: the features used in syntax must be already specified in the lexical entries.

In an obvious sense, neither "topic" nor "focus" fulfill any of the criteria mentioned in the preceding paragraph. Unlike formal features such as "wh" or "neg", and unlike semantic features such as "specific" or "distributive", they do not encode invariant properties of lexical items[2] or phrases, and unlike features such as "accented", they do not encode properties that are true relative to a certain formal dimension of comparison with clausemate elements. Whether a phrase is a topic or a focus is, in the primary interpretation of the terms, only determined relative to the "knowledge base" of the speaker and hearer, that constitutes the background of sentence interpretation.

Topicality and focality differ in yet a further and even more important way from other syntactic features: they are not projective. When a phrase is marked for "wh", "accusative", "accented", etc., it bears this feature because it contains a lexical item of which the feature is true. In a dialogue such as (1), however, only the complete object DP is the focus of the answer – none of its part is a focus (e.g., because none of the parts

2 This may seem false for overt morphemes marking "topic" or "focus" that we find in certain languages. However, I know of no example of a morpheme that fulfils the function of a "focus marker" only, in which it would have to be present in each sentence, because all sentences have a focus. Rather, "focus markers" are always linked to additional meaning components, and one can derive the relation of the marker to focus from this additional semantic content.

of the direct object answers the question, only the complete object does). Therefore, the "focus" feature violates the inclusiveness condition, and it will not help to add "focus" to one of the lexical items when it is entered into the numeration (as suggested by the anonymous reviewer), because "focus" is not a correct attribute of any of the parts of the direct object. The same is true for topics.

(1) *What did you see? I saw* [$_{DP}$ *a small yellow book*].

If information structure distinctions cannot enter the vocabulary of syntax, how can the correlation of word order with information structure be captured? Information structure correlates with prosodic properties (accent), and there can be no doubt that syntax is sensitive to prosodic properties: the syntactic behaviour of German verbal particles/prefixes depends on their prosodic properties. Focus placement can (and must) be reanalysed in terms of the placement of accented elements. The latter description seems more successful than the former, as will be discussed in the next section. As we will see there, it is the property of accentuation itself (and not "focus-marking") that drives movement in many languages. Information structure influences syntax only indirectly. The features "topic" and "focus" do not really fit into syntax. One should avoid using them there if possible.

There is a further respect in which the interaction between information structure and syntax shows surprising properties. In German, syntactic responses to information structure are always *optional*, while the prosodic encoding of information structure is mandatory. E.g., focus and topic phrases can be placed into sentence initial position (Spec,CP), but they do not have to be so. One recurrent finding of acceptability rating experiments comparing different word orders in context is that unmarked subject initial sentences are *always* at least as acceptable as sentences with marked object initial order, even in those contexts that license object fronting (Weskott et al. 2004, Fanselow et al., 2006). The situation is even worse with scrambling: apparently, scrambled OSV sentences are *less* acceptable than SOV sentences even in contexts licensing the former (Keller 2000), and there is no context in which scrambling is obligatory (see Haider and Rosengren 2003). Word order variation linked to information structure is thus not easily amenable to a treatment in terms of "feature checking" or "alignment constraints". The reordering responding to information structure has properties quite different from those typically observed with A- or A-bar movement.

German is not exceptional in this respect. That scrambling is not obligatory seems to be a widespread (if not universal) property of the construction. When we consider the placement of focal subjects and adjuncts into the preverbal focus position in SOV languages like Sinhala or Turkish, we again find optionality rather than obligatoriness (see, e.g., Morimoto 2000 for an overview). The current syntactic models certainly possess tools for the description of optionality. They might be employed when topic and focus movement is described in a given language, but this would leave it open why *wh*-movement and NP-movement do not make such a pervasive use of the optionality tools.

In certain languages, movement to focus or topic position seems indeed obligatory: Hungarian and other "discourse configurational" languages (É.Kiss 1995) such as Catalan, and Italian may be cases in point. Note, however, that, e.g., the obligatoriness of focus movement may be due to the fact that prosodic properties can only be realized in certain syntactic configurations in such languages (Zubizaretta 1998, Szendröi 2001). Under this perspective, the obligatoriness of the movement of a focus XP is not linked to information structure directly, but to prosody.

2. Focus movement is accent displacement

Conceptual reflections are of little importance when empirical facts force certain conclusions concerning the nature of grammar. The assumption of Topic and Focus heads or features may be necessary on empirical grounds, in spite of the conceptual problems they would come with. However, at least in the domain of focus movement, the empirical facts are in line with the conclusions of the previous section. Focus placement data from a wide variety of languages cannot be captured in terms of rules that refer to "focus" features. Rather, prosodic properties are relevant. This has been argued for at length in Fanselow and Lenertová (2006), and we will summarize the main point of this paper here.

The crucial observation has been made by Kenesei (1998) for Hungarian, and by Krifka (1994), Büring (1996), and Gärtner (1996) for German. A sentence such as (2) with a fronted direct object is not only compatible with a narrow focus interpretation for *ein Buch* "a book". The sentence is also appropriate as an answer to a question such as "what have you done this morning?", i.e., it can also express VP-focus.

(2) *Ein Buch hab ich gelesen.*
a book have I read
'I've read a book.'

If interpreted with a VP focus, the element *ein Buch* fronted in (2) is *not* the focus of the utterance, but rather part of the focus of the utterance. It is not an XP bearing a focus feature, but part of such an XP. The fronting in (2) cannot be captured in terms of the attraction of a phrase with a focus feature. Sentence (3) illustrates the same point from a slightly different perspective. (3) shows that parts of idioms can be moved to clause initial position. The construction again expresses VP-focus. Given that the element fronted in (3) has no meaning of its own at all (since it is part of an idiom), it is also neither the focus nor the topic of the utterance. It therefore cannot have been fronted on the basis of a pragmatically defined focus (or topic) feature.

(3) *Den Nagel hat er auf den Kopf getroffen, als er sagte, dass ...*
the nail has he on the head hit when he said that
'he found the optimal expression when he said that ...'

As Fanselow and Lenertová (2006) show, such observations are not confined to German. Rather, they are characteristic of many languages with so-called focus fronting. Thus, direct counterparts of (2) and (3) can be found in Czech, Russian, Hungarian, Estonian, Greek, and Italian.

The theoretical analysis of (2) and (3) is quite straightforward. As observed by Fanselow and Lenertová, DP- and PP-objects can be fronted in VP and IP-focus contexts, but only so if they bear an accent, and if the subject is de-accented. Quite in general, the movement illustrated in (2) and (3) cannot cross a further accented category. Since movement is in general constrained by the Minimal Link Condition[3] (Chomsky 1995), the nature of the attracted feature can be read off the crucial properties of blocking interveners. Intervening XPs block fronting when they are accented, so the movement in (2) and (3) is due to the attraction of an *accented* phrase.

Normally, more than one phrase is accented in wide focus constructions (VP- and IP-focus), and it is always the *highest* of these accented XPs that can move to Spec,CP. There is no reason to call the accent on exactly this highest XP the "focus marker" – it has no special status at all. A "focus marking feature" could be assigned to *all* accents in a wide focus VP or TP (in that case, it would only be a new label for 'accented'), but a syntactic feature of X ('is marked for focus') that is formally realized on all major categories in X is quite unusual. Thus, replacing accent attraction by the attraction of a formal "focus marking feature" is not possible.

The first step in eliminating attraction processes related to focus features thus lies in the observation that we need an attraction of accented phrases, which is able to displace parts of focused constituents as in (2) and (3). Furthermore, when an accent is attracted, larger constituents containing the accented category may be moved as well, because their pied-piping can be necessary or possible. (4b) can answer both (4a) and (4a'), and represents a case in which only the accented word is displaced. (4c) illustrates that the quantifier or determiner (or, rather, the whole DP) can (and often, must) be pied-piped when the accented noun is attracted. Pied-piping is also obligatory when the accented word is embedded in a PP, since German allows no preposition stranding, see (4d) as potential answer to (4a'). The VP may be pied piped as well (4e), both as an answer to (4a)[4] and to (4a').

(4) a. *Was hat er gekauft?*
what has he bought
a'. *Was hat er gemacht?*
what has he done
b. *Bücher hat er ein paar gekauft.*
books has he some bought
'He bought some books.'
c. *Ein paar Bücher hat er gekauft.*
d. *Unter der Brücke hat er geschlafen.*
Under the bridge has he slept
'He slept under the bridge.'
e. *Ein Buch gekauft hat er.*
a book bought has he

3 If α moves to β in order to check the formal feature f, then, according to the MLC, there can be no γ closer to β than α that also has the feature f. However, in Chomsky (2005), superiority effects (objects with feature f cannot cross subjects with the same feature when f is attracted) can no longer be derived. See Müller (2004b) for a phase-based minimalist derivation of superiority effects in which effects such as the ones described above can be captured, too. The intervention effects of the movement of accented categories may also be due to shape conservation principles as proposed by Müller (2001).

4 This judgment is not shared by everyone. For me, the dialogue (4a + 4e) sounds best when the *wh*-word gets extra stress (as in an echo question), but it is acceptable in standard situations, too.

We can conclude that the category that moves when an accented element is attracted may be smaller or larger than the focus linked to the accent. It may also happen to be identical with the phrase in focus. Therefore, we do not need an additional focus placement rule – the attraction process displacing accented words already accounts for the sentences in which the focus phrase is displaced. Focus movement thus turns out to be a special case of accent movement. Syntax makes no reference to focus.

A central notion of information structure is thus *not* accessed by the syntax, at least in the component responsible for the triggering of movement. As Fanselow and Lenertová (2006) have shown, this conclusion holds in at least those languages in which focus is expressed prosodically.

3. A position for topics in German?

Focus placement can and must be reinterpreted in terms of rules for the positioning of accents. For topic placement, a similar replacement of notions of information structure by formal or semantic features is less obvious. In this paper, we cannot discuss topic placement in general. Therefore, we set ourselves a more modest goal: we will discuss and reject the claim that German has a structural position between CP and TP that all and only the topic phrases of a clause must move to (as suggested by Haftka 1995 and Frey 2004). If such a position would exist, syntactic movement would have to be sensitive to topicality, i.e., it would involve a notion defined in informational terms.

Movement to Spec,CP has often been called "topicalization" for German, but examples such as (5) and the ones discussed above show that not only topics can appear in the left periphery of CP. Likewise, topics need not appear at the beginning of a sentence, as the position of *dieses furchtbare Gebäude* "this horrible building" in (6) shows. The syntactic term "topicalization" (=movement to Spec, CP) thus has little to do with topics in a pragmatic sense. Rather, it is an operation attracting the highest category in the functional projection directly below Comp (TP or FinP) and moving it to Spec,CP, independent of the syntactic, semantic, and pragmatic properties of that phrase (see Fanselow 2002, Frey 2004, and Müller 2004a for different versions of this idea).

(5) *Nicht einmal einen Schwan hat Derk erkannt.*
not even a.ACC swan has Derk recognized
'Derk even did not recognize a swan.'

(6) *Was gibt's Neues über das Stadtschloss?*
'Any news about the city castle?'
Laut Bürgermeister Jacobs wird man dieses furchtbare Gebäude
according to mayor Jacobs will one this horrible building
nächstes Jahr endlich abreissen.
next year at last tear down
'According to mayor Jacobs, one will finally tear down this ugly building next year.'

Frey (2004) and Haftka (1995) argue that German nevertheless has a topic position, but that it is lower than Spec,CP. Frey (2004) assumes a topic head above the sentence adverbs (which are, e.g., adjoined to TP/FinP). The specifier of such a topic head may be preceded by a further functional head, KONTRAST, which is the highest functional head below Comp. All and only the topics must move to the specifier position of the topic head, and remain there unless they are further moved to Spec,CP on the basis of the process mentioned above. (7) taken from Frey (2004) is meant to illustrate that topics must be placed in front of sentence adverbs. (7b) is a pragmatically odd continuation of the first sentence in (7), apparently because the sentence topic *Maria* has been misplaced.

(7) *Ich erzähle dir etwas über Maria.*
I tell you something about Maria
a. *Nächstes Jahr wird Maria wahrscheinlich nach London gehen.*
next year will Mary probably to London go
'Next year Mary will probably go to London.'
b. *#Nächstes Jahr wird wahrscheinlich Maria nach London gehen.*

Frey (2004) proposes a set of syntactic tests for topicality, and argues that their application always yields the same result: (aboutness) topics must move to the topic position, and only topics can appear there. (8) represents this view of German clause structure. Note that Frey (2004) uses different labels for the functional heads.

(8) $[_{CP}$ (XP) Comp ($[_{KontrastP}$ YP Kontrast) ($[_{TopicP}$ ZP Topic)($[_{TP}$ adv) $[_{TP}$ [DP T']](])(])(])]

A first difficulty of this approach lies in the fact that it is less predictive than one might think. The problematic data can be found in Frey (2004) itself: More than one sentence adverb can appear in a German clause, and XPs such as *mindestens zwei* can be placed easily *between* such sentence adverbs.

(9) *Ich denke, dass wahrscheinlich mindestens zwei leider ihren Vortrag*
I think that probably at least two unfortunately their talk
absagen werden.
cancel will
'I think that probably at least two will cancel their talks, unfortunately.'

The DP sandwiched between the two sentence adverbs is not a topic, at least not in (9), as Frey (2004) observes. If sentence adverbs are adjoined to TP (as Frey assumes), (9) illustrates that DPs can be adjoined to TP above the adverbs, as one would expect in any event given that German is a scrambling language. If sentence adverbs project their own functional phrases (see Alexiadou 1997, Cinque 1999), then (9) illustrates that DPs can be adjoined to such functional projections. *Mindestens zwei* is thus not located in a topic position in (9), so that the absence of a topic interpretation constitutes no *immediate* problem for Frey (2004).

However, this amendment necessary for (9) drastically reduces the predictive power of the topic position theory. (9) forces us to assume that DPs can *adjoin* above sentence adverbs *without being topics*. Sentences such as (10) are therefore structurally ambiguous in Frey's approach (a fact not noted in Frey 2004): *Julia* can occupy the specifier position of the topic head, or be adjoined to TP.

(10) *Ich denke, dass Julia leider ihr Hund gebissen hat.*
I think that Julia unfortunately her.NOM dog bitten has
'I think that unfortunately, her dog bit Julia.'

It thus cannot be maintained that all (non-contrastive) XPs preceding sentence adverbs are topics! Rather, due to the structural ambiguity that must be assumed because of (9), the model of Frey (2004) tolerates both topics and non-topics to the left of sentence adverbs. The model makes no testable prediction[5] concerning the pragmatic status of an XP showing up before sentence adverbs, it only makes the prediction that topics must not appear to the right of them.

Perhaps, the conclusion that the element preceding a sentence adverb need not always be a topic is not so bad after all. Frey (2004) claims that there is a contrast between (11a) and (11b) which is due to the fact that quantified expressions such as *mindestens zwei* cannot be topics. But if the sentence is elaborated by more material as in (11c), the placement of the quantified DP in front of *leider* does not sound ungrammatical at all.

(11) a. * *Während des Vortrags haben mindestens zwei leider geschlafen.*
during the talk have at least two unfortunately slept
'Unfortunately, at least two slept during the talk.'
b *Während des Vortrags haben leider mindestens zwei geschlafen.*
c. *Auch dies Jahr werden während des Vortrags wieder mal*
also this year will during the talk again once
mindestens zwei leider schlafen.
at least two unfortunately sleep
'Unfortunately, at least two will once again sleep during the talk in this year, too.'

(12) illustrates the same: non-referential quantified DPs can be placed to the left of sentence adverbs.

(12) *dass sie wen aus Hamburg wahrscheinlich/leider nicht*
that she one from Hamburg probably/unfortunately not
heiraten würde
marry would
'Probably/unfortunately, she would not marry anybody from Hamburg.'

(9)–(12) undermine the topic position model: they support the insight that XPs can appear to the left of sentence adverbs without going to a specifier position of some Topic head. But then, there is no compelling reason for why topic phrases should not also

5 If more than one non-contrastive XP is placed to the left of the sentence adverb, Frey (2004) predicts, however, that the topics among these XPs should precede the non-topics. Such a generalization also follows from the standard serialization principles of German (definite > indefinite, given > new, etc.) without the further postulation of a slot reserved for topics.

simply *adjoin* to TP/FinP just as the non-referential non-topical elements do. Topic placement would then be an instance of *scrambling*.

Frey (2004) himself notes that there are far-reaching parallels between scrambling and topic placement. With respect to binding, reconstruction, and locality, the two processes do not differ from each other at all. Frey even assumes that scrambling and topic placement behave alike with respect to the absence of a freezing effect: elements preceding a sentence adverb are transparent for movement, as (13) (=(65a) in Frey 2004) illustrates.

(13) *Was*$_i$ *hat Hans* [t$_i$ *für Leute*] *dummerweise angerufen?*
what has H. for people annoyingly called
'What kind of people has Hans annoyingly called?'

If (13) is correctly analysed by Frey as exemplifying the extraction out of a DP in a topic position, the question arises why no freezing effect blocks that movement. German has few freezing effects. In general, subjects are transparent for movement in German, a fact usually explained by assuming that subjects do not move to Spec,TP. Scrambled DPs fail to acquire island status, too. Fanselow (2001) takes this as evidence for the base-generation analysis of scrambling, while Haider and Rosengren (2003) assume that freezing arises only in the context of feature driven movement to a specifier position, and not in the context of adjunction. Neither of the two models predicts that a DP moved to the specifier position of a Topic phrase remains transparent for extraction, but they correctly characterize examples such as (13) as grammatical if topic placement is a subcase of scrambling. It is not clear what a non-stipulative model of freezing could look like that draws a distinction between the specifier of Topic (transparent) and the specifier of Tense or Comp (intransparent). Movement data such as (13) may thus turn out to be problematic for the placement of topic phrases into specifier positions.

However, (13) allows a different interpretation as well. The DP *was für Leute* can also have been adjoined to TP in front of the sentence adverb before *was* was extracted out of it. This may in fact be a more convincing analysis, since *wh*-phrases such as *was für Leute* "what kind of people" make bad topics, and therefore should not move to a topic position at all.[6] Under this analysis, (13) is just a further illustration of the fact that elements preceding the sentence adverb need not be topical.

While details thus remain a bit unclear, it seems fair to conclude that the only possible syntactic difference between scrambling and topic movement might reside in the obligatory nature of topic placement, contrasting with the optionality of scrambling. Little, however, can be concluded from the observation that a certain interpretive effect presupposes the application of some syntactic rule. Consider (14) in this respect. For many speakers of German, (14a) is unambiguous (Frey 1993), i.e., with "normal" word order, scope corresponds to c-command relations. If one wants to express an interpreta-

---

6 If it is correct that one cannot extract from topics (see, e.g., Erteschik-Shir 2006), no phrase containing a gap could ever be a topic, irrespective of position. The predictions of the different models discussed here may thus be fairly clear, yet it may turn out that they cannot be tested at all. True extraction out of DPs may be confined to very few constellations in German (see deKuthy 2000 for arguments in favour of that view) that do not involve non-contrastive topical phrases.

tion in which the universal quantifier takes scope over the existential one, the object needs to be scrambled in front of the subject, as in (14b).

(14) a. *dass mindestens ein Professor fast jeden Studenten kennt*
that at least one.NOM professor nearly every.ACC student knows
b. *dass fast jeden Studenten mindestens ein Professor kennt*
'that at least one professor knows nearly every student'

Does that mean that scrambling is *syntactically* obligatory in certain constellations, triggered by the attraction by a quantificational head? This idea is not convincing, because applications of scrambling that affect quantifier scope do not differ from the other instances in syntactic respects. Rather, there are mechanisms that govern the interpretation of syntactic structures (as proposed, e.g., in Frey 1993), and these mechanisms simply yield different results in the case of scrambled and unscrambled structures. Similarly, different portions of the sentence may be linked to different aspects of information structure, and due to this mapping, it may be that XPs with a certain pragmatic function are always found in a certain region of the clause, without this being caused by the syntactic derivation itself.

Bearing this abstract observation in mind, let us return to topic placement in German. One fact that obligatory topic movement seems to imply is the generalization exemplified by (7): topics need to precede sentence adverbs. This, however, does not capture the distribution of sentence adverbs in a completely satisfactory way. It has been noted frequently that sentence adverbs share many of the distributional characteristics of scalar particles like *nur* "only", i.e., sentence adverbs seem to be focus-sensitive operators. See Hetland (1992) and Engels (2004) for an overview.

The assumption that sentence adverbs have the syntactic properties of focus sensitive operators allows to explain sentences such as (16) easily. As Reis (2005) and Müller (2005) have shown, focus-sensitive operators combine with arguments (15) as readily as with propositional projections (VP, TP). Frey (2004) claims that each sentence adverb (being adjoined to TP) has a homophonous focus-sensitive counterpart: he proposes to distinguish the "focus-sensitive" (16) from the "neutral" use, and concedes that the generalization concerning the order of sentence adverbs and topics is true for the latter use only. However, Engels (2004) shows convincingly that the syntactic differences between "focus-sensitive" and "neutral" interpretations follow from the differences in the size of the focus constituent (wide VP/IP focus in the "neutral" use, narrow DP/PP focus in the "focus sensitive" use), so that the duplication of sentence adverbs that Frey (2004) works with is not warranted.

(15) *Nur/sogar den Fritz hat sie geliebt.*
only/even the.ACC Fritz has she loved
'She loved only/even Fritz.'

(16) *Wahrscheinlich/leider den Fritz hat sie geliebt.*
probably/unfortunately the.ACC Fritz has she loved
'Probably/unfortunately, she loved Fritz.'

What consequences does this have for topic placement? Like any other focus sensitive operator, a sentence adverb must precede the accented (focused) constituent. Focus constituents therefore follow sentence adverbs if the latter are focus sensitive operators. Furthermore, Büring and Hartmann (2001) show that focus sensitive particles should be placed into a position *adjacent* to the accented phrase if possible. Thus, when there is a narrow focus on the indirect object, it is odd to place the subject between the focus sensitive particle *sogar* "even" and the indirect object, as (17) shows. In other words, material that does not belong to the focus should be placed to the left of focus sensitive particles.

(17) a. *Gestern hat Rufus sogar [dem MÄDCHEN]$_F$ Blumen geschenkt.*
Yesterday has Rufus even the.DAT girl flowers given
'Yesterday, Rufus gave flowers to even the girl.'
b. * *Gestern hat sogar Rufus [dem MÄDCHEN]$_F$ Blumen geschenkt.*
c. *Gestern hat sogar RUFUS$_F$ dem Mädchen Blumen geschenkt.*
'Yesterday, even Rufus gave flowers to the girl.'

The relative acceptability of the data remains constant when *sogar* is replaced by *wahrscheinlich* "probably" or *leider* "unfortunately". In the spirit of Büring and Hartman, we account for (17) by assuming that *all* focus sensitive operators are preferentially placed into the lowest position in which they precede and c-command the focus. If we apply this description to (18), we can derive the word order without the assumption of a special topic position. See Engels (2004) for an elaboration of this point in an OT framework.

(18) a. *Gestern hat Rufus wahrscheinlich [dem MÄDCHEN]$_F$*
Yesterday has Rufus probably the.DAT girl
Blumen geschenkt.
flowers given
'Yesterday, Rufus probably gave flowers to the girl.'
b. * *Gestern hat wahrscheinlich Rufus [dem MÄDCHEN]$_F$ Blumen geschenkt.*
c. *Gestern hat wahrscheinlich RUFUS$_F$ dem Mädchen Blumen geschenkt.*

In general, sentence adverbs *follow* topical phrases because sentence adverbs require adjacency to the focused/accented phrase. Material that does not belong to the focus will have to be scrambled in front of the sentence adverb. This need for scrambling should not apply when the adverb is constructed with a narrowly focused argument or adjunct and when the topical element follows the narrow focus in normal word order. In this situation, the adjacency requirement for the sentence adverb and the focus is already met in normal order, because the topic does not intervene. The interpretation of sentence adverbs as focus sensitive operators predicts that topics can remain *in situ* in such a constellation, and (19) shows that this prediction is indeed borne out:

(19) *Gibt's was neues über das Stadtschloss*?
'Any news about the city castle?'
Laut dem Bürgermeister wird man wahrscheinlich in Zukunft nur
According to the mayor will one probably in future only
am SAMSTAG dieses Gebäude besichtigen können
on Saturday this building visit can
'According to the mayor, one will only be able to visit that building on Saturdays in the future.'

Topics remain *in situ* in further constellations. Scrambling is avoided (but not syntactically excluded) in those transitive sentences in which the semantic roles of the two noun phrases cannot be read off their morphological marking or semantic differences between the noun phrases. In such a situation, it seems that topics are not placed into the "topic position" either, as (20) illustrates. The two feminine noun phrases in the embedded clause in (20) are morphologically ambiguous between a nominative and an accusative interpretation, so that constituent order is the primary cue for grammatical role assignment. It would be odd to place the topical object in front of the sentence adverb, since it would precede the subject there, too, inviting a reinterpretation of grammatical roles in which the princess is the fraudulent person. Topic placement behaves like scrambling in this respect. It is avoided when the fronting of a topic would interfere with the proper identification of thematic roles. This is hard if not impossible to explain if topic placement were due to the obligatory attraction of a topic phrase to a specifier position.

(20) *Was schreiben die Zeitungen über Prinzessin Julia*?
'What do the papers write about princess Julia?'
Der Hofkurier schreibt dass leider erneut eine Maklerin
The Court Courier writes that unfortunately again a broker(FEM)
die Prinzessin um 100.000 Euro betrogen hat.
the princess by 100,000 Euro cheated has
'The Court Courier writes that unfortunately a broker cheated the princess once more out of 100,000 Euro."

(21) and (22) are further illustrations of the fact that the placement of a topic to the right of a sentence adverb is not excluded in German.

(21) *Zum Stadtschloss fällt mir nur ein, dass wahrscheinlich niemand*
To-the city castle occurs me only PRT that probably nobody
dieses Gebäude mehr liebt als der Parteichef
this building more loves than the party leader
'As for the city castle, it only occurs to me that probably nobody loves this building more than the party leader.'

(22) *Was hat er über New Orleans gesagt?*
'What did he say about New Orleans?'
Er hat gesagt, dass wahrscheinlich ein weiterer Hurrikan die Stadt
He has said that probably a further hurricane the city
endgültig zerstören wird.
finally destroy will
'He said that probably a further hurrican would finally destroy the city.'

The complement clause in (21) and the declarative in (22) are wide focus utterances, which means that the sentence adverb comes with its 'neutral' use. The most prominent accent lies on the subject, which already precedes the given and topical object, so that the object need not go to the left in order to allow the subject to be adjacent to the sentence adverb. Again, we have a constellation in which topic fronting is not mandatory.

The placement of sentence adverbs in general, and in particular relative to topics, can thus be derived from their focus sensitive nature in a satisfactory way. The postulation of an additional functional projection (Topic phrase) in German clause structure cannot be motivated on the basis of data from this domain. There is very little left then (see section 5) that could support the idea of a special topic position, and it can be explained away, too.

The reduction of topic placement rules to the focus sensitive nature of sentence adverbs is only the first step in the attempt to keep syntax free of notions of information structure. But, in the light of what we have seen above, the second and final step seems to be an easy one: the placement rules for focus sensitive adverbs and particles need not refer to the focus status of an expression in a pragmatic sense. Rather, these rules can be sensitive to the prosodic realization of the focus. E.g., we might say that a focus sensitive particle/adverb must be placed in a position adjacent to the first accented word in the category it semantically modifies. To the extent that focus sensitive adverbs can be placed in between the parts of idiomatic expressions as in (23), we can be sure that the placement rules do not refer to matters of content, but rather involve a formal property – accent.

(23) *Da wurde der Tag leider mal wieder vor dem Abend gelobt.*
there was the day unfortunately once again before the evening praised
'Unfortunately, once again, one counted one's chicken before they were hatched.'

Furthermore, the placement of topics behind sentence adverbs appears to yield considerably improved results when the topic is not expressed by a pronoun or a repetition of the proper name (or noun phrase) with which the topic was introduced. This improvement of acceptability due to lexical variation is illustrated in (24) and (25).

(24) *Was weißt Du über Josef?*
'What do you know about Josef?'
a. ?*Ich denke, dass die Polizei wahrscheinlich ihn bald verhaften wird.*
I think that the police probably him soon imprison will

b. ?*Ich denke, dass die Polizei wahrscheinlich den Josef bald verhaften*
I think that the police probably the Josef soon imprison
wird.
will

c. *Ich denke, dass die Polizei wahrscheinlich diesen miesen Verbrecher*
I think that the police probably this mean criminal
bald verhaften wird.
soon imprison will
'I think that probably the police will soon imprison him/Josef/this mean criminal.'

(25) *Was weißt Du über New Orleans?*
'What do you know about New Orleans?'

a. ?*Ich glaube, dass der Hurrikan leider New Orleans zerstört hat.*
I think that the hurricane unfortunately N.O. destroyed has

b. *Ich glaube, dass der Hurrikan leider diese schöne Stadt*
I think that the hurricane unfortunately this beautiful city
zerstört hat.
destroyed has
'I think that unfortunately the hurricane destroyed N.O./this beautiful city.'

An ongoing acceptability rating experiment carried out in collaboration with Caroline Féry will reveal whether these judgments are shared by a larger number of speakers. If real, these judgment differences are probably due to the fact that the lexical variation in the final examples in (24–25) licenses the presence of an accent placed on the aboutness topic. A repeated name/NP, on the other hand, should be deaccented. If this is the correct explanation, the data in (24)–(25) constitute further evidence for the claim that focus sensitive adverbs are placed relative to prosodic rather than pragmatic properties of their neighbours.

An influence of topicality on the placement of sentence adverbs has also been observed for English, cf. (26), and the Romance languages. Normally, sentence adverbs may precede or follow the subject in English, but when the subject is non-referential, the adverb can only precede it. Engels (2004) shows that English sentence adverbs can be adjoined to both TP and T'. Low adjunction in (26a) is excluded by the focus sensitive nature of the adverb, which requires that it precedes *nobody* (an expression that is never part of the given information). It is quite unclear how the facts in (26) could be expressed in terms of a *topic position.* English referential subjects do not target such a slot. The fact that *probably* and *wahrscheinlich* precede negatively quantified subjects finds the same explanation in English and German only if the grammatical model involves focus sensitivity.

(26) a. (*Probably*) *nobody* (**probably*) *has* (**probably*) *left.* (Belletti 1990: 51)
b. (*Probably*) *John* (*probably*) *has left.*

In this section, we have discussed three types of evidence against the existence of a topic position in the German clause. First, such an assumption cannot be reconciled easily with the possibility of having more than one sentence adverb in a clause. Second, the empirical data that might be captured by assuming a topic position (relative to sentence adverbs) are already accounted for by the focus-sensitive nature of sentence adverbs. Third, it seems that it is not the pragmatic/informational status of a phrase but rather its prosodic properties that determine its placement relative to focus sensitive adverbs.

4. Topic placement relative to sentence adverbs: an acceptability rating experiment

As we have seen in section 3, sentence topics are not always placed in front of sentence adverbs in German. One major difficulty in dealing with the issue of topic placement lies in the fact that the relevant judgments are often quite subtle and unstable, and that there is some disagreement among speakers concerning the status of the relevant examples. Therefore, systematic empirical studies concerning the acceptability of various word orders in context are called for.

Acceptability rating studies in the domain of word order variation related to information structure are confronted with at least three kinds of problems. The first problem has already been mentioned above. Object fronting to Spec, CP in proper contexts leads to structures that are perfectly acceptable, but up to now, experiments with object fronting within the middle field have always found a reduction of acceptability (as compared to SOV). This could be due to the processing problems of OSV order in the middle field, or to the low frequency of such structures, or to a pure syntactic problem, or to the fact the "truly licensing" contexts have not yet been used in experiments. Some yet unidentified factor lowers the acceptability of OSV sentences, and as long as this factor has not been identified, it is difficult to interpret the results of experiments comparing SOV and OSV.

A second difficulty stems from the fact that one knows little about the relation between acceptability ratings for single sentences, and acceptability ratings for the kind of mini-texts that one typically employs when one investigates where topics or foci have to be placed. In an experiment reported in Fanselow et al. (2006), the reaction to incoherent question-answer pairs such as *who did you see on the dancefloor – I saw a bottle on the dancefloor* was surprisingly positive. Probably, the judgment of a mini-text averages over the acceptability of the individual sentences and the degree to which they fit together. In addition, incoherent dialogues are possible to a certain extent in so far as the second sentence can reject a presupposition of the first one, etc. With the prosody of a corrective focus on the object *a bottle*, the second sentence in our mini-dialogue is in fact an appropriate reaction to the question.

The third difficulty is confined to investigations of topic placement. The part of an utterance that corresponds to the *wh*-phrase in a congruent *wh*-question is the focus of the utterance (see Büring, to appear). This definition can be used in a very direct way to force a focus interpretation of some part of an utterance in an experiment. The same is true for the notion of "givenness". In contrast, it is much harder to operationalize the notion "topic", because the idea that a sentence is "about something" is inherently vague.

The acceptability rating experiment reported here avoids the first and the second difficulty. All experimental items had subject before object order, and they consisted of a single sentence. Of course, the third difficulty cannot be circumvented in an experiment concerned with topic placement. The experiment exploited the relation between topicality and cataphoric coreference. Reinhart (1981, 1995) claimed that cataphoric pronouns can refer to topics only, so that the availability of a cataphoric reading is a positive criterion for topicality. Frey (2004) borrows this test from Reinhart, and argues that topics as identified by the cataphoric relation test have to precede sentence adverbs. However, the data do not seem compelling to us, and, in particular, we believe that sentences in which the pronoun's antecedent follows a sentence adverb improve whenever a second cataphoric relation is established in the clause:

(27) *Weil er sie sehr liebt, wird wahrscheinlich der Pfarrer*
because he her much loves will probably the priest
die Haushälterin doch heiraten.
the housekeeper PRT marry
'Because he loves her so much, the priest will probably marry the housekeeper.'

An acceptability rating experiment was designed in order to establish whether this judgment is representative.

Participants
32 students of the University of Potsdam participated in the experiment; most of them were linguistics students in their first year of education. They were not familiar with the purpose of the study.

Method and Material
The participants rated 104 sentence presented to them in a written questionnaire on a 7-point scale (1 – 7, with 1 representing the lowest degree of acceptability). There were 16 experimental items, and 88 distractor items, most of which belonged to other experiments.

As illustrated in (28), the sentences used in the experiment consisted of an adjunct clause preceding a main clause, both involving a transitive verb. In the adjunct clause, the subject and the object were pronouns. In the matrix clause, they were lexical NPs. The matrix clause also contained a sentence adverb, which could appear in front of the two NPs (28A), between them (28B), or behind them (28C). In these three experimental conditions, the gender of the pronouns was chosen such that coreference between the subjects of the two clauses would be grammatically well-formed, and the same was true for the objects. Coreference (yielding cataphoric interpretations) was also pragmatically plausible. In the fourth condition (28D), the objects of the two clauses disagreed with respect to gender, so that no cataphoric relation could be established.

(28) Condition A: no topics fronted

Obwohl	*er sie*	*liebt,*	*hat gestern*	*überraschenderweise*	*der Hans*		
although	he her	loves	has yesterday	surprisingly	the Hans		

seine Freundin verlassen.
his girlfriend left
'Although he loves her, Hans surprisingly left his girlfriend yesterday.'

Condition B: one topic fronted
Obwohl er sie liebt, hat gestern der Hans überraschenderweise seine Freundin verlassen.

Condition C: two topics fronted
Obwohl er sie liebt, hat gestern der Hans seine Freundin überraschenderweise verlassen.

Condition D: no cataphoric relation possible for object NP

Obwohl	*er*	*ihn*	*liebt,*	*hat gestern*	*überraschenderweise*	*der Hans*
although	he	him	loves	has yesterday	surprisingly	the Hans

seine Freundin verlassen.
his girlfriend left

16 sets of sentences similar to (24) were constructed. Each participant saw exactly one sentence from each of these sets, such that 4 items per condition were presented to him/her. There were 4 versions of the questionnaire, guaranteeing that each sentence was presented to 8 participants in each of the four conditions.

Results
The mean acceptability of condition A was 5.62 on the 7 point scale. For conditions B and C, mean acceptability was at 6.09, while for condition D mean acceptability was 3.25.

A computation of difference contrasts (reverse Helmert contrasts) revealed that conditions A and B(/C) differed significantly from each other: ($F_1(1,31) = 5{,}16$; $p = .03$; $F_2(1,15) = 19{,}33$, $p = .001$). Likewise, condition D differed significantly from conditions A–C: ($F_1(1,31) = 112{,}45$; $p < .001$; $F_2(1,15) = 78{,}62$, $p < .001$). No other contrasts were significant.

Discussion
The difference in acceptability between condition D on the one hand and conditions A–C on the other is crucial for the interpretability of the whole experiment. This difference shows that the participants tried to interpret the pronouns in the adjunct clause as referring to textual antecedents. In all the conditions in which the matrix clause contains NPs with which the subject and the object can co-refer, acceptability is high. When there is no plausible antecedent for the object in the matrix clause because of the gender clash, as in condition D, acceptability is fairly low. The contrast between A–C and D thus warrants the conclusion that the participants interpreted the pronouns in A–C cataphorically.

The high acceptability of condition A, in which neither of the two NPs precedes the sentence adverb (neither of them occupies the "topic position" in Frey's terms), constitutes a serious problem for the claims that (a) topics must be placed in front of sentence adverbs and that (b) cataphoric pronouns may refer to topics only. If both (a) and (b) are true, it should be impossible to assign a cataphoric interpretation to any of the pronouns in condition A, so that condition A sentences should have the low acceptability value of condition D sentences, which they do not. (a) and (b) cannot be upheld simultaneously.

Condition A sentences were rated as being slightly less acceptable than condition C sentences. If (b) holds, this could mean that a violation of the rule that topics be placed to the left of sentence adverbs induces a very mild type of unacceptability. The small size of the acceptability difference militates against the view that it is caused by the failure of carrying out an obligatory movement operation. In addition, condition B and condition C sentences cannot be distinguished from each other in terms of acceptability, in spite of the fact that one of the two NPs entering a cataphoric relation with the pronouns follows rather than precedes the sentence adverb in condition B. The optimal acceptability of condition B sentences cannot be explained by the assumption that the second NP is not a topic in the system proposed by Frey (2004), because this NP should then neither be able to appear in front of the adverb (as it does in C without inducing any unacceptability), nor be able to license a cataphoric relationship (as it does in B and C).

The contrast between A and B/C might be explained if topic placement follows the principle of "Minimal Compliance" in the sense of Richards (1997, 2001), according to which certain constraints need to be satisfied once per clause only. Other candidates for minimal compliance effects such as the superiority condition or subjacency share one further property of topic placement, viz. the relative mildness of the grammatical violation. For reasons of space, we cannot discuss this similarity here.

Except for the contrast between condition A and conditions B/C, the results of our experiment are also compatible with the view that assumption (b) only is false, i.e., the experiment could also show that cataphoric pronouns can refer to NPs that are not topics. In that case, the experimental results do not prove that topics can appear behind sentence adverbs. The experiment would, however, show that a crucial test for topicality used in Frey (2004) is not valid.

5. Concluding Remarks

In this paper, we have tried to defend the view that syntactic rules make no reference to concepts of information structure. The impact of information structure on syntax is an indirect one, mediated primarily by the prosodic manifestations of information structure, but probably also by its semantic correlates. For focus, there are very clear data (sketched in section 2) that support this view.

What we have seen in sections 3 and 4 is a first step in an attempt to also eliminate reference to the notion of "topic" from the theory of syntax. Some of the empirical observations made in section 3 need to be confirmed in acceptability rating experiments. VP-fronting is a second source of evidence for topic placement in Frey (2004) (besides the sentence adverb ordering facts discussed above). We have dealt with this domain in Fanselow (2003), but not all that was said there is compatible with what proved to be required here. Furthermore, reference to topicality may be necessary in domains other than the middle field. Left dislocation (see Grewendorf 2005b) is a case in point. Therefore, one is not yet justified to claim that information structure and syntax do not interact directly, but we have presented evidence suggesting that such a claim may not be too far off the track.

References

Alexiadou, A. 1997. *Adverb Placement: a Case Study in Antisymmetric Syntax*. Amsterdam: John Benjamins.

Arnold, J., T. Wasow, A. Losongco, and R. Ginstrom. 2000. Heavyness vs. newness: The effects of structural complexity and discourse status on constituent ordering. *Language* 76,1, 28–55.

Belletti, A. 1990. *Generalized Verb Movement*. Torino: Rosenberg and Sellier.

Büring, D. 1996. *The 59th Street Bridge Accent. On the Meaning of Topic*. Doctoral dissertation, Universität Tübingen.

Büring, D. (to appear). Semantics, intonation and information structure. In: G. Ramchand and C. Reiss (eds), *Handbook on Interface Research in Linguistics*, Oxford: Oxford University Press.

Büring, D., and K. Hartmann. 2001. The syntax and semantics of focus-sensitive particles in German. *Natural Language and Linguistic* Theory 19: 229–281.

Chomsky, N. 1995. *The Minimalist Program*. Cambridge, Mass.: MIT Press.

Chomsky, N. 2005. On phases. Ms., MIT.

Cinque, G. 1999. *Adverbs and Functional Heads: a Crosslinguistic Perspective*. Oxford: Oxford University Press.

DeKuthy, K. 2000. *Discontinuous NPs in German. A Case Study of the Interaction of Syntax, Semantics, and Pragmatics*. Doctoral dissertation, Saarbrücken.

Engels, E. 2004. *Adverb Placement. An Optimality Theoretic Approach*. Doctoral dissertation, Potsdam.

Erteschik-Shir, N. 2006. What's what? In: G. Fanselow, C. Féry, M. Schlesewsky, and R. Vogel (eds), *Gradience in Grammar*. Oxford: Oxford University Press, in press.

Fanselow, G. 2001. Features, theta-roles and free constituent order. *Linguistic Inquiry* 32, 405–437.

Fanselow, G. 2002. Münchhausen style head movement and the analysis of verb second. In: A. Mahajan, (ed.), *Proceedings of the Workshop on Head Movement*, 40–76. Los Angeles: UCLA, Linguistics Department.

Fanselow, G. 2003. Free constituent order. A minimalist interface account. *Folia Linguistica* 37, 191–231.

Fanselow, G. and D. Lenertova. 2006. Left peripheral focus: mismatches between syntax and information structure. Ms., Universities of Potsdam and Leipzig.

Fanselow, G., D. Lenertova, and T. Weskott. 2006. Studies on the acceptability of object movement to Spec,CP. Ms., Universities of Potsdam and Leipzig.

Frey, W. 1993. *Syntaktische Bedingungen für die semantische Interpretation*. Berlin: Akademie Verlag.

Frey, W. 2004. A medial topic position for German. *Linguistische Berichte* 198, 153–190.

Gärtner, H.-M. 1996. Review of "Marga Reis (ed.), *Wortstellung und Informationsstruktur* (= Linguistische Arbeiten 306). Tübingen: Niemeyer 1993." *Linguistische Berichte* 161, 90–94.

Grewendorf, G. 1980. Funktionale Satzperspektive und deutsche Wortstellung. *Linguistische Berichte* 66, 28–41.

Grewendorf, G. 2005a. The discourse configurationality of scrambling. In: J. Sabel and M. Saito (eds.), *The Free Word Order Phenomenon*, 75–135. Berlin: deGruyter.

Grewendorf, G. 2005b. The left clausal periphery: clitic left dislocation in Italian and left dislocation in German. Paper presented at the ZAS workshop *The Syntax-Semantics Interface in the CP-Domain*. To appear in the proceedings.

Grewendorf, G. and J. Sabel. 1994. Long scrambling and incorporation. *Linguistic Inquiry* 25, 263–308.

Grewendorf, G. and J.Sabel. 1999. Scrambling in German and Japanese: adjunction versus multiple specifiers. *Natural Language and Linguistic Theory* 17, 1–65.

Haftka, B. 1995. Syntactic positions for topic and contrastive focus in the German middlefield. In: I. Kohlhof, S. Winkler and H.B. Drubig (eds), *Göttingen Focus Workshop*. Arbeitspapiere des SFB 340, Bericht Nr. 69, 137–157.

Haider, H. and I. Rosengren. 2003. Scrambling: nontriggered chain formation in OV languages. *Journal of Germanic Linguistics* 15, 203–267.

Hawkins, J. 1994. *A Performance Theory of Order and Constituency*. Cambridge: Cambridge University Press,

Hetland, J. 1992. *Satzadverbien im Fokus*. Tübingen: Narr.

Keller, Frank, 2000. *Gradience in grammar: Experimental and computational aspects of degrees of grammaticality*. PhD Dissertation, University of Edinburgh.

Kenesei, I. 1998. Adjuncts and arguments in VP-focus in Hungarian. *Acta Linguistica Hungarica* 45, 61–88.

È.Kiss, K. (ed.). 1995. *Discourse Configurational Languages*. Oxford: Oxford University Press.

Krifka, M. 1994. Focus and operator scope in German. In: P. Bosch and R. van der Sandt (eds.), *Proceedings of the Conference on Focus and Natural Language Processing, Volume 1: Intonation and Syntax*, 133–152. Heidelberg: ILL.

Morimoto, Y. 2000. *Discourse Configurationality in Bantu*. PhD Dissertation, Stanford.

Müller, G. 2001. Order Preservation, Parallel Movement, and the emergence of the unmarked. In: G. Legendre, J. Grimshaw, and S. Vikner (eds.), *Optimality-Theoretic Syntax*, 279–313. Cambridge, Mass.: MIT Press.

Müller, G. 2004a. Verb-second as vP-first. *Journal of Comparative Germanic Linguistics* 7.3, 179–234.

Müller, G. 2004b. Phase impenetrability and *wh*-intervention. In: A. Stepanov, G. Fanselow, and R. Vogel (eds.), *Minimality Effects in Syntax*, 289–325. Berlin: de Gruyter,

Müller, S. 2005. Zur Analyse der scheinbar mehrfachen Vorfeldbesetzung. *Linguistische Berichte* 203, 29–62.

Pili, D. 2000. *On A and A' Dislocation in the Left Periphery*. Doctoral dissertation, Universität Potsdam.

Reinhardt, T. 1981. Pragmatics and linguistics: an analysis of sentence topics. *Philosophica* 7, 53–94.
Reinhardt, T. 1995. Interface strategies. *Technical Report OTS Working Papers*, University of Utrecht.
Reis, M. 2005. On the syntax of so-called focus particles in German. – a reply to Büring and Hartmann 2001. *Natural Language and Linguistic Theory* 23.2, 459–483.
Richards, N. 1997. *What Moves Where in Which Language?* Doctoral dissertation, MIT, Cambridge, Mass.
Richards, N. 2001. *Movement in Language.* Oxford: Oxford University Press.
Rizzi L. 1997. The fine structure of the left periphery. In: L. Haegeman (ed.), *Elements of Grammar*, 281–337. Dordrecht: Kluwer.
Weskott, T., B. Stolterfoth, I. Bornkessel, and M. Schlesewsky. 2004. The task-dependency of acceptability judgements: processing scrambling and topicalization in German. Paper presented at the 2004 annual meeting of the DGfS, Mainz.

How to Get an Object-*es* into the German Prefield*

Werner Frey

1. Introduction

For a long time, it was a standard assumption that in German, the neuter personal pronoun *es* ('it') can only be moved to the prefield[1] if it has the function of a subject; if it is an object, the prefield was assumed to be disallowed for *es* (e.g., Travis 1984, 1992, Lenerz 1993, Cardinaletti and Starke 1996). A pair like the following was supposed to demonstrate this fundamental difference:

(1) a. *Es hat das Heu gefressen.*
it has the hay eaten (it = the horse)
b. **Es hat das Pferd gefressen.*
it has the horse eaten (it = the hay)

This observation had far reaching consequences for the structure linguists proposed for German. It was one of the main points which led Travis (1984, 1992) (and similarly, e.g., Zwart 1997) to argue for the so-called asymmetric analysis of German verb-second clauses (V2-clauses). According to this theory, IPs represent V2-clauses which have the subject as the first element, whereas CPs represent V2-clauses which have a non-subject as the first element. Thus, in this theory the sentences in (1) would get the structures in (2):

(2) a. $[_{IP}$ *Es hat*$_1$ $[_{VP}$ *das Heu gefressen* $t_1]]$
b. $[_{CP}$ *Es*$_2$ *hat*$_1$ $[_{IP}$ *das Pferd* t_1' $[_{VP}$ t_2 *gefressen* $t_1]]]$

Travis (1984,119) states that unstressed pronouns may not topicalize, i.e., they may not be Ā-moved to SpecCP but can only be A-moved to SpecIP. Since it is usually assumed that a weak pronoun like *es* cannot be stressed, the data in (1) seem to follow.

However, there are examples which show that an object-*es* can in fact appear in the prefield:

(3) a. (*Ihr Geld ist ja nicht weg, meine Damen und Herren.*)
('Your money is not gone, ladies and gentlemen.')
Es haben jetzt nur andere. Lenerz (1994)
it have now just others (it = the money)
'It is just in the possession of others now.'

---

* Many thanks go to Josef Bayer, Hans-Martin Gärtner, Andreas Haida, André Meinunger and to an anonymous reviewer. Obviously, I remain responsible for all errors.

1 The term 'prefield' refers to the position in front of the finite verb in a German verb-second clause.

b. (*Hans hat im Supermarkt eine Flasche Grappa gestohlen.*)
('Hans stole a bottle of grappa in the supermarket.')
Es hatte keiner gesehen. Grewendorf (2002)
it had nobody seen (it = the stealing-event)

c. (*Wie ist denn das Kind zu dem Buch gekommen?*)
('How did the child get the book?')
Es hat ihm jemand geschenkt. Gärtner and Steinbach (2003)
it has it.DAT someone given as a present (it = the book)

d. (*Dieses Schild können sie genauso weglassen.*)
('You can just as well leave this sign out.')
Es hat ohnehin keiner beachtet. Haider (2005)
it has anyway nobody observed (it = the sign)

e. (*Gestern bin ich erstmals ohnmächtig geworden.*)
('Yesterday I became unconscious for the first time.')
Es hat zum Glück niemand mitgekriegt. Meinunger (to appear)
it has luckily nobody noticed (it = the event of becoming unconscious)

What is the difference between (1b) and the sentences in (3)? It is Meinunger (to appear) who makes the following important observation. Sentences like (1b), which disallow the object-*es* in the prefield, and sentences like the ones in (3), which allow it, differ in the referential status of their subjects appearing in the middle field.[2] (1b) has a referential subject, whereas the sentences in (3) have non-referential subjects. Furthermore, Meinunger (to appear) points out that these different subject types are likely to occupy different positions in the middle field.

Meinunger's (to appear) observation can be made a bit more precise. Consider the examples in (4):

(4) *Wo ist das Lehrbuch "Syntax"? Ich kann es nicht finden.*
'Where is the textbook "Syntax"? I can't find it.'

a. *Es hat leider jemand der Maria ausgeliehen.*
it has unfortunately someone the.DAT Maria loaned
'Unfortunately, someone loaned it to Maria.'

b. **Es hat Otto leider jemandem ausgeliehen.*
it has Otto unfortunately someone.DAT loaned
'Unfortunately, Otto loaned it to someone.'

c. ?*Es hat leider Otto jemandem ausgeliehen.*

d. (?)*Es hat leider der Maria jemand ausgeliehen.*

e. *Es hat der Maria leider jemand ausgeliehen.*

The example in (4a) is similar to the ones in (3). It also contains a non-referential subject. However, the examples in (4b,c) show that in fact it is not the referential status of the subject that really matters, but the position of the subject. The subjects in (4b,c) are

2 With regard to a V2-clause, the term 'middle field' refers to the region between the finite verb and possibly occurring non-finite verbal elements; with regard to a complementizer-introduced verb-final clause, it refers to the region between the complementizer and the verbal element(s).

identical, but their positions differ. In the bad example (4b), the subject appears to the left of a sentence adverbial, while in the much better example (4c) it appears to the right. Furthermore, it is interesting to see that the position of a further object in the middle field has no influence on whether an object-*es* may or may not appear in the prefield. (4d,e) are both fine. Thus, it seems that it is just the middle field position of the subject that determines whether an object-*es* in the prefield is possible or not.[3]

The present paper tries to explain Meinunger's observation and the data in (4). To do so, the different ingredients of the explanation have to be established. First, section 2 discusses the difference between positioning an element to the left or to the right of a sentence adverbial. Then, in section 3, different options for getting an element into the prefield will be studied. Section 4 argues that in the German middle field there exist three different regions which could be connected to different subject positions. In section 5, claims about the possibilities of the placement of personal pronouns[4] in the middle field are put forward. Section 6 combines these ingredients to deduce an explanation for the data in (4) as well as for further examples.

2. 'Middle field topics'

Many linguists subscribe to the so-called 'aboutness concept' of topics. A well-known advocate of this concept is Reinhart (1981):

(5) A topic is an expression whose referent the sentence is about. The concept 'topic' is a category of pragmatic aboutness.

In this paper, too, the aboutness concept will be assumed. What is the position of an aboutness topic in a regular German clause? Many syntacticians working on topics in German assume that a sentence internal topic has to be moved to the prefield (e.g., Vallduví and Engdahl 1996, Reis 1999):

(6) *Peter reist heute nach Italien.*
Peter goes today to Italy

It is very natural to construe *Peter* as the aboutness topic of the clause.

However, Frey (2004), for instance, argues that topics do not have to be placed in the prefield. In that paper, the following thesis is argued for (see Fanselow this vol. for a critical discussion):

3 As indicated, (4c) and (4d) are not perfect. I assume that the deviance of these sentences has nothing to do with the positioning of the object-*es* in the prefield. The degraded status of (4c) is due to the fact that a referential subject is the preferred topic of a clause, i.e., it should occur in the topic domain, which is to the left of a sentence adverbial, cf. Section 2. The degraded status of (4d) I relate to the fact that given the context in (4), scrambling of the dative object to a position below the topic domain is not really motivated.

4 In the following the term 'pronoun' will always be used as an abbreviation for 'personal pronoun'.

(7) In the middle field of the German clause, directly above the base position of sentence adverbials (SADVs), there is a designated position for topics (in the aboutness sense): all topical phrases occurring in the middle field, and only these, occur in this position.

The term 'SADV' refers to adverbials which express the speaker's estimation of the eventuality, e.g., *zum Glück* ('luckily'), *anscheinend* ('apparently'), or *wahrscheinlich* ('probably'). (7) refers to SADVs in their neutral use, in which they modify the whole proposition. In this use, the base position of SADVs is higher than the base position of any other element in the German clause (cf., e.g., Frey 2003). SADVs may also have a focus inducing use, in which the SADV relates to one narrowly focussed constituent, the rest of the clause being presupposed. In the following, the focus inducing use of SADVs is not relevant and should be disregarded.

Two kinds of evidence for (7) are given below (for further evidence, see Frey 2004):

<u>The context forces aboutness</u>

(8) *Ich erzähle dir etwas über Hans.*
I tell you something about Hans
a. *Bald wird den Hans zum Glück eine vornehme Dame heiraten.*
soon will the.ACC Hans luckily a fine lady marry
'Luckily, a fine lady will soon marry Hans.'
b. #*Bald wird zum Glück den Hans eine vornehme Dame heiraten.*

<u>Topic marking particles</u>

(9) a. *[*Die Polizei jedenfalls*] *kommt.*
the police at any rate is coming
b. *weil* [*dem Hans jedenfalls*] *zum Glück jeder gratuliert hat*
since the.DAT Hans at any rate luckily everyone congratulated has
'since luckily everyone congratulated Hans at least'
c. **weil zum Glück* [*dem Hans jedenfalls*] *jeder gratuliert hat*

The context in (8) states that the next sentence will be about Hans. Thus, it requires that if *Hans* occurs in the following sentence, it has to be an aboutness topic. The sentences (8a,b) demonstrate that if the topical *Hans* occurs in the middle field, it has to precede a SADV. With regard to (9), note that in German there are particles which signal that the phrases they extend are topics.[5] That *jedenfalls* ('at any rate') is one of these particles is shown by (9a): the subject of a thetic sentence, i.e., a non-topic, cannot be modified by *jedenfalls*. (9b,c) show that, in the middle field, a DP which is extended by *jedenfalls* has to occur in front of a SADV.

Let us draw two consequences from (7). It can easily be shown that anaphoric phrases do not have to appear in the topic position, i.e., they need not be topical:

5 This test for the claim in (7) I owe to Marga Reis (p.c.).

(10) *Gestern hat Paul [eine nette Frau]$_1$ kennengelernt. Er wird hoffentlich*
yesterday has Paul a nice woman met He will hopefully
[die Dame]$_1$ wiedersehen.
the lady see again

Next, it follows that a clause may contain more than one topic. The reason is that in the middle field, more than one phrase may precede a SADV, cf. (11), and all these phrases are topical, cf. Frey (2004).[6]

(11) *Heute wird Maria dem Hans wahrscheinlich helfen.*
today will Maria the.DAT Hans probably help

That (7) refers to a designated structural topic position and that the examples (8)-(9) do not just concern the linearization of a topic and a SADV is shown by examples like (12a,b):

(12) a. [*In Peters$_1$ Firma*]$_2$ *hat Maria überraschenderweise* t_2 *diesen Kerl$_1$*
in Peter's company has Maria surprisingly this guy
bloßgestellt.
unmasked
'In Peter's company, Maria surprisingly unmasked this guy.'
b. *[*In Peters$_1$ Firma*]$_2$ *hat Maria diesen Kerl$_1$ überraschenderweise* t_2 t_1
bloßgestellt.

In (12a), the definite description does not induce a Principle C-violation for the co-indexed *Peter*, while in (12b), it does. These data find a straightforward explanation once we take into account that in (12b) the definite description has been moved to the topic position. The sentences in (12) contain the locative adverbial *in Peters Firma*. This adverbial, be it treated as a frame adverbial or as an event-related adverbial, has its base position (indicated by t_2 in (12)) below the base position of the SADV and above the base position of the object (cf., e.g., Frey 2003). For the checking of binding principle C, the phrase containing the R-expression is reconstructed to its base position if it has been moved. The bindee is reconstructed to the highest A-position it has occupied during the derivation (cf. Frey 1993). In (12a), the definite description does not c-command the base position of the adverbial in the prefield, and, therefore, no Principle C-violation arises for *Peter*. However, due to the movement of the definite description *diesen Kerl* to the topic position in (12b), the definite description c-commands the base position of the adverbial and a Principle C-violation arises for the phrase *Peter*.

Thus, the data in (12) not only show that topic marking of a phrase happens by means of moving that phrase to a certain position, they also give first evidence that this movement is an instance of A-movement, since it is A-movement which extends the

6 For example, in the middle field inherently non-topical phrases like quantified phrases cannot appear to the left of a SADV:

(i) a. * *Während des Vortrags hat keiner glücklicherweise geschlafen.*
during the lecture has no one fortunately slept
b. * *Otto wird nur morgen leider arbeiten.*
Otto will only tomorrow unfortunately work

binding options of the moved phrase. Another piece of evidence that the movement to the topic position is A-movement is given by the fact that it is clause bound:

(13) **Seit kurzem glaubt dem Paul*$_1$ *anscheinend jeder,* [$_{CP}$ *dass*
since recently believes the.DAT Paul apparently everyone that
man t$_1$ *den Preis zuerkennen wird*].
one the prize award will
'Recently, apparently everybody believes they'll award the prize to Paul.'

(13) shows that it is not possible to move the topic of an embedded clause to the designated topic position in the middle field of a higher clause. Thus, the data in (12) and (13) constitute strong evidence that in German, topic preposing is an instance of A-movement.

The conclusion that A-movement of a phrase can be triggered by the need to license that phrase's pragmatic status of being a topic is also arrived at by Holmberg and Nikanne (2002) with regard to Finnish. Holmberg and Nikanne (2002) propose that the head of the position to which in Finnish a topic is A-moved is associated with the agreement features of the subject (Φ-features), but that these features do not define this head. On the one hand, this assumption makes it plausible that the topic position is an A-position and that a referential subject can be regarded as a kind of default topic; on the other hand, this assumption allows other categories to be moved into the topic position, i.e., to be topics. In the following, I will adopt for the structural implementation of (7) the assumptions made by Holmberg and Nikanne (2002). I will call the projection in the middle field which hosts topics TopP.

3. Movement to the prefield

Sentences like the following differ in markedness:

(14) a. *Karl hat das Paket weggebracht.*
Karl has the package taken away
b. *Das Paket hat Karl weggebracht.*
the package has Karl taken away

With a transitive verb the subject in the prefield yields an unmarked sentence, cf. (14a), while the object in the prefield yields a marked structure, cf. (14b). Travis (1984, 1992) takes these observations as further evidence for the asymmetrical analysis of German V2-clauses: a clause with a subject in the prefield is an IP, whereas a clause with another constituent in its prefield is a CP, cf. (2). However, this proposal cannot be the solution to account for the difference in (14). Certain adverbials in the prefield also give rise to unmarked structures; furthermore, the objects of certain verbs are the unmarked elements in the prefield:

(15) a. *Leider hat keiner dem alten Mann geholfen.*
unfortunately has nobody the.DAT old man helped

b. *In Europa spielen Jungen gerne Fußball.*
in Europe play boys willingly football
'In Europe, boys like to play football.'

c. *Dem Karl hat das Spiel gut gefallen.*
the.DAT Karl has the match well pleased
'Karl liked the match very much.'

The sentences (15a,b) contain in their prefields a sentence adverbial and a frame adverbial, respectively. In (15c), the prefield is occupied by the oblique object of a psych verb. Note the contrast to the marked sentences in (16), which have an event-related locative adverbial and an instrumental adverbial, respectively, in their prefields:

(16) a. *Im Görlitzer Park hat Eva den Grill aufgebaut.*
in (the) Görlitz park has Eva the barbecue set up

b. *Mit dem Hammer hat Otto das Fenster eingeschlagen.*
with the hammer has Otto the window smashed

What is the difference between the unmarked examples in (14a) and (15) on the one hand and the marked examples in (14b) and (16) on the other? The answer becomes clear if we consider the base positions the prefield constituents have in the middle field. The base positions of the prefield elements in (14a) and (15) constitute the highest positions in the middle fields of the respective sentences. For (14a) and (15c), this has been shown by Lenerz (1977), and for (15a,b) by Frey and Pittner (1998), among others. Therefore, if we transform the sentences in (14a) and (15) into sentences which have all their sentence constituents in the middle field but in the same order as (14a) and (15), we get unmarked sentences. The following sentences illustrate this for (15a,c). (17a,b) are as neutral as (15a,c) are:

(17) a. *weil leider keiner dem alten Mann geholfen hat*
since unfortunately nobody the.DAT old man helped has

b. *weil dem Karl das Spiel gut gefallen hat*
since the.DAT Karl the match well pleased has

By contrast, the base positions of the prefield elements of (14b) and (16) do not constitute the highest element in the middle field. It is uncontroversial that the base position of the object in (14b) is below the base position of the subject. Furthermore, in, for example, Frey and Pittner (1998) it is argued that the base positions of event-related locative adverbials and instrumental adverbials are between the base positions of the subject and the object of a transitive verb. Thus, the base positions of the prefield elements of (16) do not constitute the highest elements in the middle fields. This explains why the following sentences, for instance, which have their phrases in the same order as the sentences in (16a,b) but all occurring in the middle field, are marked:

(18) a. *weil im Görlitzer Park*$_1$ *Eva* t_1 *den Grill aufgebaut hat*

b. *weil mit dem Hammer*$_1$ *Otto* t_1 *das Fenster eingeschlagen hat*

These observations indicate that there exists a mechanism to fill the prefield with the highest constituent of the middle field of the same clause, which preserves the semantic/pragmatic properties of the constituent without endowing it with additional ones. Thus, this mechanism does not seem to be related to any contentful property but seems to be a purely formal one.

Therefore, Frey (2006) follows Fanselow (2002) in adopting for German a proposal originally made by Bhatt (1999) for Kashmiri. A rough characterization can be given as follows:

(19) [$_{CP}$ XP [$_{C'}$ V_{fin} [$_{IP}$ ...]]] can be derived from [$_{IP}$ XP .... V_{fin}] (by moving the finite verb to C and) by moving XP as the highest element of IP to SpecCP (Attract Closest) just in order to fulfill the EPP requirement associated with C, i.e., without any additional intonational, semantic or pragmatic effects.

In the following, I will call the operation described in (19) 'F(ormal) M(ovement)'. Note that FM can only move the highest element of the middle field to the local prefield. Since German allows long movement to the prefield, it is clear that there must exist at least one other kind of movement to the prefield besides FM. Such movement will be called 'true Ā-movement to the prefield'.

FM can be applied to a scrambled phrase if the scrambled phrase occupies the highest position in the middle field. A constituent scrambled to this position induces pragmatic markedness. If such a phrase is moved to the prefield by FM, the markedness status of the construction will be preserved. Thus, (20c) (= (16b)), for example, can be derived from (20a) by scrambling the instrumental to the highest position of the middle field followed by application of FM:

(20) a. *(dass) Otto mit dem Hammer das Fenster eingeschlagen hat* Scrambling →
b. *(dass) mit dem Hammer*$_1$ *Otto* t_1 *das Fenster eingeschlagen hat* FM →
c. *Mit dem Hammer*$_1$ *hat* t_1' *Otto* t_1 *das Fenster eingeschlagen.*

(20c) is marked because (20b) is marked. In contrast, a sentence like (21b) (= (15b)), for example, can be obtained by FM without first scrambling the constituent to be moved by FM to the highest position, because the base position of the frame adverbial constitutes the highest middle field position of the sentence:

(21) a. *(dass) in Europa Jungen gerne Fußball spielen* FM →
b. *In Europa*$_1$ *spielen* t_1 *Jungen gerne Fußball.*

(21b) is unmarked because (21a) is unmarked.

(19) also provides an explanation for (22). In (22b), the sentence adverbial is in the prefield of the clause it belongs to. The sentence is fine; however, in (22a) the sentence adverbial in the prefield of the matrix clause cannot be understood as belonging to the embedded clause:

(22) a. **Leider*$_1$ *sagte Karl, dass* t_1 *keiner dem alten Mann*
unfortunately says Karl that nobody the.DAT old man
geholfen hat.
helped has
b. *Leider*$_1$ *hat* t_1 *keiner dem alten Mann geholfen.*

A sentence adverbial cannot be topical, focal, emphasized, backgrounded, nor can it acquire any other pragmatic or semantic property. Since true Ā-movement is supposed to always have semantic or pragmatic effects (cf., e.g., Rizzi 2004), these elements cannot undergo true Ā-movement. Therefore, (22a) is ungrammatical. The assumption that elements can be moved by FM to the local prefield without any semantic or pragmatic effect makes the grammaticality of (22b) plausible.

Which options exist to fill the prefield and which effects does true Ā-movement to the German prefield have? Frey (2006) argues for the following generalization:

(23) The German prefield can be filled either
   (i) by means of FM, which may move whatever is the highest XP-constituent in the middle field – be it base-generated there or scrambled to this position – to the local prefield, and which has no intonational, semantic or pragmatic effect; or
   (ii) by means of true Ā-movement to the prefield, which may move (cyclically) any constituent in the middle field, and which always has the effect of inducing stress on the moved item and its contrastive interpretation; or
   (iii) by base generation of certain adverbials, which are licensed by C.

For evidence for the claims (23)(ii) and (iii), the reader is referred to Frey (2006). For our present purposes it is enough to consider one piece of evidence for the claim embodied in (23) that in fact there exist only the three listed options to fill the prefield.[7] In German, there are elements which cannot be scrambled:

(24) a. **dass Maria grün*$_1$ *die Tür* t$_1$ *streichen wird*
   that Maria green the door paint will
   'that Maria will paint the door green'
   b. *Julia ist Präsidentin endlich geworden.
   Julia is president finally become
   'Julia has finally become president.'

---

7 There is one option to fill the prefield not mentioned in (23). This is the so-called genuine 'Vorfeld-*es*', as it appears in the following examples:

(i) a. *Es wurde viel gearbeitet.*
   it was a lot worked
   'A lot of work was done.'
   b. *Es spielt für Sie Erwin Lehn und sein Tanzorchester.*
   it plays for you Erwin Lehn and his dance orchestra
   'And now, Erwin Lehn and his Orchestra will play for you.'

An *es* like in (i) is not an argument or an adverbial of the clause. It cannot appear in the middle field; its only function seems to be to fulfil the EPP requirement associated with a German V2-clause. Therefore, according to the standard assumption, a genuine 'Vorfeld-*es*' is base-generated in the Vorfeld as a pure expletive.

The option to fill the prefield with a genuine 'Vorfeld-*es*' is not mentioned in (23) since, at the moment, I want to leave it open whether genuine 'Vorfeld-*es*' in fact only satisfies a formal EPP property or whether it has some semantic property after all. It seems to me worthwhile to pursue the assumption that the 'Vorfeld-*es*' in (i) denotes the eventuality described by these sentences.

Thus, these elements cannot get to the highest position in the middle field. It follows that the sentences in (25) cannot be derived by FM.

(25) a. *Grün wird Maria die Tür streichen.*
b. *Präsidentin ist Julia endlich geworden.*

The prefield constituents in (25) have to be stressed and interpreted contrastively. Given (23), this follows immediately: the only option to derive these sentences is by means of true Ā-movement, which is associated with accent and a contrastive interpretation of the moved item.

True Ā-movement can be to the local prefield, or it can be long distance; FM, by definition, can only be to the local prefield. For our topic, the following data are important:

(26) a. **Es*$_1$ *glaube ich, dass* t$_1$ *das Heu gefressen hat.*
it think I that the hay eaten has
b. (*Gestern bin ich erstmals ohnmächtig geworden.*)
('Yesterday I became unconscious for the first time.')
**Es*$_1$ *glaube ich, dass zum Glück niemand* t$_1$ *mitgekriegt hat.*
it think I that luckily nobody noticed has

(26a,b) show that *es*, be it a subject or an object, cannot undergo long movement to the prefield. This means that the inherently unstressed *es* cannot get into the prefield by true Ā-movement. From this, we can conclude that the only way of getting *es* into the prefield is by FM. (See also Bayer this vol. for a discussion of data like (26).)

Therefore, as a first intermediate result we note that in all the grammatical sentences of (1), (3) and (4), *es* arrived at its position by FM. That is, at the point in the derivation of such a sentence right before FM applies, *es* must occupy the highest XP-position in the middle field.

4. Three regions in the middle field

Since Diesing (1992) at the latest, it has been a standard assumption that there exist at least two subject positions across languages. Sentences like the following constitute crucial evidence for this claim:

(27) a. *dass Jungen ja doch die Romane von Karl May lesen werden*
that boys PRT PRT the novels of Karl May read will
b. *dass ja doch Jungen die Romane von Karl May lesen werden*

The bare plural subject in (27a), which appears to the left of the particles, only has a generic interpretation. The bare plural subject in (27b), which appears to the right of the particles, allows an existential interpretation. Diesing (1992) takes the particles as marking the vP-boundary. Therefore, she assumes that the subject in (27a) occupies SpecIP while the subject in (27b) stays in SpecvP.

Kiss (1996) questions the assumption that the lower subject position is SpecvP because in English the subject in a sentence corresponding to (27b) precedes the auxiliary.

Furthermore, Kiss (1996) argues that the difference in interpretation that correlates with the different positions of the subject should be described in terms of specificity. Thus, according to Kiss, there are two subject positions outside of vP, i.e., outside of the theta-domain. The lower one is identified as SpecIP, and the higher one as the specifier of an additional projection dominating IP and dominated by CP. This projection is called Referential Phrase (RefP). Specific subjects move to SpecRefP, non-specific ones to SpecIP. With these assumptions and the assumption that sentence adverbials are adjoined to IP, Kiss accounts for the data in (28). (28a) is bad because its predicate does not allow a specific interpretation of its subject:

(28) a. ?? *Boys luckily were born.*
b. *Luckily boys were born.*

Kiss takes SpecRefP to be the position that realizes the semantic function 'subject-of-predication'. Although Kiss (1996) does not discuss the issue, it is plausible to assume that RefP does not belong to the C-domain but to what is standardly called the I-domain. Thus, the similarity between Kiss' SpecRefP and the topic position argued for in (7) is obvious.

Taking a different starting point, Bobaljik and Jonas (1996) and Holmberg and Nikanne (2002) also argue for the existence of two subject positions outside of vP. For their argumentation the Transitive Expletive Construction and the Multiple Subject Construction in Icelandic and Finnish are crucial:

(29) a. *Það lesa margir stúdentar bækur Chomskys ekki.* (Icelandic)
EXP read many students books Chomsky's not
b. *Sitä ovat nämä lapset jo oppineet uimaan.* (Finnish)
EXP have these children already learned to swim

Bobaljik and Jonas (1996) assume that in (29a), the expletive is in AgrS and *margir stúdentar* is in SpecTP. According to Holmberg and Nikanne (2002), in (29b) the expletive occupies the topic position which the authors assume for Finnish, encoding a so-called stage topic; *nämä lapset* has left the vP.[8]

Next, let us once more consider German data like the ones which motivated Diesing's assumptions:

(30) a. *Weiterhin werden wahrscheinlich ja doch* [$_{vP}$ *Jungen die Romane*
furthermore will probably PRT PRT boys the novels
von Karl May lesen]. (subj. existential or generic)
of Karl May read
b. *Weiterhin werden wahrscheinlich* [$_{TP}$ *Jungen*$_1$ *ja doch* [$_{vP}$ t$_1$ *die Romane von Karl May lesen*]]. (subj. generic)
c. *Weiterhin werden* [$_{TopP}$ *Jungen*$_1$ *wahrscheinlich* [$_{TP}$ *ja doch* [$_{vP}$ t$_1$ *die Romane von Karl May lesen*]]]. (subj. generic and topical)

8 See Guasti and Rizzi (2002) for arguments based on language acquisition that AgrSP exists as an independent projection from TP.

In (30a) the subject is in its base position, and it can get a weak reading. In contrast, the subjects in (30b) and in (30c) are in front of the particles. Thus, they have left the vP, i.e., the domain of existential closure. They are interpreted generically.[9] But indefinites which are positioned higher than the vP are still crucially different. The indefinite in (30b) is scrambled to a position below the sentential adverbial. According to (7), this indefinite cannot be a topic. Only an indefinite occurring in front of the sentential adverb is a topic. This is confirmed by tests for topichood, for instance, by the test using the topic marking particle *jedenfalls*:

(31) a. **Laut Maria spielen glücklicherweise [Fußballer jedenfalls]*
according to Maria play fortunately footballers at any rate
auf Reisen Schach.
on travels chess

b. *Laut Maria spielen [Fußballer jedenfalls] glücklicherweise auf Reisen Schach.*

In (31), the topical bare plural cannot appear below the sentence adverbial, cf. (31a). Only when it is moved to the topic position as in (31b) does grammaticality result.

The data in (30) show that in the German middle field three regions should be differentiated. Furthermore, I assume that there exists a connection between the three different subject positions presumed to exist across languages by Kiss (1996), Bobaljik and Jonas (1996) or Holmberg and Nikanne (2002), among others, and the three different regions in the German middle field. That is, I will assume that each region identified in the German middle field is made available by a projection with a special relation to the subject. The lowest region is the theta-domain, which, in the standard case, is identified with vP with the (non-ergative) subject sitting in SpecvP. That the highest region, the topic domain, is made available by a projection which has a special relation to the subject was already assumed above with reference to the proposals made by Holmberg and Nikanne (2002). Furthermore, I will assume that the medial region is made available by TP. Some authors (e.g., Bobaljik and Jonas 1996) propose that T carries the nominative case-feature (for a strong thesis about the relation between T and nominative case, see Pesetsky and Torrego 2001). However, as in the case of Top, which is assumed to be associated with the Φ-features of the subject without being defined by them, I assume that the nominative feature does not define the head T. Thus, other categories than the subject have the option of moving to the Spec-position(s) of TP.[10]

---

9 Thus, whereas in English a bare plural which, according to Kiss (1996), has left the vP can get an existential reading, cf. (28b), I assume, like Diesing (1992), that in German this option exists only for a bare plural inside vP.

10 An interesting question to be investigated concerns the relationship between Grewendorf's (2005) 'lower topic-focus field', which in his framework is positioned between the Case position of the subject and the Case position of the object, and the region provided by the 'middle' subject projection TP assumed in the present paper.

5. Options for positioning personal pronouns in the middle field

Cardinaletti and Starke (1996) put forward a very important and widely accepted proposal for the classification of pronouns:

(32) According to Cardinaletti and Starke (1996), across languages pronouns can be classified
- into strong pronouns and deficient ones.

The deficient pronouns can again be classified into
- weak pronouns and clitics.

The most important criteria this classification is based on are the following:

(33) only strong pronouns can be coordinated, can be modified and can be stressed; only deficient pronouns allow a non-human referent.

According to Cardinaletti and Starke (1996), the classes of pronouns also differ in their distributions. A strong pronoun may stay in its base position, while a deficient pronoun has to leave the θ-domain. A weak pronoun is an XP-element and appears in an XP-position; a clitic is an X^0-element and therefore has to appear in an X^0-position.

Let us consider German *es*. It can be shown that *es*, be it a subject or an object, is necessarily deficient, in contrast to the other personal pronouns (cf. Cardinaletti and Starke 1996). The masculine personal pronoun, for example, can be coordinated, modified or stressed, cf. (34); none of these options exists for the deficient pronoun *es*, cf. (35):[11]

(34) a. *Heute wird Peter ihn und Maria treffen.*
today will Peter him and Maria meet

b. *Nur er hat das Problem wirklich verstanden.*
only he has the problem really comprehended

c. *Dort drüben stehen eine Frau und ein Junge. ER hat gerufen, nicht SIE.*
over there stand a lady and a boy he has called not she

---

11 If, for instance, the masculine personal pronoun is used as a deficient pronoun, as, for example, in case its referent is non-human, it also cannot be coordinated, modified or stressed (cf. Cardinaletti and Starke 1996):

(i) a. * *Hans wird ihn und den Motorroller waschen.* (*ihn = den Wagen*)
Hans will him and the scooter wash (him = the car)

b. * *Allein er kam für Paul in Frage.* (*er = der neue VW Passat*)
only he was by Paul considered (he = the new VW Passat)

c. * *Ich habe IHN nicht gesehen, die Kuh schon.* (*ihn = den Esel*)
I have him not seen, the cow yet (him = the donkey)
'Him I didn't see, but I did see the cow.'

With regard to the examples in (26) above, it was observed that *es* cannot undergo true Ā-movement. As is to be expected, this is true for deficient pronouns in general:

(ii) * *Ihn*$_1$ *glaube ich, dass Hans bald* t_1 *waschen sollte.* (*ihn = den Wagen*)
him believe I that Hans soon wash should (him = the car)

(35) a. *Wo ist das Lehrbuch "Syntax"?*
'Where is the textbook "Syntax"?'
**Gestern hat Maria es und das Semantikbuch ausgeliehen.*
yesterday has Maria it and the semantics book loaned

b. **Nur es hat das Problem wirklich verstanden.* (*es* = *das Mädchen*)
only it has the problem really comprehended (it = the girl.NEUTER)

c. **Dort drüben stehen ein Mädchen und ein Junge. ES hat gerufen, nicht ER.*
over there stand a girl.NEUTER and a boy it has called not he

Because *es*, as other German deficient pronouns, can be positioned in the prefield (as noted above, for Cardinaletti and Starke 1996 only deficient pronouns which are subjects can occupy the prefield), Cardinaletti and Starke assume that *es* and other German deficient pronouns always have the status of a weak pronoun.

I will adopt Cardinaletti and Starke's (1996) classification, especially their assumption that *es* is always deficient. However, their claim that *es* is always weak, i.e., that it is never a clitic, will be challenged.

I will argue for the claims in (36):

(36) (i) A deficient pronoun has to leave vP. Its movement to TP or to TopP results in 'Tucking In' (cf. Richards 1997).
(ii) Between TopP and the C-domain, only clitics are possible.
(iii) Only XPs can occur in the prefield.

(36iii) will be assumed without discussion. The other two claims will be justified in the following.

In the examples in (37), the bare plural *Kollegen* has an existential interpretation. Thus, it has not left vP:

(37) a. *Ich kam gut über die eisbedeckte Straße, weil glücklicherweise* [$_{vP}$ *Kollegen mir geholfen haben*].
I came well across the iced street since fortunately colleagues me helped have

b. ?? *Ich habe mein Notizbuch wieder, weil glücklicherweise* [$_{vP}$ *Kollegen es gefunden haben*].
I have my notebook again since fortunately colleagues it found have

(37) confirms what Cardinaletti and Starke (1996), among others, observe: a pronoun like *mir*, when strong, may remain inside vP, while the deficient *es* has to leave it. (38) makes the same point:

(38) a. *Was*$_1$ *haben* t$_1$ *für Leute dir geholfen?*
what have for people you helped
'What sort of people have helped you?'

b. ?* *Was*$_1$ *haben* t$_1$ *für Leute es gesehen?*
what have for people it seen
'What sort of people have seen it?'

It is often assumed that the so-called 'was-für-split' can only occur with a phrase which is in vP (cf., e.g., Meinunger 2001). Therefore, (38b) shows that *es* may not be inside vP.

(36i) claims that the movement of pronouns results in 'Tucking In'. 'Tucking In' refers to the targets of movement in the theory of Richards (1997). Richards (1997) proposes a featural theory of movement according to which a head Z with a strong feature may have multiple Spec-positions. After a higher XP moves to SpecZP, a lower YP can move to the lower SpecZP as long as XP and YP are moving to check a strong feature in Z. Thus, in this theory multiple attraction by a single attractor is possible. Just in this case, it derives countercyclic movement. The following Bulgarian data constitute a very popular contrasting pair this theory accounts for:

(39) a. *Koj kogo vižda?*
who whom sees
'Who sees whom?'
b. **Kogo koj vižda?*

The leftmost wh-phrase in a Bulgarian question is the first to move. It has the higher base position of the two wh-phrases. The second wh-phrase has tucked in underneath the first phrase, forming the lower specifier. Richards argues that it is to be expected that the second instance of movement to a given attractor should tuck in if, in addition to the 'attractor-oriented' condition Attract Closest, movement also obeys the 'mover-oriented' economy condition Shortest Move. Shortest Move demands that the number of nodes crossed by the movement be as low as possible.[12]

Note that Richards (1997) gives evidence from different languages that Tucking In can be found with Ā-movement and with A-movement.[13]

It is an interesting fact of wh-movement in Bulgarian that if three wh-phrases are fronted, the leftmost wh-phrase must correspond to the highest element prior to movement, while the order of the following wh-phrases is free; thus, the orders wh_1 wh_2 wh_3 and wh_1 wh_3 wh_2 are both good. This is explained by Richards' (1997) 'Principle of Minimal Compliance': if the condition Attract Closest is satisfied once by the attraction of the highest element, it does not have to be satisfied any further. Thus, in the next step wh_2 or wh_3 can be attracted first.

What happens if the subject is not affected by wh-movement but lower constituents are? In Bulgarian, one gets a rigid ordering between a pair of non-subject wh-words:

---

12 A different theory to account for data like the Bulgarian examples in (39) is developed in Grewendorf (2001). This theory assumes the formation of a cluster in a low position which is afterwards moved further as a whole. For our purpose, as far as I can see, this would also be a possible way to go.

13 As evidence that instances of A-movement have the property of Tucking In, Richards (1997) refers, for example, to certain cases of local scrambling in Japanese and to object scrambling in Dutch and Icelandic.

(40) a. *Kogo kakvo e pital Ivan?*
whom what AUX asked Ivan
'Whom did Ivan ask what?'
b. ?* *Kakvo kogo e pital Ivan?*

These facts follow given Richards' (1997) assumptions for the attraction of wh-elements. In Section 6 we will consider related cases important for our concerns.

The following examples demonstrate that the movement of a deficient pronoun to the TP-region has to tuck in:

(41) a. *Ich habe mein Notizbuch wieder, weil glücklicherweise* [$_{TP}$ *Maria es*
I have my notebook again since fortunately Maria it
gefunden hat].
found has
b. **Ich habe mein Notizbuch wieder, weil glücklicherweise* [$_{TP}$ *es Maria gefunden hat*].

In the indicated analysis of (41a), the subject has left vP and has moved to SpecTP; the following pronoun *es* has also left vP, the domain in which it cannot stay. (41b) shows that the object *es* has to tuck in below the subject. It cannot occur higher in the TP-domain.

We may note in passing that several of the examples considered so far show that – at least in German – pronouns do not have to be topical. This observation is in conflict with the claim found in the literature (cf., e.g., Erteschik-Shir 1997) that pronouns are inherently topical.

In the case of the TP-domain it is not enough that a deficient object-pronoun tucks in below a subject which occurs in this domain. As Lenerz (1993, 1994) observes, a deficient object-pronoun can only occur below a sentence adverbial if the subject does too:

(42) **Ich habe mein Notizbuch wieder, weil Maria glücklicherweise es gefunden hat.*

Let us now continue to justify (36i). If we consider pronouns which are moved to the topic domain, we again find that they have to tuck in:

(43) *Die Direktorin hat nach dem Syntax-Buch gefragt.*
the director has for the syntax book asked
a. *Morgen wird Hans es der netten Dame sicherlich ausleihen.*
tomorrow will Hans it the.DAT nice lady certainly loan
b. ? *Morgen wird Hans der netten Dame es sicherlich ausleihen.*
c. ? *Morgen wird der netten Dame Hans es sicherlich ausleihen.*
d. * *Morgen wird der netten Dame es Hans sicherlich ausleihen.*

The important point is the difference between (43a,b,c) on the one hand and (43d) on the other. In (43a,b,c), the deficient pronoun has been tucked in below the subject. Although (43a) certainly sounds more natural than (43b,c), all three sentences are judged grammatical by my informants. (43d), however, is not grammatical. This demonstrates that inside the topic domain the deficient pronoun may not occur above the subject.

(43c) shows that the order of full DPs in the topic domain is not strict (cf. Frey 2004). (43c), though sounding a bit clumsy, is grammatical. Thus, we may assume either that full DPs do not tuck in or that they can be moved further inside the topic domain after having been tucked in.[14]

From the observations concerning (43), it follows that in (44), *es* must be outside of the domain made available by TopP. This is what will be argued for below: in (44), *es* is a clitic attached to the lexically realized head-position of the C-domain.

(44) *Morgen wird es Hans der netten Dame sicherlich ausleihen.*

We find the same pattern as in (43) if together with the subject and an object-*es* a low adverbial phrase is moved to the topic domain. Again *es* has to tuck in below the subject:

(45) *Das Fenster ist verklemmt. Aber ...*
the window is stuck however
a. *bald wird Hans es mit Marias Hilfe sicherlich aufbekommen.*
soon will Hans it with Maria's help certainly get open
b. ? *bald wird Hans mit Marias Hilfe es sicherlich aufbekommen.*
c. ? *bald wird mit Marias Hilfe Hans es sicherlich aufbekommen.*
d. * *bald wird mit Marias Hilfe es Hans sicherlich aufbekommen.*

(46) shows that it is not only the deficient *es* which has to tuck in, but any unstressed pronoun:

(46) a. * *Morgen wird der freundlichen Dame ihn Hans sicherlich vorstellen.*
tomorrow will the.DAT friendly lady him Hans certainly introduce
b. * *Bald wird mit ihrem Porsche ihn Maria zum Glück abholen.*
soon will with her Porsche him Maria luckily pick up

Let us now move on to (36ii). As mentioned above, Cardinaletti and Starke (1996) assume that German does not have clitics at all. The same assumption is held by, among others, Lenerz (1993, 1994).[15] However, consider the following examples:

(47) a. *Vor kurzem hat der Peter sie ihm leider weggeschnappt.*
recently has the.NOM Peter her him.DAT unfortunately snatched away
b. *Vor kurzem hat der Peter ihm sie leider weggeschnappt.*
c. *Vor kurzem hat sie ihm der Peter leider weggeschnappt.*
d. * *Vor kurzem hat ihm sie der Peter leider weggeschnappt.*

---

14 Corresponding facts hold in the TP-domain. In Richards' (1997) theory, one would assume that there exist further A-heads above Top and T, respectively, which can attract full DPs but not deficient pronouns. At the moment I have no specific suggestion to make on how to account for the permutations of full DPs inside the different domains. However, for our purposes this point is not crucial.

15 Another view is taken by Grohmann (2000,183), who assumes "that an overtly high pronoun can only be a clitic." Although there are important differences between Grohmann's approach and the present one, the cited assumption is, at least broadly, shared by the two approaches.

Let us consider (47a,b) first. The pronouns occur in different orders. It was mentioned above that Tucking In of more than two elements does not have to preserve the order of elements prior to movement. Thus, we expect that the accusative pronoun and the dative pronoun, when following the subject, may change their order. Now, compare (47c,d). Interestingly, when the pronouns precede the topical subject, only the order 'accusative – dative' is possible.

The difference between (47a,b) and (47c,d) constitutes a first piece of evidence that the status of pronouns preceding a topical lexical subject is different from the status of pronouns following a topical lexical subject. Across languages, it is a fact that clitics tend to build clusters, i.e., the privileged clitic position can contain a number of clitics, usually in a fixed order. Thus, the observed difference makes sense if we assume that the pronouns between the C-domain and the topical subject are clitics, i.e., they are X^0-elements, whereas the pronouns following the subject are XPs.[16]

Consider now (48), which has the subject below the topic domain:

(48) *Vor kurzem hat ihm sie leider jemand weggeschnappt.*
recently has him.DAT her unfortunately someone snatched away

Here, in contrast to (47d), the pronouns can occur in the order 'dative – accusative'. This is to be expected. The pronouns can be analysed as occupying different Spec-positions of TopP.

Another piece of evidence for the clitic status of pronouns occurring right-adjacent to the lexically filled head position of the C-domain ('C-position') is given by the data in (49). Cardinaletti and Starke (1996) recall the fact that, in many languages (also in unrelated ones), an accusative $1^{st}/2^{nd}$ person clitic cannot cooccur with a dative 3rd person clitic. In German, this ban holds for pronouns appearing between the C-position and a topical lexical subject, cf. (49a). It does not hold if the pronouns occur further to the right, cf. (49b):

(49) a. **Heute hat mich ihm Peter zum Glück vorgestellt.*
today has me.ACC him.DAT Peter luckily introduced
b. *Heute hat Peter mich ihm zum Glück vorgestellt.*

The fact that in front of the topical subject the order *ihm – mich* is not possible either follows from the observation we made with regard to the examples in (47).

Let us move to another difference between pronouns preceding and pronouns following a topical lexical subject. The following observation I owe to Josef Bayer (p.c.):

(50) a. ?? *weil sie, äh, ihm der Peter leider*
since her.ACC INTJ him.DAT the.NOM Peter unfortunately
weggeschnappt hat
snatched away has

---

[16] Inside the 'clitic field', the distribution of the reduced form of *es* is freer than that of the other pronouns:

(i) a. *Vor kurzem hat 's ihr Hans zum Glück gegeben.*
recently has it her Hans luckily given
b. *Vor kurzem hat ihr's Hans zum Glück gegeben.*

It is plausible to assume that this freedom is due to the special phonological status of the reduced *'s*.

b. *weil der Peter sie, äh, ihm leider weggeschnappt hat*

Although the difference in (50) is subtle, it is confirmed by informants: it is easier to interrupt with an interjection (like *äh*) (or a parenthesis, for that matter) a sequence of pronouns which follow the topical lexical subject than a sequence of pronouns which occur between the C-domain and the topical subject. This constitutes another indication that the higher pronouns build a clitic-cluster, which cannot be broken up.

A final piece of evidence that deficient pronouns right-adjacent to the C-domain are clitics builds on FM, one of the mechanisms to fill the prefield introduced in Section 3. The crucial observation is that C-adjacent pronouns do not block FM:

(51) *Wo liegt Ihrer Meinung nach Köln?*
'In your opinion, where is Cologne situated?'
a. *Köln liegt am Rhein.*
Cologne is situated on the Rhine
b. ?? *Am Rhein liegt Köln.*
c. *Am Rhein liegt 's/es.*
on the Rhine is situated it

(51a) is a natural way to answer the question. *Am Rhein* is in its base position. The subject has reached the prefield by means of FM. (51b) is awkward. The reason is that as the answering term to the wh-question, *am Rhein* cannot be scrambled (Lenerz 1977). Therefore, *am Rhein* cannot be moved to the prefield by FM but is Ā-moved and gets a contrastive interpretation (cf. (23ii) in section 3), which is not really appropriate in the given context, which is just information seeking. In the same context, however, (51c) is perfect. This means that the pronoun does not block FM. If we assume that the pronoun in (51c) is a clitic, this follows immediately. Clitics are X^0-elements, which do not interfere with FM. FM attracts the closest XP.

A confirmation that high pronouns are clitics might be given by reduction phenomena. Bayer (1999) gives the following judgements:

(52) a. *als 'm der Hans den neuen Chef vorgestellt hat*
when him.DAT the.NOM Hans the.ACC new boss introduced has
b. **als der Hans 'm den neuen Chef vorgestellt hat*
c. **als der Hans den neuen Chef 'm vorgestellt hat*

In Standard German, there are very few reduced pronouns possible. A clear case is the reduced form of *es*. For the sentences in (53), most informants have the intuitions indicated (some informants, however, accept (53b)):[17]

[17] If the sentences were transformed into my Swabian dialect, I would also have the judgements indicated in (53). With a reduced dative pronoun I have the corresponding judgements:

(i) a. *Heut hat 'm dr Kurt zum Glück gholfe.* (Swabian)
today has him.DAT the Kurt luckily helped
b. * *Heut hat dr Kurt 'm zum Glück gholfe.* (Swabian)

(53) a. *Heute hat 's Kurt gelesen.*
today has it Kurt read
b. **Heute hat Kurt's gelesen.*

So far, it has been argued that deficient pronouns occurring between the C-domain and a topical lexical phrase are clitics.[18] To conclude this section we should ask in passing whether strong pronouns may occur in this part of the clause. Consider the following sentences:

(54) a. **Heute hat IHN der Kollege anscheinend getroffen.*
today has him the colleague apparently met
b. **Heute hat ihn und Maria der Kollege anscheinend getroffen.*

A strong pronoun can be stressed and it can be coordinated. Both possibilities do not exist for a pronoun occurring between the C-domain and a topical subject. We can conclude that this part of the clause is reserved exclusively for clitics. Weak and strong pronouns are not allowed there.[19]

6. How to get an object-*es* into the prefield

In this final section, let us now try to explain the data in (4), repeated here for convenience:

(55) *Wo ist das Lehrbuch "Syntax"? Ich kann es nicht finden.*
'Where is the textbook "Syntax"? I can't find it.'
a. *Es*$_1$ *hat* t$_1$ *leider jemand der Maria ausgeliehen.*
'Unfortunately, someone loaned it to Maria.'
b. **Es*$_1$ *hat Otto* t$_1$ *leider jemandem ausgeliehen.*
'Unfortunately, Otto loaned it to someone.'
c. ?*Es*$_1$ *hat* t$_1$ *leider Otto jemandem ausgeliehen.*
d. (?)*Es*$_1$ *hat* t$_1$ *leider der Maria jemand ausgeliehen.*
e. *Es*$_1$ *hat* t$_1$ *der Maria leider jemand ausgeliehen.*

18 This paper remains neutral with regard to the question of whether there are other positions in the German middle field in which clitics occur, although I believe that the answer would be negative (cf., e.g., Uriagereka 1995 and Manzini and Savoia 2004 for claims that clitics occupy fixed positions).

19 Note that the following examples do not disprove this claim:

(i) a. *weil IHN der Kollege anscheinend getroffen hat*
since him the colleague apparently met has
b. *weil IHN und MaRIa der Kollege anscheinend getroffen hat*
since him and Maria the colleague apparently met has

In Frey (2006), it is argued extensively that true Ā-movement, introduced above in (23ii), can target an Ā-position right-adjacent to a complementizer. This position belongs to the C-domain, which according to Frey (2006) is (minimally) split in German. In (i), the phrases between the complementizer and the subject are Ā-moved to this position. Note that, as expected, these phrases are stressed and contrastively interpreted.

Frey (2006) argues furthermore that in the C-domain of German, at most one Spec-position can be filled. Thus, in the examples in (54), the phrases between the finite verb and the subject cannot be moved by true Ā-movement to the C-domain. Therefore, these sentences constitute the crucial examples to test whether strong pronouns may occur between the C-domain and a lexical topical subject.

Why is (55b) bad whereas the other examples in (55) are fine? In connection with the data in (26), it was observed that a weak pronoun like *es* can only get to the prefield by means of FM, and can never undergo true Ā-movement. This follows from the facts that weak pronouns cannot be stressed, cf. (35c) and (ic) in fn. 11, and that, according to (23ii), true Ā-movement is always associated with stress. Now, it can be shown that (55b) is the only example in (55) in which *es* cannot reach the prefield by FM.

Let us start with (55b). Recall that the attracting A-heads of the different regions in the middle field are all associated with features of the subject. Thus Top, which is relevant for the examples in (55), probes two different features. It probes the subject's Φ-features and it probes a topic feature. The topic feature, in contrast to the subject's Φ-features, has the EPP property, i.e., Top attracts a phrase by means of an EPP-feature if the phrase is a topic.[20] In the derivation, it is free which of the features of Top will be chosen first to be probed. If in (55b), Top first probes the subject's Φ-features, the topic feature of the subject will also be probed. Therefore, the subject will be the first to move to SpecTopP. If the topic feature is probed first, it is again the subject which will move first. It occupies the highest base position. It follows that the object-*es* necessarily has to tuck in below the subject. Thus, FM cannot attract *es*, the subject being closer. In order to get *es* into the prefield in (55b), true Ā-movement would have to apply to *es*, which is not possible.

Let us now consider (55a,c,d). Nothing speaks against the analyses indicated in (55a,c,d). *Es* is the only element attracted by Top. Thus, *es* has occupied the XP-position closest to the prefield. It does not matter what goes on lower in the structure. In particular, it does not matter what kind of subject occurs in the lower structure: although (55c) has the same subject as (55b), the non-topical subject of (55c) does not interfere with the topical *es*.

Let's move to (55e). Why is the indicated analysis, which allows *es* to be moved by FM, possible? In this sentence, the subject, which occupies the highest base position, is not topical, but lower constituents are. It can be shown that in such a situation, the lower constituents after being moved to the topic domain do not have to appear in the same order as they appear in the base. This is different from the related case of the Bulgarian example in (40), in which the subject is not a wh-element, but lower phrases are, and in which the movement of the lower wh-elements has to respect their order in the base. The reason for this difference is that in (40), the attracting head in the C-domain probes only one feature, whereas Top in the German middle field probes two different features. Let us see why this is so.

Richards' (1997) condition Attract Closest has to be replaced by the condition Probe Closest since according to current theorising, not all feature-licensing involves movement. Now, Top can always successfully probe the subject because Top is associated with the subject's Φ-features. Thus, if the subject's Φ-features are chosen first to be probed, the Probe Closest requirement for Top will be satisfied also in a case in which the subject is not a topic. It follows from the 'Principle of Minimal Compliance' that

---

[20] That the EPP is a property of a feature of a head – and not a property of the head itself – is also assumed by Pesetsky and Torrego (2001).

other sentential constituents can then be probed in any order.[21] Therefore, in a sentence like (55e), where the subject is not a topic, it is possible to probe the topic feature of the accusative object before that of the dative object is probed. As a result, as indicated in (55e), *es* occupies the SpecTopP-position higher than the SpecTopP-position of the dative, although in the base there is the order 'subject – dative – accusative'.[22]

Let us now investigate other examples with a weak pronoun in the prefield. Consider first the difference in (56):

(56) *Das Pferd*$_1$ *hat eine Verletzung.*
the horse has an injury
a. **Ihm*$_1$ *wird Peter zum Glück helfen können.*
him will Peter luckily help can
b. *Ihm*$_1$ *kann zum Glück geholfen werden.*
him can luckily helped be

The pronoun in (56a) cannot have a non-human referent. Thus, it is not weak. This is expected. The highest position in the middle field which *ihm* could have occupied is below the topical subject because it has to be tucked in below the subject. FM cannot catch it in this position. Therefore, in (56a) *ihm* has arrived at the prefield via true Ā-movement, which can only be possible if *ihm* is conceived as a strong pronoun. In contrast, in (56b), *ihm* can have reached the prefield via FM, because there is no interfering element. Thus, it can be a weak pronoun and may have a non-human referent.[23]

The following sentences contain an experiencer verb:

---

21 The assumption that if a head has more than one feature, the condition Probe Closest is satisfied for the head if Probe Closest is satisfied for one of its features is also held by Pesetsky and Torrego (2001).

22 In the following example, the lower arguments are probed in the other possible order, i.e., the dative object is probed before the accusative object. As a consequence, the weak pronoun *him* can get into the prefield by means of FM:

(i) *Ihm hat das Heu zum Glück bereits jemand gegeben.* (*ihm* = *dem Pferd*)
him has the hay luckily already someone given (him = the horse)

23 Meinunger (to appear) observes that the following sentence with a deficient reflexive pronoun of an inherent reflexive verb in its prefield is marginally possible:

(i) ? *sich darf nur anstellen, wer noch keine Karten hat*
REFL may only line up, who yet no tickets has
'Only those may line up who have no tickets yet.'

We may note the contrast to (ii):

(ii) * *sich darf Hans jetzt endlich anstellen*
REFL may Hans now finally line up

Given our assumptions we can explain (i) and (ii). In (i), the reflexive moves to the prefield via FM; no constituent is in its way. To account for (ii) we just have to observe that if in the middle field the lexical subject of an inherent reflexive verb is present, the reflexive either has to be a clitic as in (iiia), or it has to be c-commanded by the subject, cf. (iiib,c):

(iii) a. *weil sich jetzt endlich Hans anstellen darf*
b. * *weil jetzt endlich sich Hans anstellen darf*
c. *weil jetzt endlich Hans sich anstellen* darf

Thus, there is no way in a sentence like (ii) for the reflexive to get to the prefield via FM.

(57) a. *Ihm hat das Galoppieren anscheinend gut gefallen.* (*ihm = dem Pferd*)
him has the gallop apparently well pleased (him = the horse)
b. *Es hat dem Pferd anscheinend gut gefallen.*
it has the horse apparently well pleased

In (57a) and (57b) the pronoun in the prefield is weak. In particular, (57a) is interesting: *ihm* must have moved to the prefield via FM. Why does the topical subject not block FM of *ihm*?

The crucial difference between a standard transitive topical subject like the one in (55b), which blocks FM of an object pronoun, and the one in (57a), which does not, consists in their base positions. Whereas a standard transitive subject is generated higher than any object, it can be shown that the subject of *gefallen* is generated below the dative object (cf., e.g., Haider 1993). Let us therefore see what happens if in (57), Top probes the arguments of *gefallen*. Top carries the subject's Φ-features and a topic feature associated with the EPP property. Probe Closest can be satisfied in two different ways. In one option, the subject's Φ-features are probed first. Probe Closest is satisfied by probing the subject although the subject is generated below the dative object. The subject is the closest phrase with structural case. Since the subject is topical, this procedure also has the effect that the subject will be moved to the highest Spec-position of TopP. This structure forms the basis of (57b). Alternatively, Top first probes its topic feature. Probe Closest will first probe the topical dative object, because it is higher than the subject. The dative will be moved and will occupy the highest SpecTopP-position. This structure, then, forms the basis of (57a).

Thus, the reason why both sentences in (57) are grammatical lies in the fact that the base order of the arguments of *gefallen* is 'dative – nominative'. The base order 'dative – nominative' is also induced if a ditransitive verb is passivized. Such a verb appears in (58):

(58) a. *Ihm wurde das Heu zum Glück schon gegeben.* (*ihm = dem Pferd*)
him.DAT was the hay luckily already given (him = the horse)
b. *Es wurde dem Pferd zum Glück schon gegeben.*
it was the horse luckily already given

We find the same pattern as in (57). A weak object pronoun can occupy the prefield despite the fact that the subject occurs in the topic domain, cf. (58a), and a weak subject pronoun can appear in the prefield while the object occurs in the topic domain, cf. (58b). The explanation for (58) is the same as for (57).[24]

---

[24] Another example in which a weak object pronoun may occur higher than the subject in the topic domain is given in (ib):

(i) *Der neue 'Harry Potter' ist in England erschienen.*
the new 'Harry Potter' is in England appeared
a. *Laut FAZ werden in Deutschland die Fans ihn zum Glück*
according to FAZ will in Germany the fans him luckily
auch bald lesen können. (*ihn = der neue 'Harry Potter'*)
also soon read can (him = the new 'Harry Potter')
b. *Laut FAZ werden in Deutschland ihn die Fans zum Glück auch bald lesen können.*

Consider now (59):

(59) *(E)s hat 'r zum Glück gar net gmerkt.*
it has he luckily not at all noticed

(59) is an example in the Swabian dialect which I consider grammatical. It is of interest because it contains an object-*es* in the prefield and a subject pronoun which occurs high in the structure. The subject pronoun does not block FM of *es* to the prefield. Given our assumptions, such behaviour should be possible. In (59), the subject pronoun is treated as a clitic, i.e., as an X^0-element, which does not block FM.[25]

In the middle field, subject pronouns can occur to the right of other pronouns in some special cases.

(60) a. *Ich bin in Pauls Auto kurz ohnmächtig geworden.*
'I became briefly unconscious in Paul's car.'
Meinem Gefühl nach hat es zum Glück nur er bemerkt.
according to my feelings has it luckily only he noticed
b. *weil 's ihm ER nicht geglaubt hat*
since it him he not believed has

In (60a), the subject pronoun is part of a quantificational expression occurring below the topic domain. In (60b), the stress indicates that *er* is treated as a strong pronoun, which may occur low in the structure.

In (61), a lexical phrase and a pronoun occur in front of a topical subject pronoun:

(61) a. *da dem Kritiker sie anscheinend sehr gut gefallen hat*
since the.DAT critic she apparently very much pleased has
b. *da ihm sie anscheinend sehr gut gefallen hat*
since him.DAT she apparently very much pleased has

These examples contain an experiencer verb. Thus, Top may first probe and attract the topical dative object. Then, the topical subject pronoun is probed and tucked in below the object (cf. (57a), (58a)). The next sentence, which is based on an example of Lenerz (1993), shows that in the topic domain a high adverbial may precede a subject pronoun:

In the natural reading of the sentences in (i), *in Deutschland* is a frame adverbial. A frame adverbial has its base position above the base position of a standard transitive subject (cf., e.g., Frey 2003). Thus, the explanation for (i) is straightforward. One option for Top is to first probe the topic feature. In this case, *in Deutschland* is the first phrase to be moved to the topic domain. The other phrases can be tucked in in different orders.

25 Gärtner and Steinbach (2003) give an example in the Hessian dialect similar to (59):

(i) *'s hab isch net gewusst.*
it have I not known

(i) can be analysed like (59) since, as far as I can see, *isch* can be treated as a clitic. It should be noted, however, that Gärtner and Steinbach (2003) presumably would not agree because they argue against any special 'clitic-syntax' for the analysis of German pronouns.

(62) *weil in der gegenwärtigen weltpolitischen Situation es offensichtlich*
since in the current worldpolitical situation it obviously
von außergewöhnlicher Bedeutung ist, dass ...
of extraordinary importance is that

In this case it is the topical frame adverbial which is first probed and attracted by Top (cf. fn. 24).

In standard cases, in the middle field a topical subject pronoun of a transitive verb has to precede other pronouns (e.g., Lenerz 1993), cf. (63a); a lexical phrase to the left of a topical subject pronoun is not possible either, cf. (63b):

(63) a. **dass es sie*$_{\text{NOM}}$ *wahrscheinlich verkaufen wird*
that it she probably sell will
b. **dass das Auto sie wahrscheinlich verkaufen wird*
that the car she probably sell will

To account for sentences like (63), Hinterhölzl (2004) assumes that a 'high' subject pronoun in the middle field is realized as a clitic. As a nominative clitic it necessarily precedes other elements. However, this stipulation rules out examples like (61) and (62). Therefore, to account for (63a) I will assume that in the middle field, a sequence of deficient pronouns is treated alike, i.e., all pronouns of the sequence are treated as clitics or all are treated as XP-elements. Under this assumption, (63a) cannot be derived, since in (63a), *es* would have to be a clitic and *sie* an XP-element. However, I have to leave for further research the question as to why it should be that a sequence of deficient pronouns has to be treated in a uniform manner. To account for (63b), I assume that it is not possible to scramble a topical lexical phrase in front of a topical subject pronoun. This would have to follow from a general theory about the pragmatic effects of scrambling, which has yet to be developed.[26]

Let me finally come back to example (1b), repeated in (64):

(64) *Es hat das Pferd gefressen.*
it has the horse eaten

In Section 1, the sentence was judged ungrammatical because in Travis (1984, 1992) it was judged this way. Sometimes, however, authors are not so sure how to judge a sentence like (64). Grohmann (2000,183), for example, rates an analogous example with a question mark.

From our perspective, the sentence is bad if *das Pferd* is taken to be a topic. This is certainly a very natural thing to do. However, given the sentence in isolation, one does not necessarily have to do this. According to our assumptions, the sentence will not be judged ungrammatical if *das Pferd* is not conceived as a sentence topic.

---

26 My suspicion is that scrambling of phrases is related to the (recursive) establishing of a hierarchy of points of view from which an eventuality is reported. It might be that scrambling a phrase in front of a topical subject pronoun results in a conflict because a topical subject pronoun inherently establishes the 'highest' point of view.

References

Bayer, J. 1999. Comments on Cardinaletti and Starke "The typology of structural deficiency". In: H. van Riemsdijk (ed.), *Clitics in the Languages of Europe*, 235-241. Berlin, New York: Mouton de Gruyter.

Bayer, J. (this volume). A Note on targets of A'-movement in the Left Periphery.

Bhatt, R. M. 1999. *Verb Movement and the Syntax of Kashmiri*. Dordrecht: Kluwer.

Bobaljik, J.and D. Jonas. 1996. Subject positions and the roles of TP. *Linguistic Inquiry* 27, 195-236.

Cardinaletti, A. and M. Starke. 1996. Deficient pronouns: a view from Germanic. A study in the unified description of Germanic and Romance. In: H. Thräinsson, S. Epstein, and S. Peter (eds.), *Studies in Comparative Syntax II*, 21-65. Dordrecht: Kluwer.

Diesing, M. 1992. *Indefinites*. Cambridge, Mass.: MIT Press.

Erteschik-Shir, N. 1997. *The Dynamics of Focus Structure*. Cambridge: Cambridge University Press.

Fanselow, G. 2002. Quirky subjects and other specifiers. In: I. Kaufmann and B. Stiebels (eds.), *More than Words*, 227-250. Berlin: Akademie Verlag.

Fanselow, G. (this volume). On pure syntax (uncontaminated by information structure).

Frey, W. 1993. *Syntaktische Bedingungen für die semantische Interpretation*. Berlin: Akademie Verlag (studia grammatica 35).

Frey, W. 2003. Syntactic conditions on adjunct classes. In: E. Lang, C. Maienborn, and C. Fabricius-Hansen (eds), *Modifying Adjuncts*, 163-209. Berlin, New York: Mouton de Gruyter.

Frey, W. 2004. A medial topic position for German. *Linguistische Berichte* 198, 153-190.

Frey, W. 2006. Contrast and movement to the German prefield. In: V. Molnár and S. Winkler (eds.), *The Architecture of Focus* (Studies in Generative Grammar 82). Berlin, New York: Mouton de Gruyter.

Frey, W. and K. Pittner. 1998. Zur Positionierung der Adverbiale im deutschen Mittelfeld. *Linguistische Berichte* 176, 489-534.

Gärtner, H.-M. and M. Steinbach. 2003. What do reduced pronominals reveal about the syntax of Dutch and German? Part 2: Fronting. *Linguistische Berichte* 196, 459-490.

Grewendorf, G. 2001. Multiple *wh*-fronting. *Linguistic Inquiry*, 31.1, 87-122.

Grewendorf, G. 2002. *Minimalistische Syntax*. Tübingen: UTB.

Grewendorf, G. 2005. The discourse configurationality of scrambling. In: J. Sabel and M. Saito (eds.), *The Free Word Order Phenomenon: Its Syntactic Sources and Diversity*, 75-135. Berlin, New York: Mouton de Gruyter.

Grohmann, K. 2000. Towards a syntactic understanding of prosodically reduced pronouns. *Theoretical Linguistics*, 26.3, 175-210.

Guasti, M. T. and L. Rizzi. 2002. Agreement and Tense as distict syntactic positions: Evidence from acquisition. In: G. Cinque (ed.), *Functional Structure in DP and IP* (The Cartography of syntactic structures; Vol. 1), 167-194. Oxford, New York: Oxford University Press.

Haider, H. 1993. *Deutsche Syntax – Generativ*. Tübingen: Narr.

Haider, H. 2005. How to turn German into Icelandic – and derive the OV-VO contrasts. *Journal of Comparative Germanic Linguistics* 8, 1-53.
Hinterhölzl, R. 2004. Scrambling, optionality and non-lexical triggers. In: A. Breitbarth and H. van Riemsdijk (eds.), *Triggers* (Studies in Generative Grammar 75), 173-203. Berlin, New York: Mouton de Gruyter.
Holmberg, A. and U. Nikanne 2002. Expletives, subjects and topics in Finnish. In: P. Svenonius (ed.), *Subjects, Expletives, and the EPP*, 71-105. Oxford: Oxford University Press.
Kiss, K. É. 1996. Two subject positions in English. *The Linguistic Review* 13, 119-142.
Lenerz, J. 1977. *Zur Abfolge nominaler Satzglieder im Deutschen*. Tübingen: Narr (Studien zur deutschen Grammatik 5).
Lenerz, J. 1993. Zur Syntax und Semantik deutscher Personalpronomina. In: M. Reis (ed.), *Wortstellung und Informationsstruktur*, 117-154. Tübingen: Niemeyer.
Lenerz, J. 1994. Pronomenprobleme. In: B. Haftka (ed.): *Was determiniert Wortstellungsvarianten*? 161-173. Opladen: Westdeutscher Verlag.
Manzini, M. R. and L. M. Savoia. 2004. Clitics: cooccurrence and mutual exclusion patterns. In: L. Rizzi (ed.), *The Structure of CP and IP* (The Cartography of syntactic structures; Vol. 2), 211-250. Oxford, New York: Oxford University Press.
Meinunger, A. 2001. *Syntactic aspects of Topic and Comment*. Amsterdam: Benjamins.
Meinunger, A. (to appear): About object es in the German 'Vorfeld'. *Linguistic Inquiry*.
Pesetsky, D. and E. Torrego. 2001. T-to-C movement: causes and consequences. In: M. Kenstowicz (ed.), *Ken Hale: a Life in Language*, 355-426. Cambridge, Mass.: MIT Press.
Reinhart, T. 1981. Pragmatics and linguistics: an analysis of sentence topics. *Philosophica* 27, 53-94.
Reis, M. 1999. On sentence types in German: an enquiry into the relationship between Grammar and Pragmatics. *Interdisciplinary Journal for Germanic Linguistics and Semiotic Analysis* 4.2, 195-236.
Richards, N. 1997. *What Moves Where in Which Language?* Doctoral dissertation, MIT, Cambridge, Mass.
Rizzi, L. 2004. On the form of chains: criterial positions and ECP effects. Ms., University of Siena.
Travis, L. 1984. *Parameters and Effects of Word Order Variation*. Doctoral dissertation, MIT, Cambridge, Mass.
Travis, L. 1992. Phrase structure and V2 phenomena. In: R. Freidin (ed.), *Principles and Parameters in Comparative Grammar*, 339-364. Cambridge, Mass.: MIT Press.
Uriagereka, J. 1995. Aspects of the syntax of clitic Placement in Western Romance. *Linguistic Inquiry* 26, 79-123.
Vallduví, E. and E. Engdahl. 1996. The linguistic realization of information packaging. *Linguistics* 34, 459-519.
Zwart, J.-W. 1997. *Morphosyntax of Verb Movement – a Minimalist Approach to the Syntax of Dutch*. Dordrecht: Kluwer.

Information Structure and the Left Periphery in Dghweɗe

Katharina Hartmann and Malte Zimmermann

1. Introduction

In this article, we present some of the results of recent field research on the information structure of Dghweɗe, a VSO-language from the Chadic family, which is spoken in Northeast Nigeria.* We discuss a number of syntactic phenomena that result from the interaction with information structure in Dghweɗe. These phenomena include (i.) *wh*-question formation; (ii.) the fronting of information-structurally prominent constituents; and (iii.) the effects of fronting on the functional systems of the left periphery, in particular on the T(ense)- and C(omplementizer)-systems, which encode a clause's temporal and illocutionary properties.

As grammatical information on Dghweɗe is scarce, and given that data gathered in the field are – almost by necessity – incomplete, the discussion does not aim at an exhaustive analysis of Dghweɗe syntax and information structure.[1] Rather, we pursue the more modest goal of introducing data from a hitherto largely undocumented VSO-language that may have a bearing on the theory of the left periphery in general, and of the left periphery in VSO languages in particular. In this connection, it is remarkable that Dghweɗe shows a number of surprising similarities to the much better studied VSO-languages from the Celtic family.

In section 2, we give some background information on Dghweɗe. In section 3, the main part of the paper, we discuss the effects of information structure on the syntax of Dghweɗe. Section 4 presents a first tentative analysis of the left periphery of Dghweɗe, which is – in part – inspired by a look at the Celtic languages. Section 5 concludes.

---

* This work was carried out within the research group SFB 632 "Information Structure", which is funded by the DFG (Deutsche Forschungsgemeinschaft). We gratefully acknowledge their financial support. Many thanks to our language consultant, Malam Buba Adamu, to Jochen Zeller for comments and suggestions, and to Alhaji Maina Gimba for lending us his ear to discern subtle but decisive tonal differences in one of the interview sessions.

1 Another reason for caution stems from two facts: First, we could consult only one Dghweɗe speaker, Malam Buba Adamu, who is about 50 years old and who no longer lives in a Dghweɗe-speaking environment. Second, the elicitation language was Hausa (leaving room for potential interferences and misunderstandings). Ideally, our results should therefore be checked again with other speakers of Dghweɗe, which is not without problems, given the socio-linguistic situation of this highly endangered language.

2. Background information on Dghweɗe

Dghweɗe is a highly endangered language in the mountainous border region of Northeast Nigeria and Cameroon.[2] It is an Afroasiatic language and belongs to the Mandara/Wandala subgroup of the Central or Biu-Mandara branch of the Chadic languages. Its closest relatives are the languages in (1), taken from Wolff (2004: 46), with family relations as indicated:

(1)

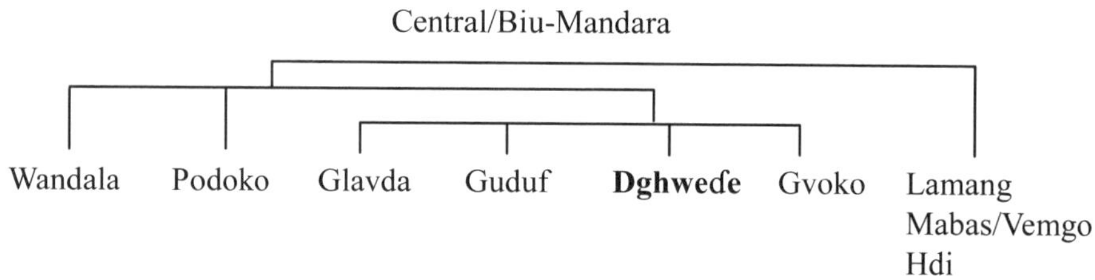

The only available grammatical information on Dghweɗe so far is found in a series of articles by Frick from the 1970s (cf. Frick 1977, 1978ab). The grammatical background information supplied in this section is taken from Frick (1978a). This background should facilitate the understanding of the empirical data in section 3.

The basic word order in neutral (= all-new) sentences in Dghweɗe is VSO – same as in Hdi (Frajzyngier 2002) and Lamang (Wolff 1983), two distant cousins. There are two real arguments, subject and object, which take no preposition, nor do they trigger extensions on the verb stem. In contrast, constituents that correspond to indirect objects semantically (in taking the semantic role of recipient or beneficient) usually trigger an extension on the verb and require the preposition *n*. We return to the syntax of Dghweɗe in section 3.

The morphological system of Dghweɗe is quite complex: the verb stem can carry a bewildering number of infixes and suffixes, which are employed to express both grammatical (e.g. argument structure) and semantic (e.g. aspect), as well as directional distinctions.[3] Aspectual distinctions, for instance, are coded by means of verbal affixes (Frick 1978a: 11-12): The non-affixed (often reduplicated) verbal stem expresses continuous aspect (2a), whereas verbs with the completive suffix *-áyà* (or with any other suffix) are marked for completive aspect (2b).

(2) a. à bá 'to put (cont.)' b. à báyà 'to put (compl.)' (Frick 1978a: 8)

It is worth pointing out that one and the same grammatical formative can have a morpho-syntactic effect in some contexts, and a semantic effect in others. For instance, the infix *-n-* functions as a transitivizer with intransitive verbs (3a), but denotes a partially completed action in connection with transitive verbs (3b) (Frick 1978a: 17).

---

2 According to Frick (1978a: 5), there were about 16000 speakers in the 1970s. It can safely be assumed that the number of speakers has declined since then.

3 Frick (1978a: 41) isolates as many as nine different infixes with more than nine different functions (which can – for the most part – co-occur) and 22 suffixes (which are mutually exclusive).

(3) a. à kə̀l-áyà 'it broke' → à kə̀lə̀-n-áyà 'he broke (it)' (Frick 1978a: 14)
b. à və̀y-áyà 'he caught (it)' → à və̀yə̀-n-áyà 'he held (it)' (Frick 1978a: 16)

The Dghweɗe phonetic inventory is special in that it has three underlying vowels (/a/, /i/, /u/), which are phonetically realized in a number of ways.[4] In addition, consecutive consonants are often linked by vocoids, the phonetic realisation of which may oscillate between [i], [ə], and [u] depending on context. It follows that the phonetic vowel quality of lexical stems may vary considerably across phonetic contexts. In our data, vowels are therefore represented phonetically as they occur in actual speech.

Finally, Dghweɗe is a tone language with two phonemic tones, H(igh) and L(ow) (Frick 1978a: 9). Apart from coding lexical distinctions, tone in Dghweɗe is used for grammatical ends, e.g. for the coding of imperative mode and for marking the argument structure in VS- and VO-sequences (see section 3.1.1). Prosodically, Dghweɗe utterances are organised in hierarchical layers, the most important of these (for grammatical purposes) being the 'pause group' (Frick 1978a: 8): Within the pause group, word-final instances of *e* are generally omitted and the completive marker *–aya* is usually shortened to *–i*. The end of each pause group is marked by polar tone, i.e. a HL- or LH-sequence. However, since a comprehensive prosodic analysis of Dghweɗe is lacking so far, we indicate only surface tones as they occur in actual speech.[5]

3. Syntax and information structure in Dghweɗe

As mentioned, the basic word order of neutral declarative clauses in Dghweɗe is VSO(X). The following examples illustrate this for the continuous aspect (4a), the completive aspect (4b), and for sentences with future reference (4c):

(4) a. (kə́nà) à zà Àdàmú àkwátì
now PRT carry Adamu box
'Adamus is carrying a box now.'
b. à v-ì Áudú plísʰè
PRT catch-COMPL Audu horse
'Audu caught a horse.'
c. də̀ skwà Hàlímá kĺfə̀ ɬ lúmà tɮá
fut buy Halima fish at market tomorrow
'Halima will buy fish at the market tomorrow.'

Example (4c) shows that locative and temporal adjuncts follow the main arguments of the verb in neutral clauses. In addition, (4a-c) show that the initial element in affirmative VSO-declaratives is not the verb, but a sentence-initial particle immediately preceding the verb, so the correct basic word order should be *PRT V S O*. As a first approximation, these sentence-initial particles seem to have temporal and/or illocutionary force:

4 In addition, the following consonants may be unfamiliar to readers without phonetic or Africanist background: bilabial (ɓ) and alveolar (ɗ) implosives (which are common throughout the Chadic languages), and voiceless (ɬ) and voiced (ɮ) lateral fricatives.

5 In addition, we only indicate tone on recorded material.

The particle *da/də* in (4c) marks future reference. In contrast, the particle *a* is found both with the perfective and the progressive aspect. As the aspectual distinction is coded on the verb by means of suffigation (cf. 2ab), the function of *a* seems to be not so much to mark a specific tense or aspect. Rather, *a* marks the status of a clause as unembedded and affirmative. Frick (1978a: 5) calls *a* an 'indicative, non-future marker'. For reasons to emerge in section 4, we prefer to use the descriptive label '*a*-marker' for the time being.[6] We will return to the syntactic distribution and nature of these sentence-initial particles in 3.4.

In Dghweɗe, information structural requirements can trigger changes in word order. In section 3.1, we show that *wh*-arguments are fronted in *wh*-questions. Likewise, both focus and topic arguments can be fronted to the left periphery of the clause, as will be shown in sections 3.2 and 3.3. In 3.4, we look at the syntactic distribution of the sentence-initial particles *da/də* and *a* in more detail. In particular, we discuss the effects of fronting operations on the presence or absence of these particles.

3.1 Wh-question formation

3.1.1 Wh-arguments

The most striking fact about the formation of *wh*-questions in Dghweɗe is that they exhibit a clear-cut asymmetry between arguments and adjuncts. To begin with, *wh*-arguments must be fronted. This is shown for the subject *wír* 'who' in (5) and for the object *ùshír* 'what' in (6):[7]

---

6 Wolff (2003) analyses a similar preverbal *a*-marker in the closely related Guduf as a marker of predicate focus, i.e. focus on the verb or on the entire proposition. We do not believe that this is the correct analysis for the *a*-marker in Dghweɗe, though, the reason being that *a* regularly co-occurs with instances of narrow constituent focus, see in particular sections 3.2.2 and 3.2.3.

7 In addition to fronting, argument questions can be (optionally) marked by a high-toned question marker *bá* in final postion, cf. (6ab) for object-*wh*-questions and (i) for subject-*wh*-questions.

(i) wìr xáb tàb(a) bá?
who drink tobacco Q
'Who is smoking a cigarette?'

The nature of *bá* as a question marker is witnessed by its final occurrence in selective Y/N-questions, as in (ii):

(ii) Mùsá ná xáb yíwè né xáb sóbò hán dì bá?
Musa DEM/TOP drink water PRT drink sobo ? ? Q
'Is Musa drinking water or is he drinking sobo?'

When *bá* is missing in *wh*-questions, its H-tone appears to be preserved and realised on the final syllable, giving the impression of a final H boundary tone in Dghweɗe *wh*-questions. Notice, however, that *bá* is also attested as a question-tag in Hausa. Its optional presence in Dghweɗe questions may therefore also arise under contact with Hausa, the dominant language in the region. The same may be true for the particle *né*, which is a focus-sensitive particle in Hausa, cf. Hartmann & Zimmermann (to appear-b).

(5) *subject wh-questions*
a. wìr tə̀g ɬùwé? continuous
who cook meat
'Who is cooking meat?'
b. wìr v-í plís^{h}è? completive
who catch-COMPL horse
'Who caught a horse?'
c. wìr də̀ tə́g gúské? future
who FUT cook chicken
'Who will cook (the) chicken?'

(6) *object wh-questions*
a. ùshír tə̀g Hàwwá (bá)? continuous
what cook Hawwa Q
'What is Hawwa cooking?'
b. ùshír skw-ì Mùsá (bá)? completive
what buy-COMPL Musa Q
'What did Musa buy?'
c. ùshír də̀ skwà Hàww(a) ft lúmà? future
what FUT buy Hawwa at market
'What will Hawwa buy at the market?'

Please observe that the future tense marker *da/də* is preserved under *wh*-fronting (5c, 6c). In contrast, the *a*-marker disappears when a *wh*-argument is fronted to the left periphery of the clause (5ab, 6ab). Also observe that the tone on the final vowel of the verb changes depending on whether the following argument is its subject (L) or its object (H), see also Frick (1978a: 10). Finally observe that *wh*-fronting does not induce a change in the form of the verb, as is the case in many other Chadic languages including Hausa (Newman 2000) and Dghweɗe's cousin Hdi (Frajzyngier 2002). In Dghweɗe, the verbal form remains constant across declarative clauses and *wh*-questions.

Optionally, instances of fronted *wh*-objects can also be followed by the particle *čə*, at least in the completive aspect, as illustrated in (7):[8]

(7) usher čə̀ zì Àdàmú ndzàyá?
what PRT carry Adamu last.year
'What did Adamu carry last year?'

Notice that the marker *čə* resembles the marker *ča/čə* in (completive) relative clauses:

8 The same particle *čə* seems to be employed in order to distinguish subject questions from object questions in clauses with two animate arguments (apart from tonal differences on the verb!). In the subject question (ia), *čə* is absent. In contrast, the presence of *čə* in (ib) indicates that the object is questioned.

(i) a. wir mbaki Musa?
who help Musa
'Who helped Musa.'

b. wir čə mbaki Musa?
who PRT help Musa
'Who did Musa help?'

As interesting as these data may be, more research is necessary in order to determine whether the presence of *čə* in (ib) is an absolute requirement, or only a preferred option.

(8) kə̀ləfe cì cà xə̀lə̀-dàwà bàngá (Frick 1978a: 21,
fish ? REL bring-from.across hyena *c* = *č* in our representation)
'the fish which the hyena had brought across'

An occurrence of *ča/če* in both *wh*-object questions and in relativisation would not be too surprising, given that both kinds of movement operations are commonly analysed as instantiating A'-movement. In addition, it may hint at the historical development of *wh*-constructions from clefts, a diachronic process which has arguably taken place in many African languages.[9]

3.1.2 Wh-adjuncts

Wh-adjunct questions differ from *wh*-argument questions in two important respects. First, adjunct *wh*-expressions are not fronted to the initial position, but occur immediately to the right of the verb.[10] This is illustrated for locative questions with *mara* 'where' in (9ab), and for temporal questions with *dua* 'when' in (10ab).

(9) a. à skwə́ márà Hàw də̀gə̀dálà? continuous
a buy where Hawwa soup-things
'Where is Hawwa buying ingredients for the stew?'
b. dà wúzà márà Ábdù màrúrà? future
FUT eat where Audu rice
'Where will Audu eat rice?'

(10) a. čà xéní dúà Mùsà m̀ Bàuché? completive
PAST sleep when Musa in Bauchi
'When did Musa sleep in Bauchi?'
b. də̀ sə̀rí dúà Ábdù? future
FUT come when Audu
'When will Audu come?'

(11) shows the same for *wh*-constituents with the semantic role of recipient or beneficient. In section 2, it was argued that such constituents have no proper argument status and that they are therefore realised as PPs (plus an extension on the verb). Same as in (9) and (10), the *wh*-adjunct occurs to the right of the verb:

(11) čə̀ ngàr-àré n wír Mùsá ráyà? completive
PAST build-IO3SG for whom Musa house
'For whom did Musa build a house?'

The *wh*-adjuncts in (9), (10), and (11) occupy the position immediately to the right of the verb, thereby preceding all the verb's arguments (if present). Recall from the discussion of (4c), though, that the immediately post-verbal position in neutral clauses is reserved for arguments. At first sight, then, there seem to be two designated positions for

9 A similar pattern is found in Gùrùntùm, where fronted focused objects are accompanied by the relative clause marker *mài* (see Haruna 2003: 121).

10 A comparable asymmetry between *wh*-arguments and *wh*-adjuncts has been observed for the closely related Guduf, see Wolff (2003) and references therein.

wh-expressions in Dghweɗe: a preverbal one for *wh*-arguments, and a position immediately to the right of the verb for *wh*-adjuncts. Nonetheless, in sections 3.2 and 4 we will provide evidence to the effect that Dghweɗe has no designated position immediately to the right of the verb to which *wh*- and focused adjuncts must move, but that all focused non-subjects occur in their base position. On this analysis, the post-verbal position of the *wh*-adjuncts in (9) to (11) is the result of an 'evacuating' movement of all intervening material to the right.

The second important difference between *wh*-arguments and *wh*-adjuncts concerns the distribution of the sentence-initial particles. (9a) shows that not only the future marker *da/də*, but also the *a*-marker is preserved in *wh*-adjunct questions. This is unlike what was found with *wh*-argument questions in 3.1.1, where it was shown that *a* cannot co-occur with *wh*-arguments in clause-initial position. The presence of *a* in *wh*-adjunct questions shows that *a* cannot be a sentence-type marker for declaratives (nor a marker of predicate focus, see fn.6).

Finally, observe that the *a*-marker is regularly replaced by the past marker *ča/čə* when the sentence has completive aspect and anterior reference. We will present more instances of the past marker *ča/čə* in section 3.2.3. There it will emerge that its presence is contingent on the presence of a *wh*-expression in the clause.[11]

Summing up, *wh*-argument questions and *wh*-adjunct questions behave differently in Dghweɗe. In argument questions, the *wh*-expression occupies a left-peripheral position and the *a*-marker disappears. In adjunct questions, the *wh*-expression occupies a position immediately to the right of the verb, and the *a*-marker precedes the verb.

3.2 Syntax and focus

Focus in Dghweɗe is realised differently on different syntactic constituents. The major split is between focused subjects on the one hand, and focus on other constituents on the other. We will discuss subject focus, object focus, adjunct focus and VP-focus in this order. We will argue (in this section and in section 4), that there is only one designated focus position, which is located in the left periphery and which typically serves as the landing site for focused subjects. In contrast, all other focus constituents can or must remain in their post-verbal base position.

---

11 That *ča/čə* indeed functions as a past marker is indicated by the Hausa translation *dâ* 'formerly', 'just now', which was volunteered by our informant. Due to the fleeting surface quality of Dghweɗe vowels (see section 2), we cannot tell for sure whether this past marker *ča/čə* is identical to the marker *če* found with some instances of *wh*-object fronting and in relative clauses, cf. (7) and (8) above. The two expressions seem to occupy the same structural position, the one immediately preceding the verb, arguing for a unified analysis. On the other hand, we have seen that the slot immediately preceding the verb can be occupied by tense markers (future *da/də*) as well as by markers indicating sentence-mood and/or non-embedding (indicative *a*), so there may well be two phonologically similar, but semantically and functionally distinct items in this position. See also Frajzyngier's (2002: 403ff.) discussion of the imperfective marker *tà* (L-tone) and the marker on preposed focused subjects *tá* (H-tone) in Hdi. Etymologically, *ča/če* and *tà/tá* seem to have the same origin.

3.2.1 Focused subjects

Focused subjects in Dghweɗe predominantly occur in preverbal position, giving rise to an SVO order.[12] This holds both for new-information focus, as found in answers to *wh*-questions, as in (12a), and for contrastive focus, as in (12b).

(12) a. Q: ùshír kd-í Ábdù? A: kúmdà kd-í Ábdù.
what kill-COMPL Audu crocodile kill-COMPL Audu
'What killed Audu?' 'A CROCODILE killed Audu.'
b. Músà v-í plís^{h}è và, Áudù vì-áyà.
Musa catch-COMPL horse NEG Audu catch-COMPL
'MUSA didn't catch the horse. AUDU caught it.'

Notice – once again – that the preverbal *a*-marker is absent when a nominal constituent occupies the sentence-initial position.

3.2.2 Focused objects

In contrast to focused subjects and to *wh*-objects, focused objects predominantly occur in their base position. This is shown for new-information focus in (13a-c).

(13) a. Q: What did Audu catch? A: à v-ì Áudú ŕdè.
a catch-COMPL Audu dog
'Audu caught a dog.'
b. Q: What did Audu catch? A: Áudú v-í plís^{h}è.
Audu catch-COMPL horse
'Audu caught a horse.'
c. Q: ùshír skw-ì Àudú ft lúmà?
what buy-COMPL Audu at market
'What did Audu buy at the market?'
A: à skw-ùd-ì Áudù gúskè ft lúmà.
a buy-?-COMPL Audu chicken at market
'Audu bought a chicken at the market

In (13a-c), a full subject NP is present in addition to the focused object: This subject NP can either occur in its base position, intervening between verb and object, and leading to the VSO order in (13a,c). Or else, the subject may move to the sentence-initial position, giving rise to the same SVO word order that is found with focused subjects, cf. (13b). Again, the preverbal *a*-marker is absent when the left-peripheral position is occupied by a – in this case discourse-old – NP. This observation is relevant, for it shows that the absence of the *a*-marker does not hinge on the focus (or *wh*-) status of the fronted constituent, but only on the initial position being filled by an NP. Finally, (13c) shows clearly that focused objects do not occur in a derived sentence-final position, but in their base position, following verb and subject, but preceding all adjuncts. From (13a) and (13c), we tentatively conclude that post-verbal focused objects do not move. Instead,

12 Although there are a few instances of VSO order with focused subjects in our corpus, the default word order with focused subjects seems to be SVO.

they seem to occur in their base position at the right edge of the inner VP or 'core sentence', which is constituted by the verb and its arguments.

(14) shows that contrastively focused objects can be realised in their base position as well. Again, the subject intervenes between the verb and the focused object.

(14) v-i Musa bungwe va, a v-i Musa plis^{h}e.
catch-COMPL Musa lion NEG a catch-COMPL Musa horse
'Musa did not catch a lion. Musa caught a horse.'

Interestingly, the *a*-marker is also absent in the negated first clause of (14). The absence of the *a*-marker in negated clauses will be crucial for its eventual analysis to be put forward in section 4.

Under certain circumstances, focused objects can also be fronted to the sentence-initial position. This is shown for new-information focus in (15a), and for contrastive focus in (15b).

(15) a. Q: ùshír wùz-ù Hàlímà? A: ngré xpùd-ù-né.
what eat-REFL Halima beans eat-REFL-?
'What did Halima eat?' 'She ate beans.'
b. bung v-i va, a v-i plis^{h}e.
lion catch-COMPL NEG a catch-COMPL horse
'He did not catch a lion, he caught a horse.'

Such fronting seems to take place in order to put an extra amount of stress or emphasis on the focused object. Again, the *a*-marker is absent after fronting (15a), and with negation (15b).

3.2.3 Focused adjuncts

Focused adjuncts behave like their *wh*-counterparts, at least when it comes to new-information focus. (16A) answers a temporal *when*-question, and the focused adjunct *txukwe* 'yesterday' occurs in an immediately post-verbal position, *preceding* subject and object.

(16) Q: čə təg-i dua Hawwa ngre na ba?
PAST cook-COMPL when Hawwa beans DEM Q
'When did Hawwa cook the beans?'
A: a təg-i txukwe Hawwa ngre na. (focused TEMP-adjunct)
a cook-COMPL yesterday Hawwa beans DEM
'Hawwa cooked the beans yesterday.'

(17A) answers a locative *where*-question, and the focused adjunct *mb raya* 'in the house' occurs in an immediately post-verbal position, *preceding* subject and object.

(17) Q: čə tәg-i <u>mara</u> Hawwa ngre na ba?
PAST cook-COMPL where Hawwa beans DEM Q
'Where did Hawwa cook the beans?'
A: a tәg-i <u>mb raya</u> Hawwa ngre na. (focused LOC-adjunct)
a cook-COMPL in house Hawwa beans DEM
'Hawwa cooked the beans in the house.'

Compare the marked position of the focused adjuncts in (16A) and (17A) with the unmarked word order of the neutral counterparts (18a) and (18b) respectively.

(18) a. à tə̀gə̀ Hàwwá ngrí txùkwé. neutral
a cook Hawwa beans yesterday
'Yesterday, Hawwa cooked beans.'
b. à tə̀gə̀ Hàwwá ngrí m̀b ràyá. neutral
a cook Hawwa beans in house
'Hawwa cooked beans in the house.'

Another important observation concerns the form of the clause-initial particle in (16) and (17). Notice that the particle *ča/čə* only occurs in the *wh*-question, while it is replaced by *a* in the corresponding answer. This suggests that *ča/čə* only occurs in completive *wh*-questions.

3.2.4 Focused VP-predicates

Let us finally look at the realisation of VP-focus as found in answers to VP-questions. From the limited data available, it seems that VP-focus is realised with VOS-order:

(19) Q: ùshír màn-ì Hàwwá bá? A: à <u>tə̀g-í ngré</u> Hàwwá.
what do-COMPL Hawwa Q a cook-COMPL beans Hawwa
'What did Hawwa do?' 'It is cooking beans that she did.'

Since (19A) is introduced by the *a*-marker, predicate fronting to a sentence-initial position does not seem to be an option here. Interestingly, though, the subject appears to be post-posed to a clause-final position. As a result, the VP appears in sentence-initial position. It seems, then, that Dghweɗe has the option of post-posing discourse-old material to a clause-final position for the sake of focused material that ends up in a more prominent position this way. In section 4, we will make use of this observation in order to account for the distribution of focused objects and adjuncts.

Summing up so far, while focused subjects are predominantly moved to a clause-initial position, focused adjuncts, *wh*-adjuncts, and focused objects are realised in a post-verbal position. Occasionally, the latter can be fronted to the clause-initial position, too. Focused (including *wh*-) adjuncts follow directly on the verb and precede the latter's arguments (if present). In contrast, focused objects can be separated from the verb by an intervening subject and occur at the end of the 'core sentence', which is made up of the verb and its arguments. The distribution of focused elements is given schematically in (20):

(20)	a.	$Subj_{Foc}$	V		Obj	(Adj)
	b.		V	Subj	Obj_{Foc}	(Adj)
	b'.	Subj	V		Obj_{Foc}	(Adj)
	b''.	Obj_{Foc}	V	Subj		(Adj)
	c.		V Adj_{Foc}	Subj	Obj	

In section 4, we will put forward some speculations as to a unified account for the post-verbal occurrence of focused adjuncts and objects. A crucial component of this unified account will be that all focused non-subjects occur in their base position – despite appearances to the contrary.

3.3 Syntax and topics

In this section, we show that topics can be fronted to the left-periphery of the clause as well, where they are incompatible with the *a*-marker. We only consider discourse topics which provide old or given information, as well as (contrastive) aboutness topics. This kind of topic fronting seems to be restricted to arguments. Argument topics are shown to differ from adjunct topics, which also occur in the left periphery, but which co-occur with the *a*-marker.

3.3.1 Topic arguments

In our discussion of object focus, we have already encountered instances of old-information subjects that are regularly fronted to the left periphery, namely in answers to object questions with post-verbal object focus. In (13b) above, as well as in (21), the subject in the answer constitutes old information and is fronted to sentence-initial position. As with fronted focus constituents, the *a*-marker is absent with fronted topics:

(21) Q: usher skw-ì Músá (bá)? A: <u>Músà</u> skw-í yíwè.
what buy-COMPL Musa Q Musa buy-COMPL water
'What did Musa buy?' 'Musa bought water.'

While fronted subject topics abound, we have found no evidence for fronted object topics in our corpus, not even in answers to subject questions, which always have the word order SVO or VSO.[13] However, old-information objects can occupy a right-peripheral position. This happens with adjunct focus, where the focused adjunct occupies the position right after the verb and precedes subject and object if present, cf. (16A) and (17A) above. (19) showed that such post-posing is an option for old-information subject topics, too.

Subject arguments can also be fronted to the left periphery when they function as (contrastive) aboutness topics. In (22A), the plural discourse topic consisting of Aishatu

13 We do not think that the absence of fronted object topics is due to structural reasons (since objects can front when focused or questioned). Instead, there may be an information-structural requirement that forbids object topics from preceding the subject, when the subject is focused and thus information-structurally more prominent. The resulting word orders are the attested SVO and VSO.

and Saratu is split up into its atomic parts, which are dealt with individually. Each of the answer sentences is about a different individual. In both sentences, the subject topic is fronted and the *a*-marker is absent.

(22)	Q:	ushir	mane	Aishatu	(s)	Saratu	ba?
		What	do	Aishatu	and	Saratu	Q
		'What are Aishatu and Saratu doing?					
	A:	Saratu	wa'a	suuta.	Aishatu	təg	kfe.
		Saratu	wash	clothes	Aishatu	cook	mush
		'Saratu is washing clothes. Aishatu is cooking mush.'					

3.3.2 Frame-setting adverbials

Adjuncts behave slightly different from arguments when they take on a topic-like function. This is the case with temporal or locative frame-setting adverbials, which locate the event or situation described by a clause in time and place. As (23) shows, frame-setting adverbials can optionally occur in a left-peripheral position as well. Unlike fronted arguments, though, left-peripheral frame-setting adverbials *do* co-occur with the *a*-marker.

(23)	və̀gə̀	ná	čə̀	dílè	à	tóxí	màrúrà.
	year	DEM	REL	pass	a	plant	rice
	'Last year, he planted rice.' (lit.: 'The year that passed...')						

The presence of the *a*-marker in (23) suggests that left-peripheral frame-setting adverbials do not occupy the same structural position as the fronted topic arguments in (21A) and (22A). This conclusion is confirmed by the following example:

(24)	Context: Two woman are at the market.						
	nʷasra	miče	na	a	skw	ɬeɬe	te.
	women	two	DEM	a	buy	eggs	3PL
	'The two woman, they are buying eggs.'						

In (24), the topicalised subject in sentence-initial position untypically co-occurs with the *a*-marker. However, unlike in (21A) and (22A), the clause contains a resumptive pronoun that refers back to the left-peripheral subject. This suggests that the sentence-initial subject in (24), and likewise the frame-setting adverbial in (23), has not moved to the initial position from within the clause, but that it is base-generated in this left-peripheral position, for instance by left-adjunction.

Summing up, then, there seem to be two left-peripheral positions for expressions with topic status. The first appears to be clause-internal and can serve as a landing site for topic arguments, in particular for subject topics. Moving the topic to this position leaves behind a trace and blocks the presence of the *a*-marker. The second position appears to be an instance of left-dislocation, where a constituent is adjoined in a clause-external position as witnessed by the presence of the resumptive pronoun in (24) and by the co-occurrence of the left-dislocated constituent and the *a*-marker in (23) and (24).

3.4 Distribution and function of the preverbal particles

In the preceding sections, we have encountered various distributional restrictions on the preverbal particles *da/də*, *ča/čə*, and *a*, which will be summed up now. This will shed more light on the nature of the structural position immediately to the left of the verb, in which these particles are located.

The future marker *da/də* obligatorily occurs in clauses with future reference. Apart from this temporal restriction, it is not restricted to a specific sentence-type, nor to a specific word order. It can occur (i.) in neutral VSO clauses (cf. 4c); (ii.) in *wh*-questions, focus constructions and topic constructions, where an argument has been fronted to the clause-initial position (cf. 5c, 6c); (iii.) in *wh*-questions and focus constructions in which the focused or *wh*-expression is located in a post-verbal position (cf. 9b, 10b). We conclude that *da/də* is a temporal marker. Its sole function is to mark the future reference of a clause.

The particle *čà/čə*, in contrast, is much more restricted in its distribution. It can only occur in completive *wh*-interrogatives (no matter whether these are argument or adjunct questions), but it does not seem to occur in declaratives. This claim is based on the examples in (16) and (17), where *ča/čə* is present in the adjunct *wh*-interrogatives, but not in the corresponding answers. (16) and (17) also show that the occurrence of *ča/čə* cannot be linked to focus, but must be linked to the clause type interrogative. From this, we conclude that the preverbal marker *ča/čə* encodes both temporal information (anteriority) as well as information concerning the illocutionary force of the clause (interrogative), i.e. information that is typically associated with the functional T- and C-system of the clause.

The feature specification of the preverbal marker *à*, finally, is not so clear. Its distribution is restricted in the following way: It is not licensed whenever the clause-initial position is filled by a fronted XP-constituent, no matter whether this constituent is a *wh*-expression, a focus expression, or a topic expression. In addition, *à* cannot occur in negated declaratives (cf. 12b, 14), and it seems to be banned from embedded clauses, too. Unfortunately, we have no data for embedded complement clauses, but the following data from Frick (1978a) show that the *a*-marker is absent from relative clauses (25a) and from narrative clauses, which do not occur as independent structures but only embedded within larger units of discourse (25b). These often come with special non-indicative verb forms, such as the subjunctive or relative aspect, in other Chadic languages:[14]

(25) a. me vákà ná gíɗə̀-dà gìɗà m̀dá (Frick 1978a: 34)
in place DEM stay-with stay 1PL
'in the place where we stayed' (lit. 'in the place with which we stayed')
b. kâ´ tə́gə̀-rè-tə́gè. (Frick 1978a: 33)
NARR cook-3PL-cook
'They are cooking.'

---

14 The absence of the *a*-marker in (25b) has nothing to do with the reduplicated verb, as *a* freely occurs together with reduplicated verbs in root clauses.

Purely descriptively, then, *a* is a marker for affirmative, non-future, VSO-root clauses in Dghweɗe. The following scheme sums up the distributional and functional differences between the three preverbal particles.

(26)

		??	*C+T*	*T*		
a.		a /	***ča** /	da	[V S O	(neutral declaratives)
b.	XP_{wh}	***a** /	ča /	da	[V … t_{XP} …	(wh-interrogatives with fronted XP)
c.	$XP_{foc/top}$	***a** /	***ča** /	da	[V … t_{XP} …	(declaratives with fronted XP)
d.		a /	ča /	da	[V XP_{wh} …	(wh-interrogatives with postverbal *wh*)
e.		a /	***ča** /	da	[V XP_{foc} …	(declaratives with postverbal focus)
f.		***a**			[V …] neg	(negated sentences)
g.	[… $[_{CP}$	***a** /	ča /	da	[V S O]]…	(embedded sentences)

In conclusion, the structural position immediately to the left of the verb hosts several head-like elements which specify temporal information (*da/də*) and illocutionary information (*ča/čə*). While most of the observed distributional restrictions on these preverbal elements (or the absence thereof) follow from their temporal or illocutionary specifications, we still lack a principled explanation for the impossibility of *a* with fronted constituents in (26bc) (or for its impossibility in negated and embedded clauses, cf. (26f,g)). In the next section we will put forward such an explanation after an illustrative look at some data from the Celtic languages.

4. Results and tentative analysis

The preceding discussion has yielded the following results concerning the effects of information structure on word order in Dghweɗe:

(27) i. The basic word order in neutral declaratives is VSO.

ii. There is a clause-initial XP-position, which can serve as the landing site:
   a. for *wh*-arguments in *wh*-questions (obligatory)
   b. for focused subjects (almost obligatory) and objects (optional)
   c. for subject topics (optional)

These expressions are usually associated with the C-domain of clauses.

iii. Focused objects mostly occur in their base-position. Focused adjuncts also occur in post-verbal position, immediately to the right of the verb. Potentially intervening discourse-old material can be postposed to the right.

iv. Dghweɗe has a clause-external left-peripheral position, in which frame-setting adverbials and topicalised nominal constituents can be base-generated.

Concerning the functional architecture of the left periphery, we obtained the following results:

(28) i. There is a structural position immediately to the left of the verb that can host various head-like elements: *a*, *ča/čə*, *da/də*.
ii. In this position, both temporal information (future, anteriority) and illocutionary information (interrogative force and (non-) embeddedness) are encoded.
iii. While the feature specification of *da/də* and *ča/čə* seems clear (and accounts for their distribution), the feature specification of the *a*-marker is not so clear: it is impossible (i.) if the clause-initial position is filled by an XP, (ii.) in negative clauses, and (iii.) in embedded clauses.

In this section, we propose a – in light of the limited amount of data available – somewhat speculative analysis of the left-periphery in Dghweɗe that would account for the observable word order phenomena, as well as for the syntactic distribution and feature specification of the left-peripheral particle *a*. Before doing so, however, we will briefly recapitulate some properties of the left periphery in another group of VSO-languages, namely the Celtic languages. This will pave the way for our analysis.

4.1 Properties of the left periphery in Celtic

As different as it may be in other respects, most of the relevant properties of the left periphery in Dghweɗe can also be found in the Celtic languages. Unless indicated otherwise, the following observations are taken from Borsley and Roberts (1996) (and the references cited there). First, the Celtic languages have VSO as their basic word order (= (27i)). Second, the Celtic languages have an XP-position before the verb. This clause-initial XP-position can serve as a landing site for constituents with marked information-structural status, such as *wh*-expressions, focus constituents, and topic constituents (= (27ii)). The postion is commonly analysed as Spec,CP (see e.g. Tallerman 1996). Third, most Celtic languages feature preverbal particles that encode temporal and/or illocutionary information (= (28i)). For instance, Welsh has root affirmative particles, a subordinating particle, and an interrogative particle, among others. Breton also has subordinating and interrogative particles (among others), whereas particles in Irish not only specify the illocutionary status of a clause, but also its temporal reference as [+/- past]. These particles are commonly analysed as being located in T or in C. For Irish particles, Chung and McCloskey (1987) even assume that they are combinations of a complementizer and an Infl-element. Fourth and last, there is a close structural affinity between the T-head and the C-head (= (28ii), even though they head separate projections. This affinity has been alternatively accounted for in terms of T-to-C-raising or C-to-T-lowering (see Roberts 2001 and references cited there).

The foregoing observations concerning the left-periphery in Celtic are summarised in the structure in (29) (neglecting functional projections below T if any):

(29) *Structure of the left-periphery in Celtic*:
a. $[_{CP}$ $[_{C}$ part] $[_{TP}$ $[_{T}$ part] V S O]] (neutral clause)
b. $[_{CP}$ $XP_{wh/foc/(top)}$ $[_{C}$ part] $[_{TP}$ $[_{T}$ part] V … t_{XP} …]] (XP-fronting)

With these structures in mind, we now proceed to our tentative analysis of Dghweɗe.

4.2 Information structure and syntax of Dghweɗe: a tentative analysis

Before we present our analysis, let the reader be reminded again that – because of the limited amount of data available – this analysis is by necessity stipulative and sketchy. Nonetheless, it accounts for *those* data that are available, and it opens up a number of specific research questions on the left periphery of Dghweɗe that can be pursued in future research and should deepen our understanding of the syntactic effects of information structure in that language. In addition, it points out parallels (as well as differences) to other VSO-languages, and as such may be fruitful for future research on the syntactic properties of VSO-languages in general.

This being said, we propose (30a) as the syntactic structure for neutral (all-new) root clauses in Dghweɗe, and (30b) as the structure for sentences in which a constituent has undergone fronting to the clause-initial position (neglecting possible functional projections below T, such as AspP, whose head may host the aspectually marked verb).

(30) *Structure of the left-periphery in Dghweɗe*:

a. $[_{CP/TP}$ $[_{C/T}$ part] V S O] (neutral clause)

b. $[_{CP/TP}$ $XP_{wh/foc/top}$ $[_{C/T}$ part] V … t_{XP} …] (XP-fronting)

As illustrated in (30ab), we assume that Dghweɗe root clauses have a combined T/C-head, which encodes both temporal and illocutionary information.[15] This double nature of the uppermost syntactic head can be captured in terms of Haider's (1988) 'matching projections', which has been proposed for the German left periphery. Two projections, originally C and I, are folded into one, while the single head of the joint projection gets the feature specifications relating to both. If correct, a major difference between the Celtic VSO-languages and Dghweɗe would be that the former do not allow for matching T- and C-projections.[16]

Because of its C-nature, the specifier of the combined T/C-head, Spec,CP/TP, can serve as the landing site for *wh*-expressions, and other constituents with a prominent information-structural status, such as focus constituents and topics (cf. 27ii). Such expressions are frequently found in the C-domain across languages. In all cases, movement of an XP to Spec,CP/TP is triggered by an (optional) C-feature [w], [foc], or [top] in the C/T-head. While the feature [w] may be spelt out as *ča/čə* in the completive aspect, the features [foc] and [top] are never morphologically realised.[17,18]

[15] Recently, various authors have argued for a tight relation between the T- and the C-domain on independent theoretical grounds, see e.g. Pesetsky & Torrego (2001 and subsequent work).

[16] Another option would be to restrict T/C-projections in Dghweɗe to root clauses, and to assume two structurally distinguished T- and C-projections for embedded clauses (see Brandt et al. 1992 on structural asymmetries between embedded and root clauses). At present, we lack the relevant data for an evaluation of this alternative.

[17] Alternatively, there may be no information-structurally motivated features [foc] and [top] at all. In this case, movement of focus and topic constituents to Spec,CP/TP would be triggered by an optional EPP-feature in C, which could be checked by moving an XP to Spec,CP/TP. See e.g. Grewendorf (2002) for discussion.

The observed restriction of Spec,CP/TP to arguments (cf. 27ii) may be due to the T-nature of the C/T-projection (Spec,TP, like its predecessor Spec,Infl, is generally thought to be an A-position because it is L-related, see Chomsky 1993). Adopting a minimalist perspective on Chomsky's Principles & Parameter framework, as explicated in Chomsky (1995), let us assume, then, that C/T contains an additional (uninterpretable) formal feature F, which must be matched against a corresponding feature in the overtly moved constituent. If no movement takes place, the F-feature apparently causes no harm (or is checked in some other way, for instance by means of the checking operation AGREE, see Chomsky 2000). But if movement takes place, both the landing site (qua its head) and the moved constituent must be specified for F. Since both subjects and objects are found in Spec,CP/TP in *wh*-questions and focus constructions, this formal feature F cannot be the nominative case feature, which is only assigned to subjects. What we need, instead, is a feature common to both subjects and objects, but not to adjuncts. At present, we see two possible candidates: Either C/T contains a D(eterminer)-feature that can be checked against the D-feature of an argument (assuming with Abney 1987 and many others that only arguments contain D-features). Or else, C/T contains a 'structural case'-feature that – again – can be checked only by arguments, since only arguments carry structural case. If an account along these lines proves feasible, all adjuncts would be blocked from moving to clause-initial position because they lack the relevant formal feature F. As a result, they must be realised elsewhere in the clause (cf. 27iii).

Turning to the temporal feature specification of the C/T-head, we assume that temporal features, like all other features in C/T, are privative features with positive values only. More specifically, we assume that C/T can contain only two temporal features, namely [future], obligatorily spelt-out as *da/də*, and [past], optionally spelt out as *ča/čə*.[19] Feature specifications such as [-future], which were attributed to the *a*-marker by Frick (1978a) are impossible on this approach. The non-future reference of clauses containing the *a*-marker simply follows from an implicature triggered by the absence of the feature [future] in C/T.

What is the nature of the *a*-marker, then? We would like to propose that *a* has no inherent feature specification of its own. Its only function is to structurally license the C/T-head, and it is therefore only inserted as a last resort operation whenever other licensing mechanisms fail. In this connection, it is highly instructive to look at the process of long head movement (LHM) in Breton, which is discussed in Borsley, Rivero and Stephens (1996). According to these authors, LHM is a last resort operation that moves a non-finite verb across an intervening auxiliary head to C in order to license Tense:

---

18 As for the question of why focused subjects almost obligatorily move to SpecCP,TP, while other focus constituents remain in situ, the special status of focused subjects can be attributed to the fact that unmarked subjects show a strong tendency to be interpreted as topics (see Givon 1976 for preverbal subjects). Therefore, if a subject is to be interpreted as focus, something special has to be done in order to indicate its marked information-structural status: In Dghweɗe, the focused subject has to be dislocated to the left-peripheral position Spec,CP/TP. On similar asymmetries between focused subjects and focused non-subjects in the Chadic languages Tangale and Hausa, see Hartmann & Zimmermann (to appear-a).

19 The optional spell-out of [past] may follow from the fact that anterior reference can in most cases be deduced from the completive aspect of the verb.

(31) lennet en deus Yann t_V al levr.
read 3SG.MASC has Yann the book
'Yann has read the book.'

Since LHM constitutes a violation of Travis' (1984) Head-Movement-Constraint, it must be subject to economy considerations, and can only apply if Tense is not licensed in some other way. Crucially for our purposes, LHM does not apply under three syntactic conditions: (i.) if Spec,CP is filled by a topic or *wh*-phrase, (ii.) in negative clauses, and (iii.) in embedded clauses. In all three syntactic contexts, the C-domain is filled by a lexical element, i.e. a fronted XP, a neg-marker, and a complementizer respectively, which can license Tense. Interestingly, these three syntactic contexts precisely match those contexts in which the *a*-marker is blocked from occurring in Dghweɗe (cf. (28iii)). We would therefore like to propose that – similar to Breton – the C/T-head in Dghweɗe is licensed (i.) if Spec,CP is filled by a topic or a focus constituent or a *wh*-phrase, (ii.) in negative clauses, (iii.) in embedded clauses, and (iv.) by an overt temporal head. If none of these licensing conditions is met, *a* is inserted in C/T as a last resort.[20] From this, it follows that even though the *a*-marker reliably indicates affirmative, non-future, VSO- root clauses, it does not do so by virtue of a particular feature specification, but only because affirmative, non-future, VSO-root clauses are the only clauses in which C/T is not licensed in some other way. This analysis of the *a*-marker as a structural licenser without an inherent feature content provides a neat account for its at first sight mysterious distribution.

The analysis proposed so far seems to capture all the relevant facts concerning the Dghweɗe left periphery, but it leaves one last question to be resolved, namely the post-verbal focus position that was observed with focused objects and adjuncts (cf. (27iii)). This is not the place to pursue this matter in great detail, but we would like to put forward at least some speculations that may point the way to an account of post-verbal focus in Dghweɗe. Recall from section 3.2 that focused postverbal objects follow the verb and the subject, whereas focused adjuncts follow the verb, but precede subject and object (if present). In our view, this difference in word order provides evidence against a designated second focus position right below the verb to which focused objects and adjuncts are moved by way of (short) leftward movement. If this was the case, both

20 A potential problem for this proposed parallelism between Breton and Dghweɗe arises from the fact that the negation element *va* in Dghweɗe occurs in clause-final position in the (completive) examples given so far (cf. (12b), (14), (15b)), unlike in Breton, where it is located in C (or Spec,CP). We see at least three alternative ways for maintaining the proposed parallelism: Either (i.) the negative element *va* occupies a right-peripheral head above CP/TP, from where it can license the empty C/T-head in negated clauses; (ii.) the negative element *va* occupies the C-position in the left periphery and the TP moves to Spec,CP adopting a Kaynean (1994) analysis (Jochen Zeller, p.c.); or (iii.) sentences in the completive aspect have a covert negation element in the left periphery. Evidence for the latter position comes from the fact that negated clauses in the continuous aspect DO have an overt element *nga* in the left periphery in addition to the clause-final marker *va*:

(i) nga Aishatu təg ngre va, təg kfe.
NEG-CONT Aishatu cook beans NEG cook mush
'Aishatu is not cooking beans, she is cooking mush.'

Our data suggest that the left-peripheral negative continuous marker *nga* must co-occur with the fronted subject in Spec,CP/TP for reasons unclear to us.

should precede the subject, or both should follow the subject, contrary to fact. Also recall from (19) in section 3.2 that discourse-old arguments can be post-posed to the right in principle. We would therefore like to propose that the immediately post-verbal occurrence of focused adjuncts is due to an evacuating movement of potentially intervening arguments to the right. This evacuating movement to the right can be likened to a similar evacuating movement to the left in German, where potentially intervening constituents between the finite verb in sentence-final position and the focus constituent can be moved out of the way for prosodic reasons (see e.g. Krifka 1998). In Dghweɗe, too, this evacuating movement may be forced by a prosodic requirement, according to which a focused non-subject constituent must be at the right edge of a particular prosodic phrase $\wp$ that is projected from the verb and subsequent material. In the default case, this prosodic phrase $\wp$ is projected from the verb and its arguments, i.e. from the inner VP or the 'core sentence'. This accounts for the fact that focused objects can remain in their base position following the subject. However, if the inner VP has been emptied of its arguments, $\wp$ can alternatively be projected from the verb and the immediately following adjunct-XP. This would account for the immediately post-verbal occurrence of focused adjuncts.[21]

Admittedly, the foregoing remarks are of a very speculative nature. We would like to point out, though, that the notion of 'core sentence' or assertion, which contains the verb and its arguments seems to play an important role both for prosodic and tonal phenomena and for the syntactic distribution of certain grammatical formatives in other Chadic and African languages (see e.g. Newman 2000: 546, Hyman 1999: 155). For this reason, we feel encouraged to pursue the prosodic structure hypothesis as a promising approach to the asymmetric distribution of post-verbal focus constituents in Dghweɗe.

5. Conclusion

In this article we have discussed a number of phenomena that have a bearing on the nature of the left periphery in the Chadic VSO-language Dghweɗe. In particular, we investigated the effects of information structure on word order in terms of fronting, and the feature specification and syntactic distribution of certain left-peripheral particles. We also proposed a tentative analysis. A particularly interesting result of the discussion was

21 Jochen Zeller (p.c.) proposes an alternative, syntactic account for the different locations of post-verbal focused objects and adjuncts, which capitalizes on the observed differences between *wh*-arguments and *wh*-adjuncts (see section 3.1). On this analysis, there would be two syntactic focus positions. The first, Spec,CP, precedes the verb and is targeted by focused arguments only. The second, call it Spec,FocP, is located below the verb and is targeted by adjuncts only. While focused arguments have the choice to remain in their base position, focused adjuncts must move to Spec,FocP, thus accounting for their obligatory occurrence in immediately post-verbal position. Notice that this analysis implies that focused adjuncts are the only constituents that must be focus-marked, for reasons unclear to us. A second remark concerns the overall architecture of the Dghweɗe clause that would be implied by the alternative analysis. Recall from (9), (10), (16) and (17) that focused adjuncts generally precede the in situ subject. Assuming that the lowest possible position for subjects is Spec,vP, this indicates that the focused adjunct must be higher than vP, which in turn implies that the finite verb may be in T (or some other functional head between Foc and T), contrary to what we assume here.

that Dghweɗe shows some surprising similarities with Celtic VSO-languages, which may open new perspectives on the study of the syntax of VSO-languages. In particular, it would be interesting to see if some of the properties of the left periphery described here can be observed in other VSO-languages, too. This notwithstanding, our discussion of information structure and the left periphery in Dghweɗe leaves a host of questions unanswered, such as (i.) In which discourse contexts do we find VOS-orders in Dghweɗe?; (ii.) What is the syntax of embedded complement clauses?; and (iii.) Is there evidence for a phonological phrase boundary at the right edge of the inner VP in neutral sentences, in sentences with a postposed subject, and in sentences with postposed arguments, e.g. in the case of adjunct focus? We hope to address these questions in future work.

References

Abney, S. 1987. *The Noun Phrase in its Sentential Aspect*. PhD dissertation, MIT, Cambridge, Mass.

Borsley, R. D. and I. Roberts 1996. Introduction. In: R. D. Borsely and I. Roberts (eds.), *The Syntax of the Celtic Languages*, 1–52. Cambridge: Cambridge University Press.

Borsley, R. D., M.-L. Rivero and J. Stephens. 1996. Long head movement in Breton. In: R. D. Borsely and I. Roberts (eds.), *The Syntax of the Celtic Languages*, 53–74. Cambridge: Cambridge University Press.

Brandt, M., M. Reis, I. Rosengren, and I. Zimmermann. 1992. Satztyp, Satzmodus und Illokution. In: I. Rosengren (ed.), *Satz und Illokution*. Vol.1, 1–90. Tübingen: Niemeyer.

Chomsky, N. 1993. A minimalist program for linguistic theory. In: K. Hale and S. J. Keyser (eds.), *The View from Building 20*, 1-52. Cambridge, Mass.: MIT Press.

Chomsky, N. 1995. *The Minimalist Program*. Cambridge, Mass.: MIT Press.

Chomsky, N. 2000. Minimalist inquiries: the framework. In: R. Martin, D. Michaels, and J. Uriagereka (eds.), *Step by Step*, 89–155. Cambridge, Mass.: MIT Press.

Chung, S. and J. McCloskey. 1987. Government, barriers, and small clauses in Modern Irish. *Linguistic Inquiry* 18, 173–237.

Frajzyngier, Z. (with E. Shay). 2002. *A Grammar of Hdi*. Berlin: de Gruyter.

Frick, E. 1977. Dghweɗe. *West African Linguistic Data Sheets*. Vol.1, 135–142.

Frick, E. 1978a. The verbal system in Dghweɗe. *Linguistics* 212, 5–43.

Frick, E. 1978b. The phonology of Dghweɗe. *Language Data – African Series* No. 11. Summer Institute of Linguistics (microfiche 1 of 1, 78-0005).

Givón, T. 1976. Topic, pronoun and grammatical agreement. In: C. Li (ed.), *Subject and Topic*, 149-188. London/NewYork: Academic Press.

Grewendorf, G. 2002. *Minimalistische Syntax*. Tübingen/ Basel: Francke.

Haider, H. 1988. Matching projections. In: A. Cardinaletti, G. Cinque & G. Giusti (eds.), *Constituent Structure. Papers from the 1987 GLOW Conference*, 101–121. Dordrecht: Foris.

Hartmann, K. and M. Zimmermann (to appear-a). In Place – out of Place: Focus in Hausa. To appear in: K. Schwabe and S. Winkler (eds.), *On Information Structure, Meaning and Form*. Amsterdam: Benjamins.

Hartmann, K. and M. Zimmermann (to appear-b). Exhaustivity marking in Hausa: A Reanalysis of the Particle *nee/cee*. In: E. Aboh, K. Hartmann, and M. Zimmermann (eds.), *Focus Strategies – Evidence from African Languages*. Berlin: de Gruyter.

Haruna, A. 2003. *A Grammatical Outline of Gùrdùŋ / Gùrùntùm*. Köln: Köppe.

Hyman, L. 1999. The interaction between focus and tone in Bantu. In: G. Rebuschi and L. Tuller (eds.), *The Grammar of Focus*, 151–177. Amsterdam: Benjamins.

Kayne, R. 1994. *The Antisymmetry of Syntax*. Cambridge, Mass.: MIT Press.

Krifka, M. 1998. Scope inversion under the rise-fall-contour in German. *Linguistic Inquiry* 29, 75–112.

Newman, P. 2000. *The Hausa Language – An Encyclopedic Reference Grammar*. New Haven & London: Yale University Press.

Pesetsky, D. and E. Torrego. 2001. T-to-C movement: Causes and consequences. In: M. Kenstowicz (ed.), *Ken Hale: A Life in Language*, 355–426. Cambridge, Mass.: MIT Press.

Roberts, I. 2001. Head Movement. In: M. Baltin and C. Collins (eds.), *The Handbook of Contemporary Syntactic Theory*, 113–147. Oxford: Blackwell.

Tallerman, M. 1996. Fronting constructions in Welsh. In: R. D. Borsley and I. Roberts (eds.), *The Syntax of the Celtic Languages*, 97–124. Cambridge: Cambridge University Press.

Travis, L. 1984. *Parameters and Effects of Word Order Variation*. PhD dissertation, MIT, Cambridge, Mass.

Wolff, H. E. 1983. *A Grammar of the Lamang Language (Gwàɗ Làmàŋ)*. Glückstadt: Augustin.

Wolff, H. E. 2003. Predication focus in Chadic languages. In: D. Ibriszimow, H. Tourneux, and E. Wolff (eds.), *Topics in Chadic Linguistics*, 137–159. Köln: Köppe.

Wolff, H. E. 2004. Segments and prosodies in Chadic: on descriptive and explanatory adequacy, historical reconstruction, and the status of Lamang-Hdi. In: A. Akinlabi and O. Adesola (eds.). *Proceedings of the 4th World Congress of African Linguistics*, 43–65. Köln: Köppe.

Old Italian Scrambling: The Low Left Periphery of the Clause

Cecilia Poletto

1. Introduction[1]

In this work I will investigate a phenomenon which has been up to now neglected in the literature on Old Romance languages, namely a set of OV (or better XPV) constructions in Old Italian (OI) which cannot be analyzed as V2 cases, because the verbal form involved is a past participle, not an inflected verb, as shown in (1):

(1) *Allora il cavalero, che 'n sì alto mestero avea la mente misa,*
then the knight, that in so high work had his mind set
(Brunetto Latini, Tesoretto, v. 1975)

I will put forth an analysis in terms of movement of the preverbal XP to a low Focus position located to the left of the low phase vP, similar to (at least one type of) scrambling studied in Germanic languages (see Grewendorf 2005). That such a low left periphery exists has been proposed in the recent literature by a number of authors (notably Jayaseelan 2001, Belletti and Shlonsky 1995, Belletti 2004 for modern Italian and Paul 2002 for Chinese among others). I will adopt this proposal and show that OI displays an interesting parallel between the high left periphery located in the CP layer at the external border of the high CP phase and the low left periphery, so that it is possible to hypothesize that the features of a functional head like Focus are parametrized as phase-independent properties. Whenever a Focus head is inserted in the syntactic structure (whether in the high or in the low phase), it is bound to be either strong or weak depending on the language, yielding uniformity of behavior across phases. If this view is correct, it predicts that also in the DP phase similar phenomena should be found, and that whenever the property is lost, all the phenomena connected to it in each phase are lost. Thus, in our case V2, scrambling and DP internal movement should be lost all together.[2] These predictions are borne out. The article is structured as follows: in section 2 I briefly sketch the peculiar type of V2 displayed by OI, in section 3 I present the phenomenon of scrambling. Section 4 contains an explanation of Egerland's (1997) generalization concerning the relation between OV and past participle agreement patterns and the core proposal that properties of functional heads must be stated phase-

---

1 It is my honor and pleasure to offer this article to Günther Grewendorf, who has been to me a master in research for his rigor and precision and a source of inspiration for his keen and far-reaching analyses.

2 As we will see in section 7 V2 is not lost at the same time for all elements. This has already been noticed by Roberts (1993) for Old French: some elements retain the possibility of the V2 construction much longer than others, and this is probably due to their intrinsic possibility of functioning as operators.

independently. Section 5 shows that the high and the low left periphery behave alike for the three phenomena already discussed for the CP in section 2. In section 6 I show that as is expected, the same type of scrambling is found also within the DP phase. Section 7 demonstrates that in accord with the predictions of the analysis, Renaissance Italian (RI) lost at the same time V2, scrambling in IP and scrambling in DP. Section 8 concludes the article.

2. The high left periphery of the clause

As is well known, OI shows some (but not all) of the typical correlates traditionally associated with the V2 property, namely subject inversion between the auxiliary and the past participle (not to be confused with free subject inversion, which is still possible in modern Italian and occurs after the past participle).[3]

(2) a. *quali denari avea Baldovino lasciati loro.*
which money had Baldovino left them
(Doc. fior. 437)

b. *...primieramente avea ella fatta a llui ingiuria,*
for first had she done to him injury
(Brunetto Latini, Rettorica, 116)

Cases like those in (2) are analysed as movement of the inflected verb to a C^0 position, as also assumed for Germanic V2.

The licensing of pro drop also shows that the inflected verb is moving higher than in modern Romance, as it is different from the usual system known from modern Romance languages in being sensitive to the main versus embedded asymmetry: pro drop is found in main clauses but not in embedded ones, where a (possibly weak) subject pronoun is realized.

(3) *E così ne provò __ de' più cari ch'elli avea.*
and so of-it tested.3SG __ of-the most dear that-he had
'So he tested some of the best friends he had.'
(oFlor.; Testi fiorentini,74)

The standard assumption is that pro can only be licensed when the verb has moved to the CP layer (cf. Benincà (1984) for OI and Roberts (1993) for Old French).

On the other hand, OI does not display any linear V2 restriction typical of V2 languages, as V3 cases are frequently attested.[4]

3 All examples are taken from the OVI data base which contains all Old Italian texts from 1200 to 1350. I follow here Lorenzo Renzi in defining Old Italian as the Florentine variety written in this period.

4 Cases of V4 and V5 are also attested. This is expected under the hypothesis that there is more than one Topic projection located higher than OpP.

(4) a. *Et dall'altra parte Aiaces era uno cavaliere franco*
and on the other hand A. was a knight courageous
(Brunetto Latini, Rettorica, p. 94, r. 7)

b. *E la reina Artemidora di Alicarnasso, che in adiuto di Serses era venuta,*
and the queen A. of Alicarnasso, who in help of Serses was come
francamente si mescolò nella battaglia
courageously herself mingles into the battle
(Bono Giamboni, Orosio, p. 92, r. 1)

I will adopt here Benincà's (2005) proposal that the verb moves to the head of an OpP projection defined as FocusP that is located in the lower portion of the CP layer and thus leaves the higher Topic positions available for other XPs yielding cases of V3.[5]

The structure of the left periphery of OI according to Benincà (2006) is the following:

(5) [Force C^0 [Rel*wh* C^0]/{$_{Frame}$ [ScSett][HT] C^0}{$_{TOPIC}$[LD] [LI] C^0}{$_{Focus}$[I Focus] [II Focus]/[Interrog*wh*] C^0}[Fin C^0] }]

The target of verb movement in OI is a projection inside the Focus field; higher positions can only be occupied by Topics (not more than one Focus can be occupied due to minimality reasons).

Another interesting peculiarity of OI is that it makes extensive use of V1 constructions (which were much more widespread also in Old Germanic languages in contrast to their modern counterparts)

(6) *Avemo ditto che è rettorica*
have told what is rhetorics
(Brunetto Latini, Rettorica, p. 5, r. 17)

No matter how these cases are analyzed, whether there is a null narrative operator (corresponding to 'then') as proposed for their Germanic counterparts or whether V1 is due to movement of the verb to the higher Topic field, it is well know that in these cases clitics are obligatorily found after the inflected verb in enclisis (this is known as a case of the Tobler-Mussafia law):[6]

(7) *Leggesi di Salamone che...*
reads of Salomone that
(Novellino, p.138, r. 1)

Therefore, OI had the following properties connected to the CP left periphery of the clause:

---

5 A number of other phenomena are connected to V2, like null topics and past participle fronting. They are left aside here because the purpose of this section is to illustrate the similarities between the CP left periphery and the low left periphery.

6 In this work I will use enclisis of the clitic pronoun as an empirical test to provide evidence for the similarity between the high and the low left periphery. I will leave aside the complex matter of deriving enclisis.

a) V2 constructions in which an XP is preposed to the inflected verb
b) V3 constructions in which a number of Topics could precede the V2 structure
c) V1 with subject inversion
d) When V1 occurs enclisis is the rule

We will see that similar properties also hold for the scrambling phenomenon.

3. XP V as scrambling to Focus

Once we factor out the cases of XP V ordering triggered by V2, a number of residual cases are still to be explained, the clearest cases being those with the object located between an auxiliary verb (in Op^0) and the past participle. Examples like those in (8) cannot be analyzed as movement of the object to the SpecOp position, which is in this case occupied by the subject preceding the auxiliary verb:[7]

(8) a. *i nimici avessero già il passo pigliato,*
the enemies had already the pace taken
(Bono Giamboni, Orosio, p. 88, r. 15)
b. *ch'egli avea il maleficio commesso*
that he had the crime committed
(Fiore di rett., p. 31, r. 12-13)
c. *dice che poi àe molto de ben fatto in guerra et in pace.*
says that then has a lot of good done in war and in peace
(Brunetto Latini, Rettorica, p. 26, r. 22)
d. *il quale da che ebbe tutto Egitto vinto,…*
whom since he had all Egypt won,…
(Bono Giamboni, Orosio, p. 83, r. 15)

On the other hand, it is not possible to assume that OI was an OV language as unmarked word order is identical to modern Italian and is typical of VO languages:

(9) a. *tenea un savio greco in pregione,*
kept a wise Greek in prison
(Novellino, p 125, r. 6)
b. *fece menare il destriere al campo*
let lead the horse to the camp
(Novellino, p.126, r. 13)
c. *Molto onoroe la donna nel parto* (Novellino, p. 234, r. 7)
A lot honoured the woman in the childbirth
d. *Torquato, consolo di Roma, fece per iustizia tagliare la testa al figliuolo*
Torquato, consul of Rome, had for justice cut the head to the son
(Fiori e vita di filosafi, p. 113, r. 2–3)

7 Given that cases with a simple verb can always be confused with V2, I will restrict the data to cases of compound tenses.

Moreover, not only direct objects can be found to the left of the past participle, but any type of internal argument (including passive subjects):

(10) a. *Ed essendo dell'unico guernimento già ispogliato,*
and being of the only ornament already stripped
(Bono Giamboni, Orosio, p. 411, r. 1)

b. *quello che per uso è già dagli antichi servato,*
what that usually is already from the ancients kept
(Bono Giamboni, Vegezio, p. 108, r. 25–26)

c. *Non crederei che fosse per voi rotto.*
(I) not believe that was by you broken
(Fiore, p. 442, r. 11)

d. *ch'elli è a fine venuto,*
that he is to end come
(Tristano Ricc., p. 397,r. 17)

e. *avegna che neuno possa buono advocato essere né perfetto*
happens that no-one can good advocate be nor perfect
(Brunetto Latini, Rettorica, p. 147, r. 1)

f. *perciò che quelli cui conviene udire sono già udendo fatigati*
for that those that have to hear are already listening tired
(Brunetto Latini, Rettorica, p. 193, r. 19)

g. *se l'avessi a mente tenuto,*
if I it-had in mind kept
(Bono Giamboni, Vizi e Virtudi, p. 16, r. 5)

h. *Assa' bene, quando sono di te acompagnata,*
very well, when I am by you accompanied
(Bono Giamboni, Vizi e Virtudi, p. 33, r. 2)

i. *E quand' ebbi cosí chiaramente a ogni cosa risposto*
and when (I) had so clearly to everything answered
(Bono Giamboni, Vizi e Virtudi, p. 37, r. 24)

j. *comandò questo giovane che fossero tutte quelle genti menate*
(he) ordered this young that were all those people led
(Novellino, p. 143. r. 24)

The same is true for (complex) adverbs and verbal modifiers in general, which in their unmarked order usually occur after the past participle:[8]

(11) a. *e holla già molte volte letta nella Bibbia*
and have-it already many times read in the bible
(Bono Giamboni, Vizi e Virtudi, p. 15, r. 22)

b. *a quelli che sono già avanti iti,*
to those that are already forward gone
(Tesoro volg. p. c350, r. 2)

---

[8] The same phenomenon is found with modal verbs and infinitival complements and with causative contructions.

c. *da tutta la gente sarai* <u>*scarso*</u> *tenuto*
of all the people will.(you).be poorly considered
(Brunetto Latini, Tesoretto, v. 1561)

d. *Poi lo fece* <u>*fuori*</u> *trarre*
then (he) him made outside take
(Novellino, p. 158, rr. 6–7)

e. *il cavaliere era* <u>*molto bene*</u> *costumato,*
the knight was so well educated
(Novellino, p. 311, r. 3)

f. *Quand' ebbero* <u>*così*</u> *ordinato*
When (they) had so ordered
(Novellino, p. 349, r. 1)

Given that this is not the unmarked word order of Old Italian, I propose that the ordering found in (8) (10) and (11) is due to a movement operation similar to scrambling that targets a Focus position located in the low IP area. Following Belletti (2004), I take this position to be in the left periphery of the low vP phase. This is clearly a position dedicated to XPs (no clitic or weak elements can occur there as they cannot be focalized), and it hosts virtually any type of constituent, all types of arguments as well as adverbials and verbal modifiers. This lack of "specialization" is a feature typical of left peripheral positions, where any type of XP can be moved.

Suppose further that the "left periphery" of each phase is construed in the same way, namely by merging a "Topic-Focus" field before the highest projection "closing up" the phase (see again Belletti (2004) for modern Italian among others).

Then, the examples in (8) (10) and (11) are to be analyzed as cases of movement to a SpecFocus position in front of the past participle but lower than the subject position (SpecAgrS or SpecT in a strict minimalist framework).

4. The relation between scrambling and past participle agreement

In his (1996) book, Egerland examines only cases of direct object scrambling and notices that the OV order is diachronically related to the possibility of past participle agreement with postparticipial objects; when postparticipial agreement is lost, OV order is lost as well:

(12) a. *quando egli avea già fatt*<u>*i*</u> *molti miracoli*
when he had already done+AGR many miracles
(Tesoro volg. p. a258, vv. 3–4)

b. *E quando il notaio ha lett*<u>*a*</u> *la proposta dinanzi a' consiglieri,*
and when the notare has read+AGR the proposal to the counsellors
(Tesoro volg. p. d335, v. 17)

c. *c'ha rifiutata la nobile città di Giadres et ha preso li marchi*
that has refused+AGR the noble city of Giadres and has taken the money
(Novellino, p. 133, r. 3)

Moreover, he shows that if the order is VO, past participial agreement is optional (as (12c) shows) while it is obligatory when the order is OV.

This observation is confirmed by modern languages, which still have OV orderings, like Friulian:

(13) a. *O ai lis sigaretis dismenteadis.*
I have the cigarettes forgotten+AGR
b. *O ai dismenteadis lis sigaretis.*
c. *O ai dismentea:t lis sigaretis.*
d. **O ai lis sigaretis dismentea:t.*

In modern Friulian only three of the four logical possibilities are attested: the sequence OV without agreement is not possible.

We can reformulate Egerland's observations in the form of two descriptive generalizations as in (14a) and (14b):

(14) a. When past participial agreement is reduced (and finally lost), OV is reduced (and finally lost) as well.[9]
b. Past participle agreement is obligatory with the order OV, but not with the VO order.

Both generalizations can be used to shed light on the account of scrambling in OI.

Generalization (14b) is to be considered a special case of a descriptive generalization proposed by Guasti and Rizzi (2002): focusing on cases of subject agreement, they note that movement and morphological richness are connected and propose the following generalization:

(15) If a feature is checked in the overt syntax, then it is expressed in the morphology.

Generalization (15) captures all cases in which a preverbal subject shows stronger agreement than a postverbal subject: when the subject moves to the specifier of AgrS thus checking the AgrS feature in overt syntax, morphology reflects this process, hence the fullest agreement pattern is always selected. This is not always the case for postverbal subjects, where there is not overt checking of the feature in the syntax, thus languages can either have full agreement, a reduced form of it, or no agreement at all. Given Egerland's generalization, confirmed by Friulian data, this is also true for objects: when the object is checking a given feature in overt syntax, this is reflected in the past participle agreement morphology.

Therefore, OV structures are to be analyzed as cases of overt movement through a position encoding strong features for object agreement[10] on their way to Focus. The derivation of the OV structure is the following:[11]

9 For data supporting this generalizations see Egerland (1996) and section 6.

(16) [$_{\text{CP}}$ *che* [$_{\text{AgrS}}$ [$_{\text{SpecAgrS}}$ *egli*] [$_{\text{AgrS°}}$ *avea*] [$_{\text{FocusP}}$ [$_{\text{SpecFocus}}$ *il maleficio*$_j$]
[$_{\text{Focus°}}$ *commesso*$_i$] [$_{\text{AgrOP}}$ [$_{\text{SpecAgrO}}$ t_j] [$_{\text{AgrO}}$ t_i] ...[$_{\text{VP}}$ [$_{\text{V°}}$ t_i] [t_j]]]]]]

VO structures in contrast are ambiguous between a movement and a non-movement analysis: when there is no agreement, the most natural hypothesis is that no movement has applied to the object DP. Given that movement to Focus is not obligatory, the object DP can stay in its base position. I assume that whenever there is object agreement, there is movement of the object DP to the SpecAgrO position. This means in turn that the past participle must have moved higher than the SpecAgrO position where the DP has moved. I assume that this position is the Focus head.

Hence, the derivation of VO orders is (17a) for non agreeing cases and (17b) for agreeing cases:

(17) a. [$_{\text{CP}}$ *quando* [$_{\text{AgrS}}$ [$_{\text{SpecAgrS}}$ *il notaio*] [$_{\text{AgrS°}}$ *ha*] [$_{\text{FocusP}}$ [$_{\text{Focus°}}$ *letto*]
[$_{\text{AgrOP}}$ [$_{\text{SpecAgrO}}$] [$_{\text{AgrO}}$ t_i] ...[$_{\text{VP}}$ [$_{\text{V°}}$ t_i] [*la proposta*]]]]]]
b. [$_{\text{CP}}$ *quando* [$_{\text{AgrS}}$ [$_{\text{SpecAgrS}}$ *il notaio*] [$_{\text{AgrS°}}$ *ha*] [$_{\text{FocusP}}$ [$_{\text{Focus°}}$ *letta*$_i$]
[$_{\text{AgrOP}}$ [$_{\text{SpecAgrO}}$ *la proposta*$_j$] [$_{\text{AgrO}}$ t_i] ...[$_{\text{VP}}$ [$_{\text{V°}}$ t_i] [t_j]]]]]]

Notice that this derives straightforwardly not only generalization (14b) but also (14a): if a strong agreement is always an instance of movement in OI, then loss of movement directly accounts for loss of strong agreement.

Alternatively, and in a minimalist framework, the agreement pattern with a postparticipial object might be derived via an Agree operation applying freely between the features of the object and those of the past participle. Note that such a hypothesis can also explain Guasti and Rizzi's generalization: given that movement is a complex operation composed of 'match', 'agree' and 'pied pipe', the fact that agreement is obligatory with movement is straightforward on this account.

The main difficulty of this approach is that the Agree operation seems too unconstrained to account for the fact that Italian has lost OV and post participle agreement at the same time (generalization (14a)) and for the fact that the agree operation is now blocked in modern Italian for postparticipial objects, but not for other types of objects like clitics or passive subjects. In other words, the operation Agree is too liberal.

One possible way to constrain the operation is to assume that there is only an indirect link between agreement and syntactic movement and to hypothesize that Agree only applies to an in situ object if the language also has a movement operation. Notice, however, that modern Italian still has agreement with object clitics and passive subjects, so we have to assume that agreement with a clitic element or agreement with a passive subject would not be sufficient to license agree with postparticipial DPs; only scrambling is. Moreover, the pattern of modern Italian, where agreement is maintained only with those elements that have been moved to the left clearly shows that movement and

10 This position could be AgrOP, as originally proposed by Kayne (1991) or another one which also has other features, as AspP. I will leave this question open here, although I use the label AgrO for the sake of concreteness.

11 In this work I assume Cinque's (1999) proposal that auxiliaries are functional heads inserted in a functional projection during the derivation and do not have a VP on their own.

agreement are directly and not only indirectly linked. Therefore I will discard this analysis here and will continue to assume that strong agreement and movement are directly related (as illustrated in (17)).

Notice that if we are on the right track with the idea that scrambling is movement to Focus there is an interesting parallel between V2 structures and OV structures: in both cases, an XP is moved to a SpecFocus and the verb is moved to the head of Focus. If both heads have the same strong property, we can hypothesize that the properties of the low Focus position in IP and in the high CP phase remain constant across phases.

It is a fairly standard assumption that the old notion of "parameter" has to be formulated in terms of properties of functional heads. This however often leads to postulating a parameter for a single syntactic construction, while the original notion of parameter has proven extremely powerful in the GB account because it accounts for a number of syntactic constructions on the basis of the same abstract property. Assuming that properties of functional heads remain constant across phases is a new way to account for different phenomena (in our case V2 and scrambling) on the basis of the same abstract feature. I will therefore assume the strongest version of this proposal, namely that a parameter corresponds to the "activation" of a given F^0 whose features must be checked in the computational component wherever it is merged.

Hence, Focus in Old Italian maintains the same property throughout all the phases where it occurs: more specifically it must be filled by a verbal head in all phases, the inflected verb fills the $Focus^0$ of the high phase, the past participle fills the $Focus^0$ of the low phase.

In the high CP-phase Focus triggers the subject inversion phenomenon typical of V2 contexts (and the other phenomena seen in section 2). In the vP phase it triggers post-participial agreement, OV, and, more generally, XP V.

This means that the high and the low left periphery should behave alike in all respects, which leads to a number of expectations concerning the low left periphery on the basis of the constructions illustrated in section 2 for the high left periphery. I will examine here three constructions: a) As there are several left dislocated items in the high left periphery, the analysis developed here predicts that more than one scrambled element above the past participle should be possible. b) Moreover, as there exists V1 in the high left periphery, this should be the case also in the low left periphery.

c) As seen in section 2, V1 constructions always trigger enclisis in the high phase. Accordingly, the same should hold in the low phase.

Two additional predictions are made by the analysis proposed here. The first concerns OI itself: if the strong feature of Focus is phase-independent we should find similar phenomena in the DP as well, given that DP is also a phase.

The second prediction is a diachronic one, namely that if V2 and XP V are effects of the same abstract property, they should be lost together. An apparent problem is raised by the equation between V2 and scrambling: if the two phenomena are always connected we should expect them to cooccur in all languages. This is clearly not the case, as there are languages with scrambling (for instance Korean) which do not display V2. The problem here is that the terms V2 and scrambling have to be qualified in terms of properties of a functional projection. The peculiar type of V2 found in OI is not the "standard" one described for Germanic languages, and as such it probably targets a much lower position in the left periphery. In other words, V2 is a complex phenomenon that always involves the left periphery but not always the same FP. The same is true of scrambling, as different types of scrambling (notably A and A' scrambling) are known to occur in different languages. Hence, given that V2 and scrambling do most probably not target the same projection in all languages, this analysis does not predict that the two phenomena always go together. If both phenomena are complex in the sense that there are some elements that are "stronger" in triggering V2 (as for instance wh-items, which still require residual V2 in English, or adverbs like "always" or "never") than others, the same could be true for scrambling, which would also progressively restrict the number of elements moved, just as we observe in the progressive loss of V2.

5. First prediction: the analogy of the high and the low phase

As seen in section 2, OI displays frequent V3 constructions, where one (or several) Topic(s) precedes the Focus element. This is, as expected, also found in the low left periphery:

(18) a. *ed ha'mi* *la cosa molte volte* *ridetta*
and has the thing many times retold
(Bono Giamboni, Trattato, p. 131)
b. *E quand' ebbi* *cosí chiaramente a ogni cosa* *risposto*
and when had so clearly to everything answered
(Bono Giamboni, Vizi e Virtudi, p. 37, r. 24)

The examples in (18) show that two elements occur to the left of the past participle. In (18a) these are the direct object *la cosa* and the adverbial XP *molte volte*, in (18b) the indirect object *a ogni cosa* and the adverbial XP *così chiaramente*. Notice that this type of adverbial XPs usually occurs after the past participle, as shown in (19), hence cases like (18) are genuine cases of multiple scrambling of an argument and an adverbial:[12]

(19) *Tullio dice, che 'l fatto è contato* *chiaramente,*
Tullio says that the deed is told clearly
(Tesoro volg. p. 138, r. 6–7)

The second type of construction frequently found in the high left periphery of the clause is V1: trivially, cases of V1 are also found in the low left periphery, where all objects and lowest adverbs[13] are located after the past participle

12 Notice that in OI *chiaramente* can only be a low manner adverb, not a high one as in modern Italian (see Poletto 2004 on this).

13 By lowest adverbs I mean here those adverbs that are usually found after the past participle both in modern Italian and OI when they are not focalized.

(20) a. *ciò che savi avevano detto intorno alla retorica*
What that the wise had said about the rhetorics
(Brun. Latini, Rett. P.7 r.19)

b. *...fue isbandito della terra*
...was banned from the earth
(Brun. Latini, Rett. P.7 r.9)

c. *poi che Tullio ae advisati li mali*
since T. has seen the evils
(Brun. Latini, Rett. P.12 r.7)

In other words, cases of V1 in the low phase are "normal" cases of VO.

The third construction found in the high left periphery is the one with V1 and enclisis of a past participle. As for the low phase, it seems difficult to test this possibility as in general it is not possible to leave the clitic inside the low phase (OI has obligatory clitic climbing to the high phase).[14]

However, absolute participial clauses of non-unaccusative verbs have been analyzed by Belletti (1990) as truncated structures corresponding to AspP in modern Italian. Hence, we can use those cases as a test for our hypothesis:

(21) a. *trovò l'arme del re Meliadus, che lli avea fatta sì bella*
found the weapons of king M. that he had done so nice
deliberanza, e donatogli: et era suo mortale nemico.
disposal, and given-to-him and was his mortal enemy
(Novellino p. 268 r. 21)

b. *Fatto ha chiamare Licomede re, e dettogli che faccia*
Made has call Licomede king, and told-him that make
chiamare le donne
call the women
(Armannino, Fioritap. 546)

As expected, in these cases enclisis is the rule just like in the high phase. We can conclude that the high and the low left periphery really behave alike with respect to the phenomena observed.

6. Second prediction: Focus in the DP phase

The second prediction made by the hypothesis that functional properties are phase-independent is that within the DP area scrambling phenomena are also possible (and are ungrammatical in modern Italian, which has also lost V2 and IP scrambling).

One interesting fact about OI is that it is possible to have modified adjectives in prenominal position (contrary to modern Italian):

---

14 There are Romance languages such as Piedmontese that leave the clitic on the past participle, but this is not the case in OI.

(22) a. *domandò se avesse più care pietre*
asked if had more valuable stones
(Novellino p. 123, 54)

b. *qual ti sembra di più ricca valuta?*
which to you seem of more rich value
(Novellino p. 127,28)

c. *Democrito fue molto grande filosofo.*
Democrito was very great philosopher
(Fiori e vita di filosafi, p. 106, r. 2)

It becomes clear how examples like (22) are to be interpreted as soon as we notice that adjectives only occurring in postnominal position in modern Italian can occur to the left of the noun in OI:

(23) a. *la quale guardava al figliuolo piccolo del morto fratello,*
whom looked at the young child of the dead brother
(Bono Giamboni, Orosio, p. 148, r. 7)

b. *e dagli usati uomini*
and from the experienced men
(Bono Giamboni, Vegezio, p. 167, r.)

c. *il ben usato cavaliere disidera battaglia*
the well behaved knight wants battle
(Bono Giamboni, Vegezio, p. 70, r. 6)

We can hypothesize that adjectives that are located in a postnominal position in modern Italian can move to the left of the noun due to a DP internal scrambling process, and that modified adjectives do the same. There is empirical evidence that the basic position of the adjective is much lower in the structure than the one where the adjective appears. Consequently, we are not dealing here with a language in which the noun raises very little and most adjectives are found to its left (like Germanic languages for instance). Consider the following examples:

(24) a. *e di gentile aspetto molto,*
and of kind appearance very
(Dante, Vita nuova, cap. 8, par. 1, v. 11)

b. *e ciò non è propia natura di cavallo*
and this not is own nature of horse
(Novellino, p. 128, r. 67)

In (24a), the adjectival modifier *molto* 'very' has remained to the right of the head noun *aspetto* 'appearance', and only the bare adjective *gentile* 'kind' has been extracted out of the complex adjectival phrase and scrambled to the left of the noun. In (24b), the portion of the AdjP to the right of the noun is its complement PP *di cavallo*. Cases like (24) clearly show that the prenominal position is not base generated, but it is the result of movement of the entire modified AdjP or of part of it.

An even more interesting case is provided by (25).

(25) a. *quando vi dissi del cavallo cosa così meravigliosa*
when I you told of the horse thing so wonderful
(Novellino p. 120 r. 14)

b. *Sì come quando ordino di ritrarre dell'antiche scritte*
So as when (I) order to draw of the ancient writings
le cose che ...
the things that
(Brun. Latini Rett. P 11 r.18)

These examples show that it is possible to move even the complement PP of a noun (in (25a) *del cavallo*, in (25b) *dell'antiche cose*) to the left. Given that in general object PPs are generated to the right of the noun (OI is not a Complement N language), (25) clearly shows that OI had a scrambling process in the DP as well as in IP, thus confirming our prediction.[15]

7. Third prediction: the loss of the medieval system

As discussed above, the hypothesis that IP-scrambling, V2, and also DP-scrambling are triggered in OI by the same abstract property relating to strong properties of a Focus head located in the left periphery of each phase entails that all these properties are lost together at the same time and at the same rate. In order to test the prediction, a 16th century text has been examined and screened for all the cases of inversion of the subject, IP scrambling, agreement with a post participial object, and DP scrambling. The text chosen is "Il Principe" by Machiavelli of which the first X chapters and chapters XX-XIII have been considered for an amount of about 120 pages.

The first phenomenon considered is V2: all cases of subject inversion and cases in which new information focus (ungrammatical in modern Italian) is found in first position have been marked.

15 A speculation left for future work concerns the possibility to connect DP scrambling to the fact that OI had very frequent determinerless DPs, in contexts which are completely excluded in modern Italian, as exemplified in (i). If the parallel between the three phases is complete, the head noun also has to raise to the head of the DP internal Focus position and through this possibly to the higher D^0 position. Evidence for N to D movement (through Focus) in OI is provided by the frequent cases of determinerless nouns with postnominal possessive adjectives, as illustrated in (ii). See on this Giusti (1992), (1996) and (2004).

(i) a. *esser figliolo di pastore* (Novellino, p. 128)
to be son of shepherd

b. *che uomo vecchio dicesse così grande villania* (Novellino, p. 129)
that man old said so big rudeness

c. *con moltitudine di gente in assedio* (Novellino, p. 130)
with lot of people in siege

d. *donami cavallo e somiere e dispendio ...* (Novellino, p. 131)
give me horse and helmet and money

e. *mandò per maestri* (Novellino, p. 127)
sent for masters

(ii) *anostra magione* (Novellino, p. 128)
to our home

In the whole sample there are only three cases of subject inversion with an auxiliary verb and eight cases with modals (six with *potere* 'can' and two with *dovere* 'must').

(26) a. *Spenti adunque questi capi, e ridotti i partigiani loro amici suoi,*
blown off then this bosses, and reduced their partisan friends,
aveva il duca gittato assai buoni fondamenti alla potenza sua (p.221)
had the duke thrown very good foundations to his power
b. *Aveva adunque Luigi fatto questi cinque errori* (p. 197)
had then Luigi made these five mistakes
c. *Mentre che durò la memoria, sempre furono i Romani incerti*
while that lasted the memory always were the Romans unsure
di quella possessione (p. 203)
of that possession

Cases with modals are the following:

(27) a. *E deve soprattutto uno principe vivere con i suoi sudditi*
and must overall a prince live with the his subjects
in modo che ... (p. 237)
so that…
b. *E con più facilità se le può un principe guadagnare* (p. 205)
and with more ease them can a prince gain

As for cases in which new information focus appears in first position, a feature typical of OI syntax which is not found in modern Italian, there exist only four cases of this type (shown in (28)) if we factor out the two adverbs *sempre* 'always' and *mai* '(n)ever' which are very frequently placed in first position even in contexts where they do not seem to be contrastively focalized (in which case they are still possible in modern Italian as well):

(28) a. *Confido assai che per sua umanità gli debba essere accetta* (p. 173)
(I) trust a lot that for his humanity to-him must be accepted
b. *Più facilmente si tiene una città usa a vivere libera*
more easily one keeps a city adapted to live free
con il mezzo dei suoi cittadini, che in altro modo (p. 204)
with the means of its citizens that in other way
c. *Come di sopra si disse* (p. 213)
as above was said
d. *Ordinare che da Firmani fosse ricevuto onoratamente* (p. 234)
order that from the Firmani was received with honour

(29) a. *Sempre si trova dei malcontenti* (p. 202)
always one finds someone grudging
b. *Mai si troverà ingannato da lui* (p. 243)
never himself-will-find cheated by him

Notice that in non-V2-languages like Spanish and Catalan, adverbs corresponding to OI *sempre* and *mai* are always found to the left of the inflected verb and are probably located in a dedicated position inside the Focus layer (see Grava (2005)), so they cannot be taken as real indications of a V2 grammar.

Relative clauses deserve particular attention, as instances of V2 embedded under the relative complementizer or wh-item are much more frequent than relative clauses that lack the V2 effect.

(30) a. *Quelli i quali per vie virtuose simili a costoro diventano*
those who along ways winding similar to them become
principi (p. 210)
prince

b. *Quelli che di sua qualità gli avevano invidia* (p. 211)
the ones that of his qualities to-him had envy

c. *Quelle armi che vicine lo potevano offendere* (p. 223)
those weapons that close him-could offend

Apparently, the context of relative clauses is special in maintaining V2 longer than other clause types, and this is probably due to the fact that in these cases, the CP layer is already activated by the wh item sitting in a higher relative position (see Rizzi (1997) on the position of relative elements).[16]

Therefore, the V2 syntax seems partially accessible still at this stage, although in a restricted way and under special conditions (as in relative clauses).[17]

Egerland (1996) proposed that at this stage two grammars were available, one simulating the OI syntax, which is less frequently used, and the new one, in which V2 has already been lost. If we are on the right track, it seems that the V2 grammar still available is subject to specific conditions and very limited in its use.

As for IP scrambling, we have a parallel situation: very few cases of scrambling are found in the sample and they are all restricted to compound tenses with the auxiliary *essere* 'be' or copular clauses.

(31) a. *Non è, oltre a questo, la provincia spogliata da' tuoi ufficiali* (p. 187)
not is, beyond this, the province stripped by your officers

b. *E benché dai Cartaginesi fusse due volte rotto* (p. 232)
and although by C. was twice broken

c. *Da coloro che saranno in quella malcontenti* (p. 190)
by those who will-be in that unhappy

---

16 One might wonder why this is not the case in interrogative clauses, where a wh-item activated the CP layer. However, as already shown by Rizzi (1997), interrogative wh-items are located low in the CP structure, in the same Focus field where in OI V2 occurs. Therefore, interrogative wh-items and the focalization of an XP are not compatible.

17 Notice that past participle agreement here originates from movement of the wh-item in the lower phase through an AgrOP and possibly through the low SpecFocus projection. The fact that the activation of the left periphery in the CP phase is mirrored by the activation of the low left periphery of the vP phase is expected in the framework.

d. *Non consentono che sia tra gli eccellentissimi uomini*
(they) not consent that is among the best man
celebrato (p. 233)
celebrated

e. *Che senza l'una e l'altra fu da lui conseguito* (p. 233)
that without the-one and the-other was by him achieved

Two cases with *have* have been found, both in a construction with a bare noun:

(32) a. *Che esso abbia con loro obbligo* (p. 246)
that it has with them obligation

b. *Altrimenti non ha nelle avversità rimedio* (p. 240)
otherwise not has in the misfortune remedy

Most relevant examples involve modals (recall that the same observation has been made for subject inversion in the case of V2) or causative constructions:

(33) a. *Lo può con grandissima difficoltà perdere* (p. 187)
it can with greatest difficulty loose

b. *Si possono con più difficultà corrompere* (p. 202)
one can with much difficulty bribe

c. *Tanto potette in su tale fondamento edificare ogni edificio* (p. 212)
so-much could on this foundation build every building

d. *Fece da' suoi soldati uccidere tutti i senatori e i più ricchi*
made by his soldiers kill all the senators and the richest
del popolo (p. 232)
of the people

e. *Non abbiano ancora mai potuto ne' tempi pacifici mantener*
not have yet ever could in peaceful times keep
lo Stato (p. 235)
the state

f. *Che non fa i fondamenti prima, li potrebbe con una gran virtù*
that not does the foundations before, them could with great virtue
fare dipoi (p. 213)
do afterwards

g. *Per potere con quelli tenere il papa in freno* (p. 224)
to can with those keep the pope at bay

Postparticipial agreement is also extremely rare, only five cases are found in the sample:

(34) a. *Per aver tenuta più lunga possessione in Italia* (p. 93)
for have kept+AGR more long possession in Italy

b. *Basta avere spenta la linea del principe che*
enough have turned+AGR off the line of the prince who
li dominava (p. 86)
them dominated

c. *Se egli avesse osservate le regole sopradette e tenuti*
if he had observed+AGR the rules aforementioned and kept+AGR
sicuri e difesi tutti quelli suoi amici (p. 195)
safe and protected all those his friends

d. *Arebbero sempre tenuti gli altri discosto dalla impresa*
had always kept+AGR the others away from the enterprise
di Lombardia (p. 197)
of Lombardy

e. *Si vedrà lui aversi fatti gran fondamenti*
One will-see him have done+AGR grand foundations
alla futura potenza (p. 213)
to future might

Interestingly, past participle agreement is extremely frequent in relative clauses, that is exactly the same construction in which V2 is still quite consistently found, as discussed above:

(35) a. *La quale opera io non ho ornata né ripiena di*
the which deed I not have adorned+AGR nor filled+AGR of
clausole ample (p. 173)
big clauses

b. *E quelli fondamenti che gli altri hanno fatti avanti che*
and those foundations that the others have done+AGR before that
diventino principi (p. 213)
become princes

c. *Di quelli cardinali che lui avesse offesi* (p. 229)
of those cardinals that he had offended+AGR

We can conclude that scrambling and post participial agreement are also extremely restricted and are more frequently found in the same contexts that still trigger V2, namely modals and relative clauses.

DP scrambling, as expected, is also quite restricted but still present:

(36) a. *Ed uno de maggiori rimedii e più vivi* (p. 187)
and one of the main remedies and more powerful

b. *Troverà difficultà grande* (p. 202)
will.find difficulty great

Therefore, we can conclude that the Renaissance system exhibits a consistent behaviour of all the phenomena that we attributed to one and the same abstract property relating to Focus heads. The phenomena are all still present, though apparently only in some constructions, which, at least for V2, scrambling and past participle agreement, seem to be the same.

8. Conclusion

In this work, I have tried to show that OI is not only a V2 language, but also has scrambling to an IP internal Focus position. Scrambling is both diachronically and synchronically connected to past participle agreement: when scrambling occurs, agreement is obligatory, while only optional when the object occurs in a post participial position; moreover, the two phenomena disappear at the same time. I have adopted an analysis of past participial agreement in terms of syntactic movement (and rejected an analysis in terms of the operation agree with in situ objects) of the DP to a SpecAgrO position and further movement of the past participle to Focus.

This analysis of OV orders opens up new perspectives for a unitary treatment of functional projections that occur in different phases: if Focus is marked strong in OI independently of the phase where it is merged, V2 and scrambling are two sides of the same coin, one occurring in the high, the other in the low phase. We have seen that the two phenomena have a number of properties in common, for example the possibility of V1 and V3 on a par with V2. The hypothesis of feature uniformity in all phases further predicts that the same type of reordering phenomena is also found within the DP, which is confirmed by the cases of prenominal adjectives and PPs present in the sample. The last prediction is that all reordering phenomena (i.e. V2, IP and DP scrambling) and past participle agreement are lost at the same time. A close examination of a Renaissance text shows that they all occur mainly in restricted contexts, involving relative clauses and modals and only very sporadically in other contexts.

Such a far-reaching hypothesis as the one assuming the uniformity of features across phases must clearly be tested in other domains, and in other languages. The next domain for which one should test whether the connection between V2 on the one hand and IP and DP scrambling on the other really holds are other Old Romance languages. It is important to keep in mind that both V2 and scrambling are complex phenomena; they have been analyzed as targeting different projections in different languages (as the well known distinction between asymmetric and symmetric V2 languages or the distinction between A-scrambling and A'-scrambling attest) which do not always involve the same head in all languages. This means that we should not compare different types of V2 and different types of (IP and DP) scrambling before we are sure that they really target one and the same head in different languages. This is a very vast field of inquiry that cannot be tackled in this work, because it is clear that a confirmation of the proposal concerning the parallel between the CP, vP and DP phases requires detailed and careful analysis that will show whether the phenomena compared here are really determined by the same functional head.[18]

[18] In a recent work by Ledgeway, he notes that Old Neapolitan has the same features that OI displays, and Old French seems to confirm the connection as well.

References

Belletti, A. 1990. *Generalized Verb Movement.* Turin: Rosenberg and Sellier.

Belletti, A. 2004. Aspects of the low IP area. In: L. Rizzi (ed.), *The Structure of CP and IP*, 16–51. New York and Oxford: Oxford University Press.

Belletti, A. and U. Shlonsky. 1995. The order of verbal complements: a comparative study. *Natural Language and Linguistic Theory* 13, 489–526.

Benincà, P. 1984. Un'ipotesi sulla sintassi delle lingue romanze medievali. *Quaderni Patavini di Linguistica* 4, 3–19. Reprinted in Benincà (1994).

Benincà, P. 1994. *La Variazione Sintattica.* Bologna: Il Mulino.

Benincà, P. 2001. Syntactic focus and intonational focus in the left periphery. In: G. Cinque and G. Salvi (eds.), *Current Studies in Italian Linguistics Offered to Lorenzo Renzi*, 39–64. Amsterdam: North-Holland.

Benincà, P. 2006. A detailed map of the left periphery of Medieval Romance. To appear in: R. Zanuttini, H. Campos, E. Herburger, and P. Portner (eds.), *Negation, Tense and Clausal Architecture: Cross-linguistics Investigations*. Washington. Georgetown University Press.

Cinque, G. 1999. *Adverbs and Functional Heads.* Oxford: Oxford University Press.

Egerland, V. 1996. *The Syntax of Past Participles*. Lund: Lund University Press.

Giusti, G. 1992. *La sintassi dei sintagmi nominali quantificati*. PhD thesis, University of Venice.

Giusti, G. 1996. Is there a FocusP and a TopicP in the noun phrase structure? *UVWPL* 6.2, 105–28.

Giusti, G. 2004. At the left periphery of the Romanian noun phrase. To appear in: M. Coene (ed.), *On the Expression of Time and Space*. Bucarest: Clusium.

Grava, R. 2005. *Estudio Comparativo entre Italiano Espanol y Catalan sobre la posicion de los denominados adverbios "bajos"*. Tesi di Laurea, University of Venice.

Grewendorf, G. 2005. The discourse configurationality of scrambling. In: J. Sabel and M.Saito (eds.), *The Free Word Order Phenomenon: Its Syntactic Sources and Diversity*, 75–135. Berlin: de Gruyter [Studies in Generative Grammar 69].

Guasti, M. T. and L. Rizzi. 2002. Agreement and Tense as distinct syntactic positions: evidence from acquisition. In: G. Cinque (ed.), *Functional Structure in DP and IP*, 167–194. New York and Oxford: Oxford University Press.

Kayne, R. 1991. Romance clitics, verb movement, and PRO. *Linguistic Inquiry* 22, 647–686.

Jayaseelan, K. 2001. IP-internal Topic and Focus phrases. *Studia Linguistica* 55, 39–75.

Ledgeway, A. (to appear). Old Neapolitan word order: Some initial observations. In: A. L. Lepschy and A. Tosi (eds.), *Histories of the Italian languages*. Oxford: Legenda.

Paul, W. 2002. Sentence internal topics in Mandarin Chinese: the case of object preposing. *Language and Linguistics* 3.4, 695–714.

Poletto, C. 2004. La struttura della frase in Italiano antico. Ms., University of Padua.

Rizzi, L. 1997. The fine structure of the left periphery. In: L. Haegeman (ed.), *Elements of Grammar*, 281–337. Dordrecht: Kluwer.

Roberts, I. 1993. *Verbs and Diachronic Syntax: a Comparative History of English and French*. Dordrecht: Kluwer.

Selective Residual V-2 in Italian Interrogatives

Luigi Rizzi

1. Introduction

Bare *wh* elements in Italian interrogatives must normally be adjacent to the inflected verb. If the subject remains in its canonical preverbal position intervening between the *wh* element and the inflected verb, the structure is ungrammatical, as in (1a), (2a), (3a); if the subject is null or clause final, the structure is fine, as in b:

(1) a. **Che cosa Gianni dice?*
'What Gianni says?'
b. *Che cosa dice (Gianni)?*
'What does Gianni say?'

(2) a. **Dove Gianni va?*
'Where Gianni goes?'
b. *Dove va (Gianni)?*
'Where does Gianni go?'

(3) a. **Come Gianni parla?*
'How Gianni speaks?
b. *Come parla Gianni?*
'How does Gianni speak?'

A notable exception among bare *wh* elements is *perché* (why), which does not require inversion:

(4) a. *Perché Gianni parla?*
'Why Gianni speaks?'
b. *Perché parla Gianni?*
'Why speaks Gianni?

Another exception to the adjacency requirement is that lexically restricted, D-linked *wh* elements do not require adjacency, at least not as forcefully as bare *wh* elements. A minimal pair:

(5) a. **Dove Gianni ha messo il libro?*
'Where Gianni has put the book?'
b. *In quale cassetto Gianni ha messo il libro?*
'In which drawer Gianni put the book?'

(5b) sounds fully acceptable, if slightly marked. In other cases, e.g. when the *wh* phrase is the direct object, lack of adjacency is somewhat degraded, but not as much as with a bare *wh* element:

(6) a. **Che cosa Gianni ha buttato via?*
'What Gianni threw away?'
b. ?*Quale dei tuoi libri Gianni ha buttato via?*
'Which one of your books Gianni threw away?'

Focusing on the basic facts of (1)-(3), it is natural to put forth the hypothesis that the a examples are excluded for the same reason for which the English glosses are excluded: Italian appears to pattern with most of the Germanic and Romance languages in requiring adjacency between the *wh* element and the inflected verb in root questions. This is the approach adopted in Rizzi (1991). But a unified analysis of the adjacency requirement must deal with the language-specific exceptions we have just noted, and with other syntactic differences between English and Italian (and, more generally, with the language-specific manifestations of the adjacency requirement).

2. A parametric approach

In order to get started, let us first try to identify some basic properties that a comparative approach to interrogatives should specify. Some languages require inversion in main interrogatives, other languages do not. Compare, e.g., English and Brazilian Portuguese:

(7) *Who did she meet?*

(8) *Quem que ela encontrou?* (BP)
'Who that she met?'

English must make the bearer of the verbal inflection adjacent to the (non-subject) *wh* element, and this is achieved by moving it to the C-system; BP does not have this requirement, and apparently it never moves a verb to the left periphery (the fact that BP expresses an element corresponding to *that* in this construction may not be crucial, as other languages not requiring inversion, e.g. the Semitic languages, do not have this property). The idea put forth in Rizzi (1991) was that the verbal inflection bears a featural specification (*wh* or, more perspicuously, Q) marking the clause as interrogative, and that this specification must be brought to the C system to create a local configuration between the *wh* operator and this featural specification. The principle requiring such a local configuration was called a Criterion, the *Wh* or Q Criterion.

Generalizing somewhat this approach, we may now assume that scope-discourse semantic properties (Chomsky 2000, 2001, 2004, 2005) are expressed in the syntax by a family of Criteria (Rizzi 2003): natural language syntax disposes of a set of scope-discourse, or criterial, features (Q, Top, Foc,….) expressed in the functional system of the clause, and a given expression receives the relevant scope-discourse property (as an interrogative operator, a topic, etc.) by being moved to a position local to the relevant feature, where the typical case of locality may be expressed by the Spec-head configuration (but perhaps not only: Rizzi and Shlonsky 2005). In this view, a scope-discourse feature transparently signals to the external interpretive systems such interpretively relevant articulations as operator – scope, focus – presupposition, topic-comment, and possibly subject-predicate (Rizzi, op. cit. Rizzi and Shlonsky, op.cit.). Back to inversion in (7), in this perspective its functional role can be understood as a necessary step for the satisfaction of the Q Criterion: if T possesses the Q specification in English, but the Spec of T is already taken up by an s-selected temporal adverbial (Cinque 1999), and/or is involved in the functioning of the Case-agreement system (and under the assumption that a free proliferation of specifiers is disallowed, possibly for locality reasons: a lower specifier would always intervene between a higher specifier and the head, in violation of locality), the only option for permitting the satisfaction of the Q Criterion is that T bearing Q moves to a higher position in the C-system, where the required local configuration with a *wh* operator can be formed.

In a strict "movement as last resort" approach, the relevant application of T to C movement may be thought of as triggered by a system of abstract uninterpretable features, in a probe-goal framework (Pesetsky and Torrego 2001), or by an abstract affix. An alternative may be offered if one takes seriously Chomsky's (2004, 2005) idea that movement is just a subcase of merge, internal merge, and is as freely available as external merge (possibly submitted to milder versions of "last resort" requirements, such as the requirement that the head of a chain must yield an interface effect), with problems of look-ahead limited to the computation of the current phase in a "Derivation by Phase" model (Chomsky 2001, 2005). I will not try to address the question of the formal implementation of movement any further here.

If the Q feature is needed for the proper interpretation of main questions in an English-type language, what about languages of the non-inverting type? Under the assumption of the fundamental uniformity of the interpretive systems, it is natural to assume that a Q feature is involved in the latter case as well. Then, the option that comes to mind is that the structural position in which Q may be inserted in the structure may vary. If inverting languages typically have Q generated in the inflectional system, non-inverting languages may have it directly generated in the C system (Poletto 2000). If T is the carrier of Q in English-type languages, it is natural to think that the carrier of Q in the C-system of non inverting languages is the T-related head of the C-system, Fin(iteness) in the system of Rizzi (1997). So, consider the following parameter distinguishing inverting and non-inverting languages:

(9) Q is in T/Fin.

Now, back to Italian. The word order in (1a), etc. can be excluded as a violation of the Q Criterion if Italian, like English, has Q in T. What about the acceptable b examples with inversion? The assumption in Rizzi (1991) was that the case is exactly parallel to English inversion, with T to C head movement carrying Q to the C-system for criterial satisfaction.

One immediate problem with this idea is that in case of a complex verbal structure the parallel with English is broken: the subject appears in between the functional verb and the lexical verb in English, while it cannot appear in this position in Italian, where it must be clause-final:

(10) *What has John said?*

(11) a. **Che cosa ha Gianni detto?*
'What has Gianni said?'
b. *Che cosa ha detto Gianni?*
'What has said Gianni?'

Rizzi (op.cit.) adopted an idea borrowed from Rizzi and Roberts's (1989) analysis of French inversion, according to which I to C movement in Italian destroys the environment for nominative Case assignment/checking, so that (11a) is excluded because the subject DP ends up in violation of Case requirements. But this analysis is not easily compatible with current assumptions on Case checking, which is assumed to take place derivationally when the appropriate Agree relation is established.

Here I would like to consider the following alternative. In inverting languages Q is generated in T, but the languages may exploit two devices for bringing the feature to the C system, corresponding to the two fundamental types of movement: head movement of T, or phrasal movement of the projection of T. The former case yields the word order of English (10), the latter yields (11b).

More specifically, consider the following intermediate level of the derivation of the Italian example:

(12) *Gianni* X [ha_Q *detto che cosa*]

where X may be the head Subj, discussed in Rizzi (2003) based on Cardinaletti (2004). At this point Fin is merged, and the Q feature must be brought to the C-system. Suppose that head movement of ha_Q to Fin and further is not available (possibly due to the minimality effect induced by X, or for a primary parametric choice of the language: head vs XP movement), and then the language resorts to phrasal movement of [...Q...] to the Spec of Fin. This may be sufficient to endow the C-system with the required Q feature (it may also be that further movement in the left periphery of the carrier of Q is required, for instance it could be that, after [...Q...] has moved to the Spec of Fin, Fin acquires the Q feature via Spec-head agreement from the [...Q...] constituent, and then Fin moves on to the head hosting the *wh* phrase -- Foc under our assumptions --; but these specific mechanical details are not of immediate relevance here). So, after *wh* movement of *che cosa* to Spec of Foc (on which see below), the derived order is (11b), and (11a) is correctly underivable.

Many questions are raised by this approach, but let us adopt it for concreteness and address the question of the exceptions to the inversion requirement mentioned in the introduction.[1]

---

1 An anonymous reviewer points out that, given the proposed derivational mechanism of inversion, one would expect only high adverbials to occur in between the clausal constituent preposed to Spec Fin and

3. *Perché* and the Q Criterion

Why can *perché* questions survive without inversion? Here I will rely on the analysis given in Rizzi (2001a), which is summarized in what follows. In fact, *perché* differs from other *wh* elements also in other respects. It can be preceded and followed by a topic, expressed by Clitic Left Dislocation (Cinque 1990), which I will assume to involve movement to the left periphery in Romance (as may be the case for dislocation more generally: Grewendorf 2002) while other bare *wh* elements can only be preceded by a topic:

(13) a. *A Gianni, perché gli avete parlato ieri?*
'To Gianni, why did you speak yesterday?'
b. *Perché, a Gianni, gli avete parlato ieri?*
'Why, to Gianni, did you speak yesterday?'

(14) a. *A Gianni, che cosa gli avete detto ieri?*
'To Gianni, what did you say yesterday?'
b. **Che cosa, a Gianni, gli avete detto ieri?*
'What, to Gianni, did you say yesterday?'

Moreover, while other bare *wh* elements are incompatible with a contrastive focus, in any order, *perché* is compatible with focus in the fixed order *perché* Foc:

(15) a. **A GIANNI che cosa avremmo dovuto dire (, non a Piero)?*
'TO GIANNI what we should have said (, not to Piero)?'
b. **Che cosa A GIANNI avremmo dovuto dire (, non a Piero)?*
'What TO GIANNI we should have said (, not to Piero)?'

(16) a. *Perché QUESTO avremmo dovuto dirgli (, non qualcos'altro)?*
'Why THIS we should have said to him (, not something else)?'
b. **QUESTO perché avremmo dovuto dirgli (, non qualcos'altro)?*
'THIS why we should have said to him (, not something else?)'

The incompatibility illustrated by (15) is accounted for in Rizzi (1997) by assuming that normal *wh* elements target the Spec of Focus, so that a contrastive focus and a *wh* element compete for the same position, hence are in complementary distribution. By putting this together with what has been proposed in section 2 we end up with the

---

the subject, i.e., adverbs that are able to occur in between the Fin and the Subj layers; but in fact, various lower adverbs can intervene in examples like the following:

(i) *Che cosa ha detto subito Gianni?*
'What has said immediately Gianni?'

How can this order arise? Remember that Italian is a Null Subject Language, in which the thematic subject does not have to raise to Spec Subj: this position can be filled by expletive *pro*, with the thematic subject remaining in a lower position (in particular, in a lower Top or Foc position, according to the analysis of Belletti 2004a). So, the order in (i) may arise through this option (presumably with *Gianni* in the lower topic position, as the clausal focus is taken up by the *wh* element: this is also suggested by the fact that (i) is optimally acceptable with a slight pause between the adverb and the subject), with the whole clausal constituent *ha detto subito Gianni* moved to Spec Fin.

following picture: normal *wh* operators target the focus position, and for the construction to be properly interpreted as a question its C-system must be marked with the Q feature (which, in English type languages, is done through inversion).

This granted, we are led to the conclusion that *perché* occupies a distinct position, necessarily higher than Focus, so that the strict order of (16a) follows. In Rizzi (2001b) the position hosting *perché* in its Spec is identified as Int(errogative), the position normally hosting the yes-no complementizer *se* (if) in indirect questions, which has the same word-order properties. So, Int can be preceded and followed by topics, while it necessarily precedes focus; this leads to the following partial map (irrelevant positions are omitted: see various papers in Belletti, ed. 2004b, Rizzi, ed. 2004b for discussion):

(17) Force ... Top ... Int ... Top ... Foc ... Fin ...

In order to account for the lack of inversion with *why* in Italian-type languages, we are led to assume that the following holds in Italian:

(18) Int has Q.

Notice that Int (or the Q feature *per se*) cannot be an attractor of movement, otherwise any *wh* element could be moved to Spec Int without the necessity of triggering inversion. So, *wh* movement (in questions) is movement to Focus in the core case (but see below). Int can only s-select in its Spec a high clausal operator, like a yes-no operator or a reason adverbial, high enough to be compatible with external merge in such a position; in this case its s-selection and scope-discourse positions coincide, and no movement is needed.

Given that *perché* can occupy a higher position than other *wh* elements, we may expect, contrary to fact, that *perché* and another *wh* element could be allowed to cooccur in a fixed order, much as *perché* and focus in (16a):

(19) **Perché che cosa avremmo dovuto dirgli?*
'Why what should we have said to him?'

Clearly, clause-internal recursion of Q must be blocked. There may be a simple interpretive reason for that: the interpretive import of Q is to require its Spec to be interpreted as an interrogative operator, and its complement as the scope domain of the operator (much as Top defines the topic-comment articulation, etc.). Now, the scope domain of an interrogative operator must be a proposition, not another question (i.e., we want to know the value of the variable which makes the following clause a true proposition); if so, the clause internal recursion of Q is excluded, even if more then one possible site for it is syntactically available (I'm following here a line of argument akin to the one which plausibly excludes recursion of focus in the left periphery: Rizzi 1997; of course, languages allowing multiple *wh* elements to occur in the left periphery must dispose of a mechanism of absorption which overcomes this problem).

4. D-linked *wh* phrases

Why is it that D-linked *wh* phrases are at least marginally compatible with lack of inversion in Italian interrogatives (5a-b)?

Revising and updating the approach to *wh* D-linking in Rizzi (2001c), I would like to consider the idea that D-linked *wh* phrases can target a position distinct from the one targeted by bare *wh* elements (except *why*). That this may be the case is directly suggested by the observation that, in multiple *wh* movement languages, D-linked *wh* phrases can often subvert the basic order of elements, which must be preserved by bare *wh* phrases. Consider the following Rumanian data taken from Alboiu (2000):

(20) a. *Cine ce a dat lui Mihai?*
'Who what gave to Mihai?'
b. **Ce cine a dat lui Mihai?*
'What who gave to Mihai?'

(21) a. *Cine cu care candidat a votat?*
'Who for which candidate voted?'
b. *Cu care candidat cine a votat?*
'For which candidate who voted?'

This is immediately explainable if the D-linked phrase has at least the option of an independent landing site, and the order preservation effect is restricted to positions of the same kind, whatever the exact nature of the mechanism (Richards 2001, etc.; this is only a rough first approximation: see Krapova and Cinque 2004 for a detailed analysis of the ordering constraints in Bulgarian, and Laenzlinger and Soare 2005 on Rumanian).

What kind of independent landing site? A D-linked *wh* is one in which the range of the variable is familiar from the discourse context (Pesetsky 1987). This interpretation is generally favored by the presence of a lexical restriction, enforced by an explicit partitive form, and sensitive to the choice of the *wh* element (*what* vs *which*) and the contextual conditions. This basically coincides with the notional characterization of topics: a topic generally highlights a referent familiar from context, and D-linking has explicitly been mentioned as a constitutive characteristic of topics (Grewendorf 2002). The similarity is strong enough to encourage the hypothesis that the special position targeted by D-linked *wh* elements be some kind of topic position. This is not unproblematic. Grewendorf (2003) observes that *wh* is not compatible with topichood in Japanese as *wh* elements (D-linked or not) typically do not carry the topic-marker *wa*, which casts doubts on the possibility of a full assimilation. Nevertheless, Benincà and Poletto (2004), among others, argue that different kinds of topics should be distinguished. Perhaps, the weakest notion, which seems to be the one expressed by Italian Clitic Left Dislocation, is the notion of partitivity topic: the topic picks up a referent belonging to a set which has already been introduced in discourse, even if the referent itself has not (see Ionin 2003 for a recent discussion of this notion,). This weak condition allows for indefinite topics as well. For instance, the following is felicitous in Italian:

(22) *Un libro, lo ho già letto.*
'A book, I have read.'

provided that the book belongs to a set preestablished in discourse, e.g. in a context in which a set of books for the preparation of an exam has been mentioned (see Rizzi 2005 for discussion, and Grewendorf 2002 for the attempt of connecting the weakness of the topichood requirement in Italian with other properties of the Italian construction, such as the full range of reconstruction effects). Different topic constructions in different languages pick different notions of topicality, possibly more restrictive than the weak partitivity notion. For instance, partitivity is sufficient for Clitic Left Dislocation, but not for Romance Hanging Topic, which requires familiarity of the particular referent, hence is inconsistent with indefiniteness:

(23) a. *Quanto a questo libro, lo ho già letto.*
'As for this book, I have already read it'
b. **Quanto a un libro, lo ho già letto.*
'As for a book, I have already read it'

So, if the particular topichood expressed by *wa* in Japanese is more restricted than just partitivity, the hypothesis that D-linked *wh* may target a topic position may be consistent with Grewendorf's observation.

Pursuing this hypothesis, I will assume that D-linked *wh* phrases can target a Top position (in the weak sense just discussed of a position expressing partitivity) in the left periphery of the clause. Of course, the targeted position is not just a topic position. In the transparent view on the syntax – interpretation mapping that we are adopting, the syntax must express the fact that D-linked questions are *wh* questions, operator-variable structures. If the crucial signal to the interpretive systems to activate the interpretive procedure for questions is given by the Q feature appearing on some clausal head, this feature must be involved in D-linked *wh* questions as well. Putting together this consideration and the previous remarks on the parallel D-linking - topicality, I will assume that the position targeted by D-linked *wh* phrases is characterized as +Top, +Q. If Q is typically expressed in T in Italian, one possible derivation will involve inversion, as in ordinary non-D-linked questions, with movement of the clausal constituent including T to Spec Fin, and perhaps successive movement of Fin endowed with the Q feature via Spec-head agreement to a Top position, thus forming a +Q, +Top head. This kind of head can host a D-linked *wh* phrase, where D-linking justifies the targeting of a Top position, and the *wh* specification satisfies the Q Criterion.

The composite character of the hosting head accounts for two important differences between D-linked *wh* questions and simple topic constructions. First, while topics can freely proliferate in the Romance Clitic Left Dislocation construction, there can be a single D-linked *wh* in the left periphery of a question. In this respect, D-linked and non D-linked *wh* questions pattern alike:

(24) a. *Di questa storia, a Gianni, (gliene) ho parlato ieri sera.*
'Of this story, to Gianni, I spoke (of it to him) yesterday evening.'
b. **Di quale storia a quale ragazzo hai parlato ieri?*
'Of which story to which boy did you speak yesterday?

Second, Clitic Left Dislocation requires a clitic if the topical element is a direct object (and is anyway compatible with a clitic in cases of PP topicalization), while a D-linked wh, also a direct object, is fine with a gap, and permits a clitic only very marginally:

(25) a. *Questo libro, vorrei darlo /* dare a Gianni.*
'This book, I would like to give it / give to Gianni.'
b. *Quale di questi libri vorresti dare / ?? darlo a Gianni?*
'Which one of these books would you like to give / to give it to Gianni?'

As for the uniqueness of D-linked wh, what has been said of non-D-linked *wh* carries over to this case: the complement of Q must be a proposition, hence recursion of Q is forbidden. On the other hand, no interpretive problem is raised by recursion of Top: the comment of a higher Top can in turn express the topic-comment articulation (the notion of "comment" is very weak, and doesn't preclude this possibility).

As for the requirement that object topics require a resumptive clitic (as in (25a)), Cinque (1990) has proposed that this is due to the fact that a gap in object position could not have a well-formed status in this case, in particular it could not be a variable, as topics in Romance do not have the syntactic status of operators: so, a clitic is required in (25a) to permit a well-formed A'-dependency. In the case of the D-linked *wh* phrases, the operator status of the phrase is ensured by the *wh* morphology and by the Q specification on the complex attracting head, so that a gap is well-formed as a legitimate variable, and the clitic has the marginal status that is characteristic of a language which has not grammaticalized the resumptive strategy for operator-variable dependencies. It should be noticed nevertheless that the occurrence of a resumptive clitic is clearly less degraded in (25b) than in the case of a non-D-linked construction like the following:

(26) *Che cosa vorresti dare / *darlo a Gianni?*
'What would you like to give / to give it to Gianni?'

(and the variant of (25b) with the clitic becomes almost fully acceptable if the *wh* phrase is followed by an intonation break *Quale di questi libri, vorresti darlo a Gianni?*, which underscores the similarity with clitic Left Dislocation). This can be understood by observing that the clitic satisfies at least part of the requirement of the construction expressed by the composite head (the Top part in +Top +Q in (25b)), while in (26) the *wh* element, targeting a purely quantificational +Foc, +Q position, is fully incompatible with clitic resumption.

This analysis has a number of significant consequences. For instance, the insensitivity of D-linked *wh* elements to weak islands (Comorowski 1989 and much subsequent literature) could be made to follow from the composite character of the attracting head: an element targeting a +Top +Q position is able to skip an intervening purely quantificational position in an approach to Relativized Minimality effects like the one in Rizzi (2001a, 2004a) (see Starke 2001 for a detailed analysis along such lines). The partial insensitivity of D-linked question operators to Superiority and Weak Cross-over effects should also be related to this approach. I will not pursue these important issues here.

5. Lack of inversion with D-linking

Back to the fact that D-linked *wh* phrases can, at least marginally, survive in non-inverted interrogatives in Italian; why is it so? The idea that the *wh* element in these interrogatives targets a Top-like position does not automatically yield this result:

(5) a. * *Dove Gianni ha messo il libro?*
'Where Gianni has put the book?'
b. *In quale cassetto Gianni ha messo il libro?*
'In which drawer Gianni put the book?'

Notice that D-linked *wh* elements also differ from ordinary *wh* elements in that they are more easily compatible with a lower topic, even though (27b) remains quite marked:

(27) a. * *Dove, questo libro, lo potremmo mettere?*
'Where, this book, we could put?'
b. *In quale cassetto, questo libro, lo potremmo mettere?*
'In which drawer, this book, we could put?'

Moreover, if compatibility with a lower focus remains quite difficult, in terms of relative acceptability the sequence D-linked *wh* – Focus, illustrated on (29a) is the least deviant:

(28) a. * *A chi QUESTO vorresti dire?*
'To whom THIS you would weant to say?'
b. * *QUESTO a chi vorresti dire?*
'THIS to whom you would want to say?'

(29) a. ?? *A quale dei tuoi amici QUESTO vorresti dire?*
'To which one of your friends THIS you want to say?'
b. * *QUESTO a quale dei tuoi amici vorresti dire?*
'THIS to which one of your friends you want to say?'

Lack of inversion, compatibility with a lower topic, and (at least marginal) compatibility with a lower focus make the positional properties of D-linked *wh* akin to those of *perché*. A partially unified analysis thus seems to be in order.

At first sight, it is not clear what D-linked *wh* and *perché* might have in common. In other respects the two elements behave in a sharply different way: for instance while D-linked *wh* can easily jump across weak islands, *perché* is probably the most sensitive element to weak islands, extraction of *perché* from an indirect question being sharply excluded (but not from a declarative, so that extraction of *perché/why* from a lower clause must be possible in general).

A possible partial unification is suggested by the cartographic structure, repeated here for convenience:

(17) Force ... Top ... Int ... Top ... Foc ... Fin ...

Suppose that, as has been proposed in section 3, Int can be endowed with Q in Italian. Still, we don't want to say that Int can directly attract D-linked *wh* phrases, which would be a completely arbitrary assumption. But there is another possibility. Suppose that Int endowed with Q, much as T endowed with Q, can move to a higher head capable of functioning as an attractor. An attractor higher than Int is Top in the cartographic representation (17). So, Int_Q moves to Top, thus creating the complex head Int_Q+Top, which, due to its feature constitution, is able to attract a D-linked *wh* element (attractable to Top because of its topic-like properties). So, we end up with a representation like the following:

(30) *In quale cassetto* Int_Q+Top [t_{Int} ... [Fin [*Gianni ha messo il libro* ...]]]

This structure satisfies the Q Criterion without requiring inversion, as the Q feature generated in Int, hence already in the C-system, is exploited. The same trick is not applicable in the case of a non-D-linked question given the cartographic structure (17): Int is necessarily higher than the Foc position in (17), hence it cannot be lowered to this position to create Q+Foc; therefore non-D-linked *wh* elements, which must move to Foc, always require inversion to bring the Q feature (in T) to Foc, through the mechanism envisaged in section 3. The only attractor that Int can move to, given the no-lowering constraint and structure (17), is Top, but then the mechanism permitting questions without inversion is restricted to *wh* elements suitable to be attracted to Top, D-linked *wh* elements.

References

Alboiu, G. 2000. *The Features of Movement in Romanian.* PhD Thesis, University of Manitoba, Winnipeg.

Baltin, M. and Ch. Collins (eds.). 2001. *A Handbook of Syntactic Theory.* Oxford: Blackwell.

Belletti, A. (2004a). Aspects of the low IP area. In Rizzi (ed.) (2004b), 16-51.

Belletti, A. (ed.). 2004b. *Structures and Beyond: The Cartography of Syntactic Structures, Vol. 3.* New York: Oxford University Press.

Beninca, P. and C. Poletto. (2004). Topic, focus, and V-2: defining the CP sublayers. In Rizzi (ed.) (2004b), 52-75.

Cardinaletti, A. 2004. Towards a cartography of syntactic positions. In Rizzi (ed.) (2004b), 115-165.

Chomsky, N. 2000. Minimalist inquiries: the framework. In R. Martin, D. Michaels, and J. Uriagereka (eds.), *Step by Step – Essays in Minimalist Syntax in Honor of Howard Lasnik*, 89-155. Cambridge, Mass.: MIT Press.
Chomsky, N. 2001. Derivation by phase. In M. Kenstowicz (ed.), *Ken Hale: A Life in Language*, 1-52. Cambridge, Mass.: MIT Press.
Chomsky, N. 2004. Beyond explanatory adequacy. In A. Belletti (ed.) (2004), 104-131.
Chomsky, N. 2005. On phases. Ms., MIT.
Cinque, G. 1990. *Types of A' Dependencies*. Cambridge, Mass.: MIT Press.
Cinque, G. 1999. *Adverbs and Functional Heads: A Cross-linguistic Perspective*. Oxford: Oxford University Press.
Comorowski, I. 1989. Discourse-linking and the Wh-Island Constraint. In J. Carter and R.-M. Déchaine (eds.), *Proceedings of NELS 19*, 78-96. University of Massachusetts, Amherst, Mass: GLSA.
Grewendorf, G. 2002. Left dislocation as movement. In S. Mauck and J. Mittelstaedt (eds.), *Georgetown University Working Papers in Theoretical Linguistics* 2, 31-81.
Grewendorf, G. 2003. Improper remnant movement. *Genko Kenkyu – Journal of the Linguistic Society of Japan* 123, 47-94.
Ionin. T. 2003. *Article Semantics in Second Language Acquisition*. PhD dissertation, MIT, Cambridge, Mass.
Krapova, I and G. Cinque. 2004. On the order of wh-phrases in Bulgarian multiple wh-fronting. Ms., University of Plovdiv, University of Venice.
Laenzlinger, C. and G. Soare. 2005. Intervention effects in French and Rumanian. Ms., University of Geneva.
Pesetsky, D. 1987. Wh in situ: movement and unselective binding. In E. Reuland and A. ter Meulen (eds.), *The Representation of (In)definiteness*, 98-129. Cambridge, Mass.: MIT Press.
Pesetsky, D and E. Torrego. 2001. T-to-C movement: causes and consequences. In M. Kenstowicz (ed.), *Ken Hale: A Life in Language*, 355-426. Cambridge, Mass.: MIT Press.
Poletto, C. 2000. *The Higher Functional Field. Evidence from Northern Italian Dialects*. New York: Oxford University Press.
Richards, N. 2001. *Movement in Language: Interactions and Architectures*. Oxford: Oxford University Press.
Rizzi, L. 1990. *Relativized Minimality*, Cambridge, Mass.: MIT Press.
Rizzi, L. 1991. Residual verb second and the Wh Criterion. *Technical Reports in Formal and Computational Linguistics* 2. Department of Linguistics, University of Geneva, republished in Rizzi (2000), 213-240.
Rizzi, L. 1997. The fine structure of the left periphery. In L. Haegeman (ed.), *Elements of Grammar*, 281-337. Dordrecht: Kluwer.
Rizzi, L. 2000. *Comparative Syntax and Language Acquisition*. London: Routledge.
Rizzi, L. 2001a. Relativized Minimality effects. In Baltin and Collins (eds.) (2001), 89-110.
Rizzi, L. 2001b. On the position Int(errogative) in the left periphery of the clause. In G. Cinque and G. Salvi (eds.), *Current studies in Italian syntax. Essays offered to Lorenzo Renzi*, 287-296. Amsterdam: North Holland.

Rizzi, L. 2001c. Extraction from Weak Islands, reconstruction, and agreement. In G. Chierchia, M.-T. Guasti, and C. Cecchetto (eds.), *Semantic Interfaces. Reference, Anaphora and Aspect*, 155-176. Stanford, Calif.: CSLI Publications.

Rizzi, L. 2003. On the form of chains: criterial positions and ECP Effects. To appear in L. Cheng, N. Corver (eds.), *On Wh Movement*. Cambridge, Mass.: MIT Press.

Rizzi, L. 2004a. Locality and left periphery. In Belletti (ed.) (2004), 223-251.

Rizzi, L. (ed.). 2004b. *The Structure of CP and IP – The Cartography of Syntactic Structures Vol. 3*. Oxford: Oxford University Press.

Rizzi, L. 2005. Some remarks on subjects and topics. To appear in *Proceedings of the XXX Incontro di Grammatica Generativa*, Venice.

Rizzi, L. and I. Roberts. 1989. Complex Inversion in French. *Probus - An International Journal of Latin and Romance Linguistics* 1.1, 1-30.

Rizzi, L. and U. Shlonsky. 2005. Strategies of subject extraction. Ms., University of Siena, University of Geneva.

Starke, M. 2001. *Move Dissolves into Merge*. Doctoral dissertation, University of Geneva.

Impossible Infinitival Interrogatives and Relatives

Joachim Sabel

1. Introduction

Infinitives in the languages of the world behave differently in several respects. One difference relates to their potential to be used in the formation of infinitival questions. In English, for example, they are possible and are realized as *wh-* and *yes-/no-*questions, as shown in (1). In other languages such as German, infinitival questions are impossible, as illustrated in (2).[1]

(1) a. *Lena has decided* [*who to visit* ___].
    b. *Lena has decided* [*whether to visit him or not*].

(2) a. **Lena hat entschieden,* [*was Tom* ___ *zu sagen*].
    Lena has decided what Tom(DAT) to say
    'Lena has decided what to say to Tom.'
  b. **Lena hat entschieden,* [*ob Tom etwas zu sagen*].
    Lena has decided whether Tom(DAT) something.ACC to say
    'Lena has decided whether to say something to Tom.'

This paper discusses why embedded infinitival questions are excluded in some languages and are acceptable in others. First, it will be shown that the impossibility of *wh*-infinitives is rooted in syntax rather than the lexicon. The syntactic account given is based on a comparative analysis of different European languages which examines those languages which do and do not allow for *wh*-infinitives. Following this analysis, additional phenomena which correlate with the observed differences are discussed. It turns out that *wh*-infinitives are impossible in languages in which the left periphery of the infinitive cannot be occupied with an infinitival complementizer, an infinitival marker, or more generally, with a base-generated phonetically realized element. The same correlation can be observed with respect to the impossibility of infinitival relatives in different languages. I propose that languages without *wh*-infinitives have a "defective" infinitival C-system.

---

1 The interpretation of *wh*-infinitivals and infinitival relatives involves "modality." This means that (1a) is understood as *Lena does not know* [*who she should/could visit* ___], see Bhatt (1999) for discussion.

2. The impossibility of infinitival questions: syntax vs. lexicon

The impossibility of (2) is not due to a lexical gap in German, i.e. because German does not have verbs that select interrogative complements and *zu*-control infinitives. Both options are found, with verbs such as *entscheiden* 'decide', for example:

(3) a. *Frank hat entschieden, [den alten Wagen zu verkaufen].*
Frank(NOM) has decided the.ACC old car to sell

b. *Frank hat entschieden, [ob er heiraten wird].*
Frank(NOM) has decided whether he.NOM marry will

c. *Frank hat entschieden, [wen er ___ heiraten wird].*
Frank(NOM) has decided who.ACC he marry will

In English, verbs such as *decide, forget* and *tell* take all four different complement types with respect to the properties [± wh, ± fin]. In German only three of the four possibilities are attested for these verbs, i.e. [–wh, –fin], [–wh, +fin] and [+wh, +fin] but not [+wh, -fin]. Given this pattern, it would be ad hoc to argue that the absence of *wh*-infinitives in German has lexical reasons. The following list contains further German verbs that select infinitives as well as interrogative complements.

(4) *berichten* 'tell/report', *besprechen* 'discuss', *entscheiden* 'decide', *erkennen* 'realize', *erklären* 'explain', *mitteilen* 'inform', (*sich*) *überlegen* 'consider', *vergessen* 'forget', ...

A further reason for assuming that (2) is not excluded for lexical reasons has to do with another construction involving operator movement inside an infinitival clause, i.e. with infinitival relative clauses. In German – in contrast to English – infinitival relative clauses are impossible, as can be seen from (5a) vs. (6a). As with indirect *wh*-questions, only the finite variant of (6a), illustrated in (6b), is possible.

(5) a. *This is a topic [about which to argue ___].*

b. *This is a topic [about which one can argue ___].*

(6) a. **Maria sucht das Messer [mit dem Käse ___ zu schneiden].*
M.(NOM) seeks the-ACC knife with the.DAT cheese to cut
'Maria seeks the knife with which to cut the cheese.'

b. *Maria sucht das Messer [mit dem man Käse ___ schneidet].*
M.(NOM) seeks the-ACC knife with the.DAT one.NOM cheese cuts
'Mary seeks the knife with which one cuts cheese.'

Relative clauses are not lexically selected. Obviously, the fact that *wh*-movement may not end up in the left periphery of an infinitive in German but may do so in English, must be traced back to a systematic difference in the syntax of these languages.[2, 3]

3. Infinitival questions in other languages

Let us now turn to other European languages to investigate whether they allow for infinitival interrogatives. In languages such as Swedish (7), Norwegian (8) and Danish (9) *wh*-infinitives are excluded, as in German.

(7) **Jag vet inte* [*wart* [*att gå* ___]]. (Holmberg 1983)
I know not where to go

(8) **Det er uklart* [*hva* [*å gjøre* ___]]. (Christensen 1984)
it is unclear what to do

(9) **Han har glemt* [*hvad* [*at købe* ___]]. (Per Bærentzen, p.c.)
he has forgotten what to buy

Other languages besides English that allow for *wh*-infinitives are Polish (10), French (11), Spanish (12), European-Portugese (13), Italian (14) and Dutch (15):

(10) *Janek nie wie* [*gdzie* [*skukać Marka* ___]]. (Zabrocki 1981: 53)
Janek not know where to-seek Marka

(11) *Je lui ai dit* [*où* [*aller* ___]]. (Kayne 1984:104)
I him have said where to-go

(12) *No se* [*qué* [*decirle* ___]]. (LaPolla 1988)
not (I-)know what to-say-him

(13) *Nós não sabemos* [*quem* [*convidar* ___ *para o jantar*]]. (Raposo 1987)
we not know who to-invite for to-eat

(14) *Non so* [*dove* [*andare* ___]]. (Napoli 1976)
not (I-)know where to-go

---

2 The infinitivals that I mainly discuss in this paper are control infinitivals. Further infinitivals like raising and bare infinitivals that appear as arguments of modals and ECM (i.e. AcI-) verbs in German will not be discussed. The reason is that the latter infinitivals are not selected by verbs that allow for interrogative complements. Furthermore, consider (i), an embedded infinitival without *zu*:

(i) *Ich weiß nicht was tun.*
I know not what do

The construction in (i) is not productive in German. For more discussion of these infinitivals as well as of the corresponding matrix *wh*-infinitivals such as *Was tun?* 'What to do?' see Reis (1985, 2003), Giusti (1989), Wilder (1989) und Sabel (1996, chapter 8).

3 Several authors have analyzed control infinitives as smaller categories than CPs (see, for example, Wurmbrand 2000, Cinque 2004). A discussion of the various analyses and arguments for the CP-analysis (adopted here) can be found in Sabel (1996, chapter 4).

(15) *Ik weet niet* [*wie* [___ *te bezoeken*]].
I know not who to visit

Differences can sometimes be observed regarding the possibility of *wh*-infinitives in one and the same language. We saw already that English allows for *wh*-infinitives and infinitival relatives with *to*-infinitives (cf. (1) and (5a)). However, indirect questions and relative clauses in connection with another infinitive, the gerundive construction, are excluded:

(16) a. **I wonder* [*where* [*going* ___]].
b. **The table* [*on which* [*putting your coat* ___]] *is in the house.*

In trying to ascertain why infinitival questions and infinitival relative clauses are ruled out in some languages (or with certain infinitival constructions, as in English), it is enlightening to look at a phenomenon that correlates with the possibility of infinitival questions and relatives, a phenomenon that, like operator-movement, relates to the C-system of the relevant infinitives. The question that I address in the next section is whether overt subordinating elements like complementizers are possible in German control infinitives with *zu*.

4. The nature of the left periphery and the impossibility of infinitival interrogatives

4.1 Prepositions, complementizers, and infinitival markers

In this section, I will illustrate that a close connection exists between the potential presence (or absence) of overt subordinators in the infinitival C-system and the possibility of question- and relative clause formation with infinitives: languages which do not have phonetically realized complementizers with certain infinitives do not allow for infinitival questions and relative clauses with these infinitives.

Let us consider first the status of the infinitival marker *zu* in German and *to* in English. Infinitives in Indo-European languages have developed from verbal nouns. As a verbal noun, the infinitive was governed by a preposition, for example, by the preposition *to* in Old English and *zi* in Old High German. In modern European languages the infinitive has become a verb and *zi* and *to* have lost their prepositional categorial nature. They have become "infinitival markers." This is the case with the control infinitives of all modern Germanic languages. For several reasons (discussed in Sabel 1996 chap. 1; chap. 8, 2000a), *zu* has to be analyzed in a way similar to *to*, i.e. as a verbal particle in T^0 that has the distribution of an auxiliary, although *zu* differs from *to* in so far as *zu* is a bound and *to* is a free morpheme. Neither *to* nor *zu* occupy a position in the infinitival C-system (see also Beukema and den Dikken 1989, IJbema and Abraham 2000 for discussions).

Are there any subordinators or complementizers in German *zu*-infinitives that might be argued to be base-generated in the C^0 or Spec CP position of infinitives? In fact, there are none: phonetically realized complementizers are excluded in German infinitival complement clauses ((17) = (2b)):

(17) **Lena hat entschieden* [$_{CP}$ *ob* [$_{TP}$ *Tom etwas zu sagen*]].
L.(NOM) has decided whether T.(DAT) something.ACC to say

(18) **dass sie versuchte* [$_{CP}$ *um* [$_{TP}$ *das Buch zu lesen*]].
that she.NOM tried COMP the book to read

In German, we find adjunct clauses that are selected by prepositions such as *ohne* 'without', *um* 'in order', and *anstatt* 'instead of'. Some of these prepositions alternatively select finite CPs introduced by the complementizer *dass* 'that':

(19) a. *Maria hat kritisiert, dass Hans Bücher kauft,*
Maria(NOM) has criticized that Hans(NOM) books buys
[$_{PP}$ *ohne/anstatt* [$_{CP}$ *sie zu lesen*]].
without/instead-of them to read
b. *Maria hat kritisiert, dass Hans Bücher kauft,*
Maria(NOM) has criticized that Hans(NOM) books buys
[$_{PP}$ *ohne/anstatt* [$_{CP}$ *dass er sie liest*]]
without/instead that he.NOM them reads

(20) *Tom und Lena gingen ins Museum* [$_{PP}$ *um* [*die*
Tom and Lena went into-the museum in-order the.ACC
alten Meister anzuschauen]].
old masters to-take-a-look-at
'Tom and Lena went to the museum in order to take a look at the old masters.'

Note that German *zu*-infinitives differ in this respect from the infinitives in the two Germanic languages with *wh*-infinitives mentioned above, i.e. from *to*-infinitives in English (1) and from *te*-infinitives in Dutch (15). In *to*-infinitives, the prepositional complementizer *for* is realized in the C-system:

(21) a. *I want* [$_{CP}$ *for* [$_{TP}$ *Chris to win*]].
b. *There is someone* [$_{CP}$ Op *for* [$_{TP}$ *me to talk to* ___]].

Dutch has the *om-te* infinitive. This construction is similar to the German *um-zu* infinitive in (20), in so far as *om*, like its German counterpart *um*, appears in adjunct clauses:

(22) *Bernard ging naar Amerika* [$_{PP}$ *om* [*beroemd te worden*]].
'Bernard went to America, in order to become famous.'

But in contrast to *um, om* may already appear as a complementizer in infinitival complement clauses ((18) vs. (23)):[4]

---

4 A further difference between the *om-te* infinitive in (23) and *um-zu* infinitives is illustrated in (i). In contexts in which German *um* does not occur, *zu* likewise cannot appear. Compare (20) with (i). Obviously, the absence of *um* is not a result of complementizer deletion in (ia), as it is the case with the absence of *om* in (23).

(i) a *Tom und Lena gingen ins Museum* [*die alten Meister anschauen*].
Tom and Lena went into-the museum the old masters take-a-look-at

(23) *dat zij probeerde* [$_{CP}$ (*om*) [$_{TP}$ *het boek te lezen*]].
that she tried in-order the book to read.

I conclude from these observations that prepositional elements like *ohne*, *um* and *anstatt* are not generated in the C-system of the *zu*-infinitives. They are prepositions, structurally realized outside the infinitival clause, i.e. as [$_{PP}$ P [$_{CP}$ infinitive]]. Further confirmation for this structure comes from language change facts. Lightfoot (1979: 196) mentions that the construction illustrated in (19a), (20) and (22) existed in Old-English as well ([$_{PP}$ *for* [PRO *to leave*]]) and that the preposition has recently become a complementizer in Modern English. (The older construction still exists in Belfast English.) Old English seems to be a language similar to Modern German with respect to the non-realization of infinitival complementizers. This similarity is not astonishing. Van Kemenade (1985) notes further parallels between infinitives in Old English and Modern German: control infinitives in Old English and Modern German display the same restructuring phenomena. Finally data from creolization provide evidence for the fact that a category change may take place by which infinitive-embedding prepositions become infinitival complementizers.

To sum up, in languages without *wh*-infinitives, such as German, we do not find infinitival complementizers, in contrast to languages with *wh*-infinitives such as English and Dutch. Infinitival complementizers are not found since (i) the infinitival marker is not located within the CP and (ii) the category change from preposition selecting infinitive to infinitival complementizer has not taken place.

4.2 The *Wh*-Infinitive-/Infinitival Relative-Correlation

As already mentioned in section 3, *wh*-infinitives are also excluded in Swedish (25), Norwegian (26) and Danish (27), as well as with gerunds in English (28).

(24) **Lena hat entschieden* [*was Tom* ___ *zu sagen*].
Lena(nom) has decided what Tom(DAT) to say

(25) **Jag vet inte* [*wart* [*att gå* ___]].
I know not where to go

(26) **Det er uklart* [*hva* [*å gjøre* ___]].
it is unclear what to do

(27) **Han har glemt* [*hvad* [*at købe* ___]].
he has forgotten what to buy

(28) **I wonder* [*where* [*going* ___]].

As was observed with respect to German *zu*-control infinitives, it is generally assumed that the C-system of gerunds in English, and control infinitives in Norwegian, Danish (see, for example, Holmberg 1986:154ff, Sigurðsson 1989:52, Thráinsson 1998) and

b. * *Tom und Lena gingen ins Museum* [*die alten Meister anzuschauen*].
Tom and Lena went into-the museum the old masters to-take-a-look-at

Swedish (Johnson und Vikner 1994, Thráinsson 1998) cannot be introduced by a phonetically realized base-generated element, i.e by a conjunction that takes the form of a preposition or by an infinitival marker that introduces an infinitival clause of the relevant type. *At* in Danish, *å* in Norwegian and *att* in Swedish are infinitival markers that – like German *zu* and English *to* – do not occupy a position in the C-system but a position inside TP.

Let us next consider whether the C-system of control infinitives can be occupied by a phonetically realized conjunction, with an overt complementizer, or an overt infinitival marker in the other languages with *wh*-infinitives discussed in section 3. Control infinitives in Polish (29), French (30), Spanish (31), European-Portugese (EP) (32), Italian (33), the *te*-infinitive in Dutch (34) and the *to*-infinitive in English (35) differ in this respect from the languages/constructions in (24)-(28). Here the relevant infinitival C-system may be filled with a base-generated, phonetically realized element. I leave open the matter of in which position the relevant element is located in the C-system, i.e. whether it is in C^0, in Spec CP or adjoined to CP (see Sabel 1996: 221-224 for discussion).

(29) *Chciałem* [$_{CP}$ *żeby* [*aprosić Kasię*]]. (Zabrocki 1981:69)
I-wanted to-invite K.

(30) *Il a oublié* [$_{CP}$ *de* [*nettoyer la chambre*]]. (Long 1976)
he has forgotten to-clean the room

(31) *Acabamos* [$_{CP}$ *de* [*ofrecer se los*]]. (Luján 1980)
we-have-just to-offer-him-them
'We have just offered them to him.'

(32) *Não sei* [$_{CP}$ *se* [*ir o não ir*]]. (Kayne 1991)
not (I-)know whether to-go oder not to-go

(33) *Gianni non sa* [$_{CP}$ *se* [*andare al cinema*]]. (Kayne 1991)
Gianni not know to-go to cinema

(34) *dat zij probeerde* [$_{CP}$ *om* [*het boek te lezen*]]. (Koster 1987)
that she tried the book to read

(35) *I want* [$_{CP}$ *for* [*Chris to win*]].

To sum up, in the languages in which non-finite relative clauses and *wh*-infinitives are possible, the C-system of the relevant infinitives may host a phonetically realized base-generated element, i.e. a phonetically realized subordinating conjunction, for example, an infinitival marker or a (prepositional) complementizer. This gives rise to the following empirical generalization:

(36) *The Wh-Infinitive-/Infinitival-Relative-Correlation*
If a language possesses *wh*-movement to Spec CP in infinitives, then this language possesses the option of filling the C-system of this (type of) infinitive with a base-generated overt element.

Let us call the C-system of infinitives that cannot be contructed as indirect questions or relative clauses "defective", where defective will be understood for the moment only as a descriptive term. I return in due course to a definition in the technical sense with respect to defective C^0. Thus, the German control infinitive, for example, has a defective C-system. Operators may not end up in the Spec CP position in infinitival relative clauses and infinitival interrogatives. The discussed control infinitives in Norwegian, Swedish, Danish, and English Gerunds likewise have a defective C-system. This is summarized in (37a) below.[5]

It is expected that further languages exist that conform to generalization (36). For example, Icelandic does not possess *wh*-infinitives. According to Thráinsson (1993, 1998) and Christensen (2005), the infinitival marker *að* in control infinitivals occupies a position below the C-system and prepositional complementizers do not exist. Thus Icelandic control infinitives seem to have a defective infinitival C-system (37a) and seem to confirm (36).

(37)	a. Defective infinitival C-system Danish, German, Norwegian, Swedish, English (Gerunds), ...	b. Non-defective infinitival C-system Dutch, English (*to*-infinitive), European-Portugese, French, Italian, Polish, Spanish, ...

In languages and constructions in which the infinitival C-system is not defective, *wh*-infinitives and infinitival relatives are possible. This situation arises with *to*-infinitives in English, with *te*-infinitives in Dutch ((38) = (15)) and with the other languages mentioned in (37b).

(38) *Ik weet niet* [$_{CP}$ <u>*wie*</u> [___ *te bezoeken*]].
I know not who to visit

(39) ... *en bal* [$_{CP}$ Op *om* [*mee* ___ *te spelen*]].
a ball (which) with to play

Further languages or dialects of type (37b) should exist. Brandner (2004) and Gärtner (2006), for example, have discovered that in certain dialects of German, embedded infinitival *wh*-questions are possible, i.e. in Reichenau German and Pennsylvania German. Brandner (2004) argues that in Reichenau German, the inflected preposition *zum* ('to-the.DAT') has turned into an infinitival complementizer. According to Gärtner (2006), similar assumptions might be made about *fer* ('for') in Pennsylvania German. Thus, both Reichenau German and Pennsylvania German seem to have non-defective infinitival C-systems and seem to confirm generalization (36).

Further systematic differences between the control infinitives in (37) should exist. For example, in German, *wh*-infinitives with *zu* do not appear as root expressions **Wen zu besuchen?* 'Who to visit' in contrast to Dutch *wh*-infinitives with *te*, cf. *Wie te bezoeken?* Whether this correlation holds for all the languages in (37) still needs to be in-

5 As pointed out in Sabel (1996, chapter 8), in some languages, the possibility of realizing infinitival relatives and *wh*-infinitives may differ in accordance with (36).

vestigated. Furthermore, Gärtner (2006) observes that languages that allow certain *wh*-words to be used as indefinites have no *wh*-infinitives.

We now arrive at the issue of how the generalization in (36) can be derived. I will outline three possible accounts.

1. One possibility seems to be that the control infinitives in the languages/constructions in (37a) have not developed a C-system at all. This would explain why *wh*-infinitives, infinitival relatives and infinitival complementizers are excluded or possible at the same time in the languages in (37). The explanation for (36)-(37) then is due to the conditions under which C-systems, i.e. their different layers,[6] evolve.

2. A second potential account is based on the idea that control infinitives are CPs. The idea is that the C-head of a defective infinitival C-system has the parametric property of obligatorily triggering head movement into its C-system (Agr^0-to-C^0 within the sentence structure $[_{CP}\ [_{ArgP}\ [_{TP}\ [_{vP}]]]]$). Now the observed facts follow by assuming that *wh*-infinitives and infinitival relatives are excluded for the same reason as V-to-C movement is excluded in finite indirect questions and relative clauses. The impossibility of V-to-C in these contexts is then explained in terms of the *Wh*-Criterion (Rizzi 1996) and selectional violations (Rizzi and Roberts 1996). The *Wh*-Criterion states that (a.) each X^0-[+*wh*] must be in a Spec-head relation with a *wh*-phrase and that (b.) each *wh*-phrase must be in a Spec-head relation with an X^0-[+*wh*]. In indirect *wh*-questions a [+*wh*]-feature is introduced on the embedded C^0 as a result of selection by the matrix predicate. V-to-C-movement overwrites the feature [+*wh*] in C^0 or, alternatively, forms a complex head adjunction structure that conflicts with selectional requirements. The derivation results in a violation of a CI-interface condition. This explanation can be extended to infinitive relative clauses. The [+*pred*]-feature in C^0 obeys the same restrictions as the [*wh*]-feature.

3. A final alternative rests likewise on the assumption that the control infinitives in (37) are all CPs but that the C-system of the control infinitives in (37a) is defective in terms of its featural make-up. As pointed out in Chomsky (2000, 2001, 2005), Case is normally assigned to the subject of a control infinitive via T^0. Control T^0 is non-defective. Control CPs, for example English *to*-infinitives, are phases. The difference between (37a) and (37b) might result from the following parametric property. Control C^0 is defective in the languages in (37a) but not in the languages in (37b), "defective" being understood here in analogy to defective T^0. T_{def} only bears a subset of φ-features in raising- and ECM-infinitives. C_{def} in control infinitives likewise cannot bear the complete range of C-features. In the languages in (37a) it bears a full set of φ-features and Tense-features that is transferred to T^0 (assuming the technology in Chomsky 2005), but it lacks the possibility of being endowed with a [*focus*]-/[*wh*]- or with a [*topic*]-/[*pred*]-feature. (I have argued elsewhere (see Sabel 1998, 2000b, 2000c) that both a [+*wh*] and a [+*focus*] feature are universally involved in any *wh*-question formation (cf. also Gre-

6 Rizzi (1997, 1999) has shown that the C-system is more articulated than illustrated in the examples above. It consists of different layers such as (at least) ForceP (for the representation of illocutionary force), TopP (for topicalized elements), FocP (for focused material), and FinP (for (in-) finiteness). Under this analysis it would not be surprising to find that other kinds of A'-movement into the left periphery of the (same) infinitvals in the languages in (37a) are attested. For ease of presentation, I have used the more general concept of the C-system.

wendorf 2005); likewise two features, i.e. [*topic*]- and [*pred*]-features, seem to be involved in relative clause formation.) The consequence is that (similar to an NP in the edge of T_{def}) a *wh* or a relative operator may move to Spec CP due to the possibility of C_{def} being endowed with an EPP-feature, but it may not remain in the edge of C_{def}, i.e. in a position in which it cannot be properly interpreted. This analysis derives the fact that Spec C_{def}P and Spec T_{def}P are only intermediate landing sites. (The operator in situ variant is likewise excluded.)

All of the three proposals are in line with the common assumption that functional categories are the locus of parametric variation. However, proposal one has the disadvantage of being incompatible with the evidence discussed in (for example) Sabel (1996; chapter 4 and 8, 2001), where it is shown that German control *zu*-infinitives have the distribution and categorial status of CPs. Let us next turn to proposal two according to which *wh*-infinitives and infinitival relatives are excluded in the languages in (37a) for the same reason as why verb-movement to C is excluded in finite indirect questions and relative clauses. Recall that in Rizzi's (1996) analysis, C^0 acquires its [*wh*]-feature in matrix questions as a result of I-to-C movement. Assuming I- (or Agr-) to-C in infinitives in (37a), proposal three, but not proposal two, offers an "automatic" explanation for the above-mentioned fact that *wh*-infinitivals in German are also impossible with matrix *zu*-control infinitivals, as well as for the fact that partial *wh*-movement in German (triggered by a [*focus*]-feature in the analysis of Sabel 1998, 2000b, 2000c) to the initial position of a [–*wh*] infinitival clause is ungrammatical (McDaniel et al. 1995).

(40) a. * *Was glaubst du* [$_{CP}$ <u>*wen*</u> [PRO *wahrscheinlich* ___ *zu treffen*]]?
what believe you who.ACC probably to meet

b. *Was glaubst du* [$_{CP}$ <u>*wen*</u> [*du* *wahrscheinlich* ___ *triffst*]]?
what believe you who.ACC you.NOM probably meet
'Who do you believe to meet probably?'

Assuming that *wh*-infinitives do not appear as root expressions not only in German but (with the relevant infinitives) in all the languages in (37a), I assume that proposal three is on the right track as an explanation of (36)-(37).

References

Beukema, F. and M. den Dikken. 1989. The position of the infinitival marker in the Germanic languages. In: D. Jaspers et al. (eds.), *Sentential complementation and the lexicon. Studies in honor of Wim de Geest*, 57–75. Dordrecht: Foris.

Bhatt, R. 1999. *Covert Modality in Non-Finite Contexts*. Ph.D. dissertation, University of Pennsylvania.

Brandner, E. 2004. Microvariation within German *to*-Infinitives. Ms., University of Konstanz.

Chomsky, N. 2000. Minimalist inquiries: the framework. In: R. Martin, D. Michaels, and J. Uriagereka (eds.), *Step by Step. Essays on Minimalist Syntax in Honor of Howard Lasnik*, 89–155. Cambridge, Mass.: MIT Press.

Chomsky, N. 2001. Derivation by phase. In: M. Kenstowicz (ed.), *Ken Hale: a Life in Language*, 1–52. Cambridge, Mass.: MIT Press.
Chomsky, N. 2005. On phases. Ms., MIT.
Christensen, K. K.. 1984. Infinitival (pseudo-) complementation of noun phrases in Norwegian. In: W. de Geest and Y. Putseys (eds.), *Sentential Complementation*, 75–82. Dordrecht: Foris
Christensen, K. R. 2005. The infinitival marker across Scandinavian. Ms., University of Aarhus.
Cinque, G. 2004. Restructuring and functional structure. In: A. Belletti (ed.), *Structures and Beyond. The Cartography of Syntactic Structures, vol. 3*, 132–191. New York: Oxford University Press.
Gärtner, H.-M. 2006. From German quirk to universal tendency: a speculation on (the absence of) *wh*-infinitives. Ms., Zentrum für Allgemeine Sprachwissenschaft (Center for General Linguistics), Berlin.
Giusti, G. 1989 *Zu*-infinitivals and the structure of IP in German. Ms., University of Geneva.
Grewendorf, G. 2005. The asymmetry of short and long wh-movement in German. In: H.-G. Obenauer (ed.), *L'Architecture Propositionelle. La Syntaxe de la Périphérie Gauche* (*Recherches Linguistiques de Vincennes* 33), 35–54.
Holmberg, A. 1983. Infinita Frågebisatser I Finlanfssvenk dialekt. Ms., University of Stockholm.
Holmberg, A. 1986. *Word Order and Syntactic Features in the Scandinavian Languages and English.* Doctoral dissertation, University of Stockholm.
IJbema, A. and W. Abraham. 2000. Die syntaktische Funktion des infinitivischen *zu*. In: R. Thieroff et al. (eds.), *Deutsche Grammatik in Theorie und Praxis*, 123–137. Tübingen: Niemeyer.
Johnson, K. and S. Vikner. 1994. The position of the verb in Scandinavian infinitives: in V^0 or in C^0 but not in I^0. *Working Papers in Scandinavian Syntax* 53, 61–84.
Kayne, R. 1984. *Connectedness and Binary Branching*. Dordrecht: Foris.
Kayne, R. 1991. Romance clitics, verb movement, and PRO. *Linguistic Inquiry* 22, 647–686.
Kemenade, A. van. 1985. Old English infinitival complements and West-Germanic V-raising. In: R. Eaton (ed.), *Papers from the 4th International Conference on English Historical Linguistics*, 73–84. Amsterdam: John Benjamins.
Koster, J. 1987. *Domains and Dynasties. The Radical Autonomy of Syntax.* Dordrecht: Foris.
LaPolla, M. V. 1988. Clitic Movement in Spanish and the Projection Principle. In: D. Birdsong and J.-P. Montreuil (eds.), *Advances in Romance Linguistics*, 217–231. Dordrecht: Foris.
Lightfoot, D. W. 1979. *Principles of Diachronic Syntax.* Cambridge: Cambridge University Press.
Long, M. E. 1976. French infinitival complementizers and their place in Generative Grammar. In: M. Lujàn and F. G. Hensey (eds.), *Current Studies in Romance Linguistics*, 205–220. Washington, DC: Georgetown University Press.
Lujàn, M. 1980. Clitic promotion and mood in Spanish complements. *Linguistics* 18, 381–484.
Napoli, D. J. 1981. Semantic interpretation vs. lexical governance: clitic climbing in Italian. *Language* 57, 841–887.

Raposo, E. 1987. Case theory and Infl-to-Comp. *Linguistic Inquiry* 18, 85–109.
Reis, M. 1985. Satzeinleitende Strukturen im Deutschen. In: W. Abraham (ed.), *Erklärende Syntax des Deutschen*, 271–311. Tübingen: Niemeyer.
Reis, M. 2003. On the form and interpretation of German wh-infinitives. *Journal of Germanic Linguistics* 15, 155–201.
Rizzi, L. 1996. Residual verb second and the Wh-Criterion. In: A. Belletti and L. Rizzi (eds.), *Parameters and Functional Heads. Essays in Comparative Syntax*, 63–90. New York, Oxford: Oxford University Press.
Rizzi, L. 1997. The fine structure of the left periphery. In: L. Haegeman (ed.), *Elements of Grammar*, 281-337. Dordrecht: Kluwer
Rizzi, L. 1999. On the position "Int(errogative)" in the left periphery of the clause. Ms., Università di Siena.
Rizzi, L. and I. Roberts. 1996. Complex inversion in French. In: A. Beletti and L. Rizzi (eds.), *Parameters and Functional Heads. Essays in Comparative Syntax*, 91–116. New York, Oxford: Oxford University Press.
Sabel, J. 1996. *Restrukturierung und Lokalität. Universelle Beschränkungen für Wortstellungsvarianten*. Berlin: Akademie Verlag.
Sabel, J. 1998. *Principles and Parameters of Wh-Movement.* Habilitation thesis, Universitat Frankfurt/Main.
Sabel, J. 2000a. Das Verbstellungsproblem im Deutschen: Synchronie und Diachronie. *Deutsche Sprache* 28, 74–99.
Sabel, J. 2000b. Expletives as Features. In: R. Billerey & B. Lillehaugen (eds.), *WCCFL 19 Proceedings*, 411–424. Somerville: Cascadilla Press.
Sabel, J. 2000c. Partial *wh*-movement and the typology of *wh*-questions. In: U. Lutz, G. Müller & A. v. Stechow (eds.), *Wh-Scope Marking*, 409–446. Amsterdam: John Benjamins.
Sabel, J. 2001. Das deutsche Verbum infinitum. *Deutsche Sprache* 29, 148–175.
Sigurðsson, H. 1989. *Verbal Syntax and Case in Icelandic*. Doctoral dissertation, University of Lund.
Thráinsson, H. 1993. On the structure of infinitival complements. *Harvard Working Papers in Linguistics* 3, 181–213.
Thráinsson, H. 1998. Infinitival complements in some Old and Modern Scandinavian languages. In: J. O. Askedal (ed.), *Historische germanische und deutsche Syntax. Akten des internationalen Symposiums anlässlich des 100. Geburtstages von Ingerid Dal*, 335–363. Frankfurt: Peter Lang.
Wilder, C. 1989. *The Syntax of German Infinitives*. Doctoral dissertation, University of London (UCL).
Wurmbrand, S. 2000. *Infinitives: Restructuring and Clause Structure*. Berlin and New York: Mouton de Gruyter.
Zabrocki, T. 1981. *Lexical Rules of Semantic Interpretation. Control and NP-Movement in English and Polish*. Poznan.

Expletive Replacement Reconsidered: Evidence from Expletive Verbs in Japanese

Mamoru Saito

1. Introduction*

It is argued in Hoshi 1995, and Saito and Hoshi 2000 that the Japanese expletive verb *su* shares the basic properties with the English expletive *there*. These works suggest further that its distribution can be accounted for by the expletive replacement analysis proposed in Chomsky 1986. The aim of this paper is rather modest: it is to confirm these conclusions by further developing the analysis of the Japanese expletive verb.

It has been known that the English existential construction, exemplified in (1), exhibits properties that indicate that the indefinite (associate) NP occupies the subject position in place of the expletive *there*.

(1) *There were linguists in the room.*

Thus, the indefinite NP *linguists* participates in subject-verb agreement exactly as in (2).

(2) *Linguists*$_i$ *were* t_i *in the room.*

The parallelism between (1) and (2) goes further. For example, when NP-movement as in (2) is illicit, the corresponding existential sentence is ungrammatical. This is illustrated in (3) and (4).

(3) a. **Linguists*$_i$ *seem to* t_i *that Warlpiri is the most fascinating language.*
b. **There seem to linguists that Warlpiri is the most fascinating language.*

(4) a. **Linguists*$_i$ *seemed that it was likely* t_i *to be in the room.*
b. **There seemed that it was likely linguists to be in the room.*

Given these observations, Chomsky (1986) proposed that the indefinite NP moves to the subject position and replaces the expletive at LF. The failure of expletive replacement, according to him, results in a violation of Full Interpretation, which requires that every element receive interpretation at the interface levels, LF and PF. Since *there* is void of meaning, its presence at LF is illicit. The expletive replacement is itself NP-movement

---

* While my indebtedness to Günther Grewendorf in my research on scrambling is obvious, I have also benefited much from discussions with him on the topic of complex predicates. I am happy to be able to contribute a paper on this topic to the volume in his honor. The material in this paper is based on joint research with Hiroto Hoshi over the years. His contributions, which are evident in the pages to follow, are gratefully acknowledged. I would also like to thank an anonymous reviewer for detailed, helpful comments on the earlier version of the paper.

and hence, (4b) is ruled out due to the SSC effect, precisely as (4a). Chomsky further proposed Last Resort in this context as a principle to exclude (3a) and (3b). The principle states that movement can take place only to satisfy a morphological requirement of the moved item. It prohibits the movement of *linguists* in (3a-b) since the NP is checked for Case at its base position and there is no need for this NP to move to the subject position. (3b), then, violates the Last Resort Principle if expletive replacement takes place, and if not, it is excluded by Full Interpretation.

The analysis just described played an important role in the development of the syntactic theory toward Minimalism. The Last Resort Principle provided the incentive to pursue the economy of derivation, and Full Interpretation is just another name for the economy of representation. However, as the Minimalist model was developed, the Last Resort Principle lost its place within the theory and was eliminated in favor of a more refined theory of feature-checking in Chomsky 1995. The purpose of this paper is neither to discuss this development nor to examine the analysis of the English existential construction. Instead, I will show that the analysis in terms of Last Resort and Full Interpretation successfully extends to expletive verbs in Japanese. This not only adds to the data to be considered in the analysis of expletives but also suggests that there is an insight behind expletive replacement that must be captured even in a more refined analysis.

The following section concerns the distribution of the expletive verb *su* in the Japanese light verb construction. I will first go over the analysis of the construction presented in Saito and Hoshi 2000. Then, I will revise the analysis of one of the constraints on the construction and argue that it can be derived from the Last Resort Principle and Full Interpretation. In Section 3, I will discuss another construction in which this expletive verb appears, i.e., the one in which the topic marker *wa* or a focus particle like *sae* 'even' is attached to the regular verb. After updating Hoshi's (1995) analysis of the distribution of the expletive *su* in this construction, which is in fact in terms of the Last Resort Principle and Full Interpretation, I will show that it has further desirable empirical consequences. Section 4 contains a summary and a brief remark on a consequence of the proposed analysis within the derivational model of syntax.

2. Expletive verbs in the light verb construction

In Section 2.1, I will go over the properties of the Japanese verb *su* and discuss Sells' (1989) argument that it can function as an expletive verb. Then, in Section 2.2, I will present and develop the covert head movement analysis proposed in Saito and Hoshi 2000.

2.1 *Su* as an expletive verb

The verb *su* mentioned above appears in contexts such as those in (5).

(5) a. *Hanako-ga yama-nobori-o sita.* (*sita* = *su* + *ta* (past))
Hanako-NOM mountain-climbing-ACC did
'Hanako did mountain-climbing.'
b. *Hanako-ga Taroo-ni toti-o zyooto-sita.*
Hanako-NOM Taroo-to land-ACC giving-did
'Hanako gave a piece of land to Taroo.'
c. *Hanako-ga Taroo-ni* [$_{\text{NP}}$ *toti-no zyooto*]*-o sita.*
Hanako-NOM Taroo-to land-GEN giving-ACC did
'Hanako gave a piece of land to Taroo.'

In (5a), it is used as the main predicate, very much like the main verb *do* in English. (I will henceforth refer to this *su* as the main verb *su*.) In this case, it takes an agentive subject and an accusative object that typically refers to some action. It may optionally take other arguments such as the goal phrase in (6).

(6) *Taroo-ni sore-o suru koto-wa yurus-are-nai.* (*suru* = *su* + *ru* (present))
Taroo-to it-ACC do fact-TOP permit-PASS-not
'It is not permissible to do that to Taroo.'

In (5b), *su* is a category-changing suffix that turns a noun into a verb. This *su* can attach to an unaccusative, unergative, or transitive noun to create the corresponding verb, as shown in (7).[1]

(7) a. *Mizu-ga zyoohatu-sita.* (unaccusative)
water-NOM evaporation-did
'The water evaporated.'
b. *Taroo-ga sanpo-sita.* (unergative)
Taroo-NOM taking.a.walk-did
'Taroo took a walk.'
c. *Hanako-ga Taroo-o hihan-sita.* (transitive)
Hanako-NOM Taroo-ACC criticism-did
'Hanako criticized Taroo.'

(5c) is an example of what is called the 'Japanese light verb construction' and represents the case where *su* is used as an expletive verb. As discussed in Grimshaw and Mester 1988, *su* can be void of meaning and the accusative noun *zyooto* 'giving' can serve as the predicate in this example. It is pointed out by Terada (1990) and others that *su* in examples of this kind can plausibly be analyzed as a main verb because the main verb *su* can sometimes take a goal argument as mentioned above. However, Sells (1988) persuasively argues that there is indeed an expletive *su* based on examples such (8).

---

1 See Miyagawa 1989 and Tsujimura 1990 for detailed discussion on the properties of this *su*.

(8) ?? *Hanako-ga* *Taroo-ni* *toti-o* *zyooto-o* *sita.*
Hanako-NOM Taroo-to land-ACC giving-ACC did
'Hanako gave a piece of land to Taroo.'

This example is degraded because it violates the ban on multiple accusative phrases, known as the 'double-*o* constraint'. It is shown in Harada 1973 and Shibatani 1973 that this constraint has two separate subtypes. When there are two accusative argument NPs in a simple sentence, the result is hopeless as in (9a).[2]

(9) a. * *Hanako-ga* *Taroo-o* *sake-o* *nom-aseta.*
Hanako-NOM Taroo-ACC sake-ACC drink-made
'Hanako made Taroo drink sake.'
b. ?? *Hanako-ga* *Taroo-o* *hamabe-o* *hasir-aseta.*
Hanako-NOM Taroo-ACC beach-ACC run-made
'Hanako made Taroo run on the beach.'

On the other hand, the result is only marginal when one of the two accusative NPs is a non-argument. In (9b) *hamabe* 'beach' is a locative adverbial and hence, the example is much better than (9a). What Sells points out is that (8) has the grammatical status of (9b) and not of (9a). This implies that one of the accusative NPs in this example is a non-argument. If *su* is a main verb and assigns θ-roles to all arguments, then both of the accusative NPs would be arguments and hence, we would expect a strong violation as in (9a), contrary to the fact. On the other hand, if *su* is an expletive verb and *zyooto* 'giving' serves as the predicate of the sentence, the marginal status of the example is correctly predicted. In this case, the only accusative argument in the sentence is *toti* 'land'. Thus, examples like (8) indicate that *su* indeed can be an expletive verb. And if (8) contains an expletive *su*, we expect this expletive verb to be able to occur in (5c) as well.[3]

The remaining task is to provide an analysis for the light verb construction. If a head can assign θ-roles only to phrases within its maximal projection, the construction is quite interesting because the θ-role assigning noun heads the accusative object NP and (some of) its arguments appear outside this NP in the surface configuration. The analysis, then, must explain how θ-role assignment is possible in the construction. In particular, it must show how the nominal head *zyooto* 'giving' assigns θ-roles to the clausal arguments in (8) and (5c).

---

2 In the Japanese causative construction, the causee can be marked either dative or accusative as shown in (i).

(i) *Hanako-ga* *Taroo-ni/-o* *hasir-aseta.*
Hanako-NOM Taroo-DAT/-ACC run -made
'Hanako made Taroo run.'

The examples in (9) are perfect if the causee *Taroo* is marked dative instead of accusative.

3 If Terada (1990) is correct, (5c) is ambiguous. That is, the verb *su* in this example may be a main verb or an expletive verb.

2.2 Expletive verb replacement in LF

Given this background, it was proposed in Saito and Hoshi 2000 that the θ-role assigning noun in the light verb construction covertly moves to the position of the expletive verb and discharges its θ-roles from that position. Let us consider the structure of (8), shown in (10).

(10)

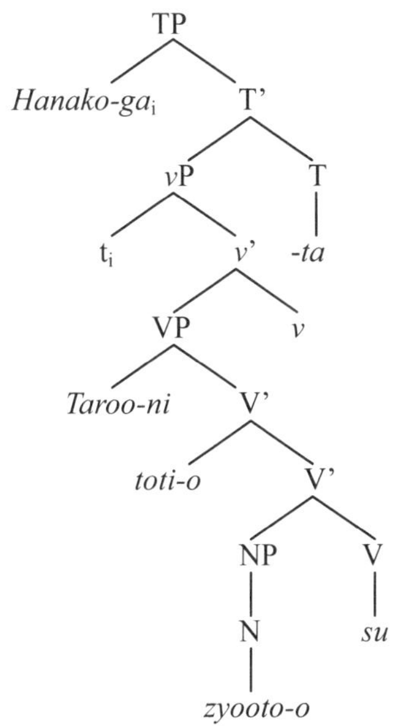

According to this analysis, the noun *zyooto* 'giving' moves to the position of *su* in LF, and discharges its theme role to *toti* 'land' and its goal role to *Taroo*.[4] Note that the expletive verb is replaced (or adjoined to) as a result and does not appear in the LF representation. Although Saito and Hoshi did not consider this crucial, the analysis is thus consistent with Full Interpretation.

One piece of evidence for this analysis is that the θ-role assigning noun resists any kind of overt movement, a fact reported in Grimshaw and Mester 1988. The cleft sentences in (11) illustrate the generalization.

(11) a. [$_{CP}$ Op$_i$ [$_{IP}$ *Mary-ga* *John-ni* t$_i$ *zyooto-o* *sita*] *no*]-*wa* *toti-o*$_i$ *da*.
Mary-NOM John-to giving-ACC did COMP-TOP land-ACC is
'It is a piece of land that Mary gave to John.'
(*Lit.* 'It is a piece of land that Mary did giving to John.')

b. *[$_{CP}$ Op$_i$ [$_{IP}$ *Mary-ga* *John-ni* *toti-o* t$_i$ *sita*] *no*]-*wa* *zyooto-o*$_i$ *da*.
Mary-NOM John-to land-ACC did COMP-TOP giving-ACC is
(*Lit.* 'It is giving that Mary did a piece of land to John.')

---

4 Saito and Hoshi (2000) actually do not assume the *v*-projection and place the subject at the specifier position of VP. Hence, according to their analysis, the raised noun assigns the agent role to *Hanako* as well. I will come back to this point later in this section when the choice becomes relevant.

(11a-b) are based on the multiple accusative sentence (8). In (11a), the theme argument *toti-o* 'land-ACC' is focused. The example is better than (8) as one of the two accusative NPs is dislocated.[5] (11b), on the other hand, is derived by focusing the θ-role assigning noun *zyooto-o* 'giving-ACC', and the result is hopeless. This readily follows from the covert head movement analysis. The θ-role assigning noun would have to move sideways in order to discharge its θ-roles in this example. Hence, the required θ-role assignment fails to take place.

Similarly, scrambling of the θ-role assigning noun results in ungrammaticality, as shown in (12).

(12) a. **Hanako-ga* *zyooto-o*$_i$ *Taroo-ni* *toti-o* t_i *sita.*
Hanako-NOM giving-ACC Taroo-to land-ACC did
'Hanako gave a piece of land to Taroo.'
b. **Zyooto-o*$_i$ *Hanako-ga* *Taroo-ni* *toti-o* t_i *sita.*
giving-ACC Hanako-NOM Taroo-to land-ACC did
'Hanako gave a piece of land to Taroo.'

These examples can be explained in basically the same way as (11b).

This account of (12) is consistent with the (radical) reconstruction properties of scrambling. As shown in detail in Tada 1993 and Nemoto 1993, VP-internal scrambling exhibits strict A-properties and is not subject to LF reconstruction. Thus, (13b) contrasts sharply with (14b).

(13) a. *Hanako-ga* (*kinoo*) *karera-ni* *otagai-o* *syookaisita.*
Hanako-NOM yesterday they-to each other-ACC introduced
'Hanako introduced them to each other yesterday.'
b. **Hanako-ga* *otagai-o*$_i$ (*kinoo*) *karera-ni* t_i *syookaisita.*
Hanako-NOM each other-ACC yesterday they-to introduced
c. **Otagai-o*$_i$ *Hanako-ga* (*kinoo*) *karera-ni* t_i *syookaisita.*
each other-ACC Hanako-NOM yesterday they-to introduced

(14) a. *Karera-ga* *otagai-o* *semeta.*
they-NOM each other-ACC blamed
'They blamed each other.'
b. *Otagai-o*$_i$ *karera-ga* t_i *semeta.*
each other-ACC they-NOM blamed

It was proposed in Saito 1989 on independent grounds that scrambled phrases can be placed back in their initial positions at LF. This offers a possible account for the grammaticality of (14b). But then, the ungrammaticality of (13b) shows that VP-internal scrambling (as opposed to scrambling across the subject) is not subject to this LF reconstruction. Similarly, the ungrammaticality of (13c) indicates that scrambling out of VP proceeds through the edge of VP (or *v*P), and reconstruction applies only to the movement originating from this position. Then, *zyooto-o* 'giving-ACC' in (12) is at the edge of

5 As noted in Harada 1973 and Shibatani 1973, this kind of improvement is observed with the weak type of "double-*o*" effect represented by (9b) but not with the strong type shown in (9a).

VP (or *v*P) at LF, and hence, it must lower to the position of *su* in order to discharge its θ-roles.

I have so far introduced the covert head movement analysis of Saito and Hoshi 2000. There is another argument presented in favor of this analysis in the paper, and that is where refinement is necessary. I will now turn to this argument.

Grimshaw and Mester (1988) note the following as one of the peculiar properties of the Japanese light verb construction:

(15) At least one internal argument of the θ-role assigning noun must be realized outside the NP it projects.

The examples in (16) illustrate this generalization.

(16) a. *Hanako-ga* *Taroo-to* *kekkon-o* *sita.*
Hanako-NOM Taroo-with marriage-ACC did
'Hanako married Taroo.'
b. ? *Hanako-ga* [$_{NP}$ *Taroo-to-no* *kekkon*] *-o* *sita.*
Hanako-NOM Taroo-with-GEN marriage -ACC did

(16a) receives a straightforward interpretation as a regular instance of the light verb construction with the expletive *su*. (16b), on the other hand, is marginally allowed but only with the main verb interpretation of *su*. It roughly means that there is a specific act of getting married with Taroo and Hanako did it. Grimshaw and Mester, hence, consider (16b) to be ungrammatical as an instance of the light verb construction. This example contradicts the generalization in (15) since only the external argument *Hanako* is realized outside the NP headed by *kekkon* 'marriage'.

Saito and Hoshi (2000) argued that the generalization in (15) follows from the Last Resort Principle. The structures they posit for (16a-b) are shown in (17a-b) respectively.

(17)
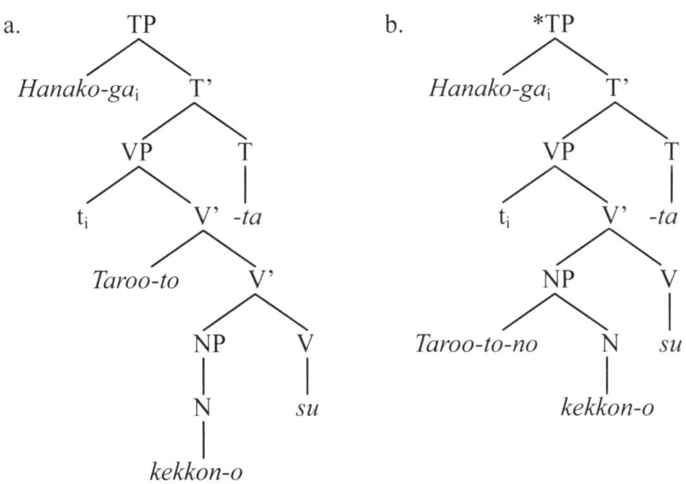

It is assumed here that the subject *Hanako* is generated in VP Spec. In (17a), *kekkon* 'marriage' moves covertly to the position of the expletive verb and assigns θ-roles to *Taroo* as well as to *Hanako*. In (17b), on the other hand, *kekkon* discharges its internal θ-role to *Taroo* within the NP. In this case, there is no motivation for the θ-role assigning noun to move to the position of *su* because a noun, as opposed to a verb, only optionally assigns its external θ-role. The movement is, therefore, excluded by the Last Resort Principle and consequently, the subject *Hanako* fails to receive a θ-role.

This analysis, however, cannot be maintained under the hypothesis that *v* assigns the external θ-role. The structure of (16b) would then be as in (18).

(18)

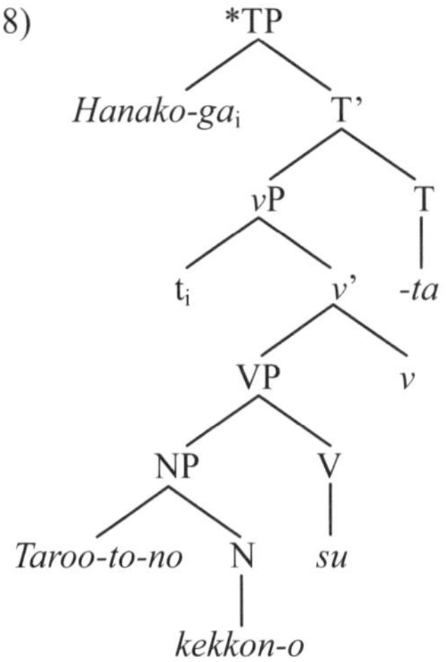

In this structure, *Hanako* and *Taroo* receive θ-roles in their base positions from *v* and *kekkon* 'marriage' respectively, and hence, θ-role assignment takes place properly even in the absence of the covert movement of the θ-role assigning noun. In Saito 2001, I tentatively assumed that the θ-role assigning noun assigns the external θ-role together with *v*. Then, *kekkon* must and hence, can move to the position of *v* in order to discharge the external θ-role, but this movement violates the head movement constraint. Although this account provides a technical solution to the problem, it begs the question because it after all assumes that an external θ-role is assigned by a lexical head.

There is an obvious, straightforward alternative analysis for the illicitness of (18). Since all arguments are successfully assigned θ-roles, there is nothing wrong with the θ-role assignment. But because the θ-role assigning noun discharges its θ-role at the base position, there is no reason for it to replace the expletive verb. Consequently, the expletive verb remains at LF in violation of Full Interpretation. According to this slightly modified analysis, (16b) is excluded in precisely the same way as the English (3b), repeated below as (19).

(19) **There seem to linguists that Warlpiri is the most fascinating language.*

Since the indefinite NP *linguists* is checked for Case at its base position, the Last Resort Principle prevents it from replacing the expletive *there*. In the case of (16b), since the θ-role assigning noun *kekkon* 'marriage' discharges its θ-role at its base position, again, the Last Resort Principle prohibits it from replacing the expletive verb *su*. As the result, both (19) and (16b) are excluded by Full Interpretation.

The grammatical (16a) parallels (1), repeated in (20).

(20) *There were linguists in the room.*

In this example, *linguists* moves to the position of the expletive *there* in order to be checked for Case. Similarly, in (16a), *kekkon* moves to the position of the expletive verb *su* in order to discharge its θ-role. Because of these independently motivated movements, the expletives are successfully replaced in both (20) and (16a).

If the slight revision of the analysis of (16b) presented above is correct, then the Japanese light verb construction provides evidence not only for the Last Resort Principle but also for the necessity of expletive replacement to satisfy Full Interpretation. In the following section, I will consider another construction with the expletive *su* that points to the same conclusion.

3. Expletive verbs in the VP focus construction

The verb *su* occurs also when the regular verb is followed by the topic marker *wa* or a focus particle like *mo* 'also' and *sae* 'even'. This is illustrated in (21b-d).

(21) a. *Hanako-ga Taroo-o nagutta.*
Hanako-NOM Taroo-ACC hit
'Hanako hit Taroo.'

b. *Hanako-ga Taroo-o naguri-wa sita.*
Hanako-NOM Taroo-ACC hit-TOP did
'Hanako did hit Taroo.'

c. *Hanako-ga Taroo-o naguri-mo sita.*
Hanako-NOM Taroo-ACC hit-also did
'Hanako also hit Taroo.'

d. *Hanako-ga Taroo-o naguri-sae sita.*
Hanako-NOM Taroo-ACC hit-even did
'Hanako even hit Taroo.'

Kuroda (1965), who first discussed this phenomenon, postulated a rule of *si*-insertion, similar to *do*-support in English. Hoshi (1995), on the other hand, reconsiders the phenomenon based on the properties of the verb *su* discussed in the preceding section. He proposes in fact to explain the distribution of the expletive *su* in this construction in terms of the Last Resort Principle and Full Interpretation. In Section 3.1, I will discuss the basic properties of this construction, focusing on the particle *sae* 'even', and introduce Hoshi's analysis. Then, in Section 3.2, I will present independent evidence for the analysis and thereby confirm the conclusion that Last Resort and Full Interpretation both play crucial roles in the explanation of the phenomenon.

3.1. The illicitness of VP-scrambling with expletive *su*

Given that *su* can function as a main verb or an expletive verb, as was shown in the preceding section, it is tempting to analyze the occurrences of *su* in (21) as instances of these. The analysis in terms of the main verb *su* seems straightforward. (21d), for example, can have the following structure:

(22)

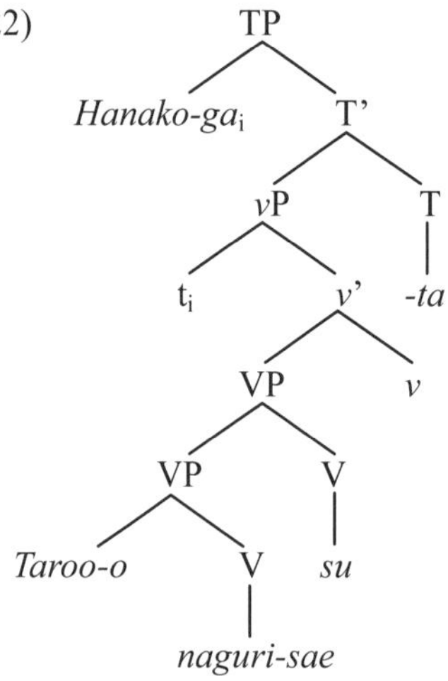

This is identical to the structure of the sentences with the main verb *su* discussed in the preceding section except that the complement of *su* is a VP instead of an NP.[6]

Examples such as (21b-d) have drawn much attention in part because they allow VP-scrambling. Thus, the VP headed by *naguri-sae* 'hit-even' in (21d) can be preposed as shown in (23).

(23) [$_{VP}$ *Taroo-o* *naguri-sae*] *Hanako-ga* t_{VP} *sita.*
Taroo-ACC hit-even Hanako-NOM did
'Hanako even hit Taroo'

Nothing seems to prevent this VP-scrambling if (21d) has the structure in (22).

Hoshi (1995) argues that there is another possible analysis for examples like (21b-d), that is, that the *su* in these examples may be an expletive verb. Given that *su* can function as an expletive verb in the light verb construction, this would be the null hypothesis. More specifically, he proposes that (21d) may have the structure in (24) with the expletive *su*.

6 The examples in (21b-d) are ungrammatical without the particles *wa*, *mo*, and *sae*. I assume that this is due to a property of *su* that it can only be merged with NPs and phrases with topic/focus particles. This suggests that those particles are nominal in some sense.

(24)

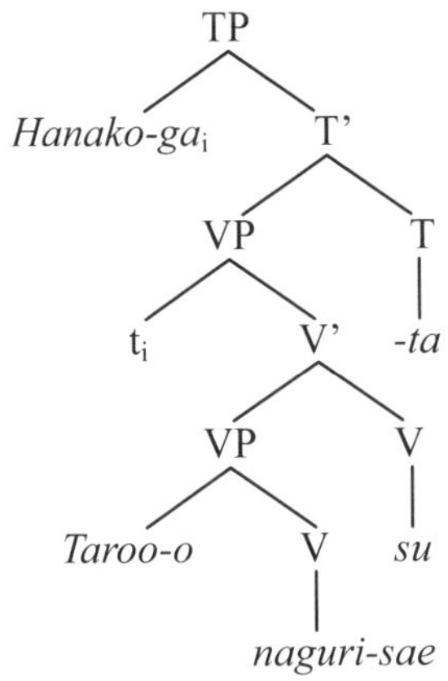

Hoshi (1995), like Saito and Hoshi (2000), assumes that external θ-roles are assigned by lexical heads and not by *v*. The verb *nagur* 'hit' assigns its theme role to *Taroo* in its base position and covertly moves the position of *su* in order to discharge its agent role to the trace of *Hanako*.[7]

Hoshi specifically argues that (25) is not a possible structure with the expletive *su*.

(25)

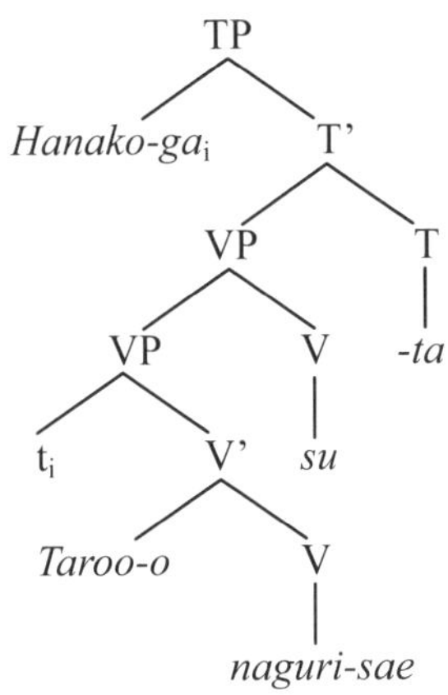

Since *nagur* 'hit' assigns all of its θ-roles in the base position, there is no reason for it to move to the position of *su*. Hence, the Last Resort Principle prohibits the movement, and the expletive *su* remains at LF. Hoshi proposes that the structure is consequently excluded by Full Interpretation.

The analysis introduced above makes an interesting prediction, as Hoshi notes. If the structure in (25) were possible, VP-scrambling should be allowed even with the expletive *su*. Nothing seems to prevent the scrambling of the lower VP in (25). On the other hand, if (24) is the only option with the expletive *su*, VP-scrambling should be impossible. If the lower VP in (24) is scrambled, the verb *nagur* 'hit' is no longer in the

7 Note that the structure in (24) may be possible also with the main verb *su*. In that case, the verb *su* assigns the agent role to the subject.

c-command domain of the expletive *su*. Hence, it would have to move sideways to the position of *su* in order to assign its external role to the trace of *Hanako*. Hoshi's analysis predicts, then, that although examples like (21d) are ambiguous between the main verb and the expletive verb interpretations of *su*, once VP-scrambling applies, only the former interpretation obtains. Hoshi states that this prediction is indeed borne out. That is, the main verb interpretation of *su* is forced in (23) but not in (21d).

Although Hoshi's analysis is complete, one drawback is that it is somewhat difficult to distinguish the main verb and the expletive verb interpretations of *su* in examples like (21d) and (23). It is therefore desirable to examine examples where the distinction comes out more clearly. In the following subsection, I will consider the illicit cases of VP-scrambling discussed in Hoji, Miyagawa and Tada 1989, and argue that they constitute evidence for Hoshi's analysis. But before I move on to this, I will briefly update the analysis assuming that *v* is the external θ-role assigner.

The required modification is straightforward. If *v* assigns the external θ-role, the structure in (25) would be as in (26).

(26)

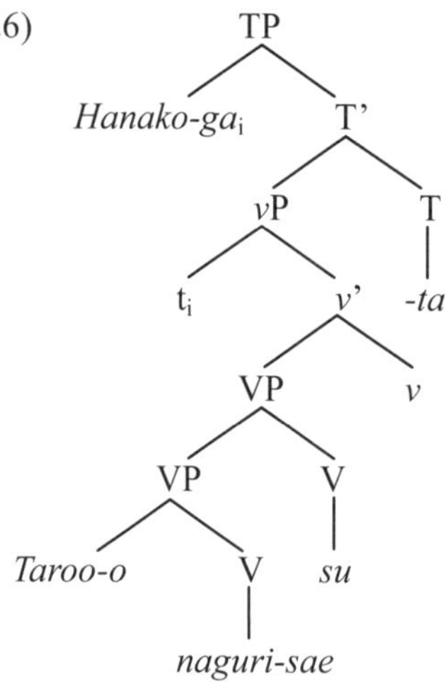

This is disallowed according to Hoshi's analysis. The verb *nagur* 'hit' discharges its theme role at its base position and has no reason, therefore, to move to the position of *su*. The expletive verb remains at LF and the structure is ruled out by Full Interpretation.[8]

It is then necessary to place an internal argument within the projection of the expletive verb *su*, exactly as in the case of the light verb construction. The structure is shown in (27).

8 Note that (26) is identical to (22), the structure of (21d) with the main verb interpretation of *su*. If *su* is the main verb, it satisfies Full Interpretation and need not be replaced. Hence, there is nothing wrong with the structure in this case.

(27)

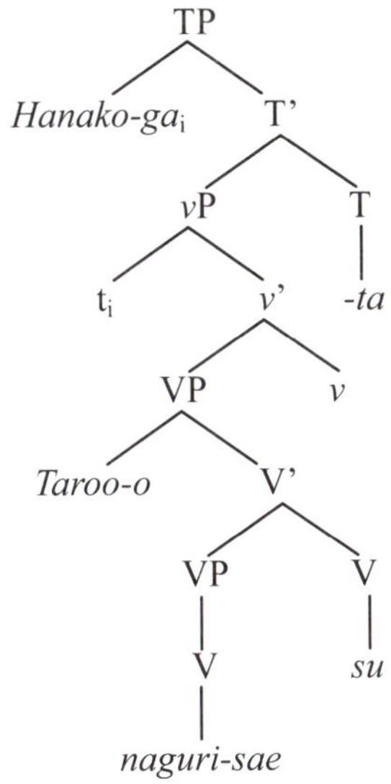

Here, the verb *nagur* 'hit' must and hence, can move to the position of *su* in order to discharge its theme role, and as a result, the expletive verb is successfully replaced. (21d), thus, has the structure in (22) with the main verb *su* and that in (27) with the expletive *su*. The appropriate structure depends on which *su* is included in the numeration.

The prediction that VP-scrambling is illicit with the expletive *su* remains the same. In the case of (27), *Taroo-o* and *naguri-sae* 'hit-even' do not form a constituent. But since multiple scrambling is allowed in Japanese, the object NP and the VP dominating *naguri-sae* could scramble to the sentence-initial position separately. This is ruled out because the verb would have to lower to the position of *su* in LF in order to discharge its theme role. Consequently, the object fails to receive a θ-role and the expletive verb fails to be replaced. Thus, VP-scrambling in this configuration is ruled out in exactly the same way as the scrambling of the θ-role assigning noun in the light verb construction. A parallel example of the light verb construction, (12b), is repeated below as (28).

(28)	**Zyooto-o*$_i$	*Hanako-ga*	*Taroo-ni*	*toti-o*	t$_i$	*sita.*
	giving -ACC	Hanako-NOM	Taroo-to	land-ACC		did

'Hanako gave a piece of land to Taroo.'

3.2 VP-scrambling in unaccusative and passive sentences

Clear evidence for Hoshi's analysis discussed above can be found in Hoji, Miyagawa and Tada 1989 (henceforth, Hoji, et al.). They discuss illicit cases of VP-scrambling such as those in (29)-(31) and propose to rule them out by the proper binding condition.

(29) a. *Hanako-ga Taroo-o naguri-sae sita.*
Hanako-NOM Taroo-ACC hit-even did
'Hanako even hit Taroo.'
b. **Naguri-sae*$_i$ *Hanako-ga Taroo-o* t_i *sita.*
hit-even Hanako-NOM Taroo-ACC did

(30) a. *Ame-ga huri-sae sita.*
rain-NOM fall-even did
'It even rained.'
b. **Huri-sae*$_i$ *ame-ga* t_i *sita.*
fall-even rain-NOM did

(31) a. *Hanabi-ga utiage-rare-sae sita.*
firework-NOM set.off-PASS-even did
'They even set off fireworks.'
b. **Utiage-rare-sae*$_i$ *hanabi-ga* t_i *sita.*
set.off-PASS-even firework-NOM did

It looks like the verbs are preposed by themselves in the ungrammatical examples in (29)-(31). But the examples can be derived by VP-scrambling, as Hoji, et al. point out. A possible derivation of (29b) is shown in (32).

(32) [$_{VP}$ t_j *naguri-sae*] [*Hanako-ga* [*Taroo-o*$_j$ t_{VP} *sita*]]
hit-even Hanako-NOM Taroo-ACC did

The object *Taroo-o* is first scrambled out of the VP headed by *naguri-sae*, and then, the remnant VP is scrambled to the sentence-initial position. Similarly, (30b) can be derived as in (33).

(33) [$_{VP}$ t_j *huri-sae*] [*ame-ga*$_j$ [t_{VP} *sita*]]
fall-even rain-NOM did

Since *hur* 'fall' is unaccusative, Hoji, et al. assume that the subject *ame* 'rain' originates in its complement position and moves to the specifier position of TP. After this NP-movement applies, the VP headed by *huri-sae* is scrambled over the subject.

Although the ungrammatical examples in (29)-(31) can be derived by VP-scrambling, the preposed VPs contain unbound traces as can be seen in (32) and (33). The VP in (32) contains a trace of scrambling and that in (33) a trace of NP-movement. Hoji, et al. then argue that the examples violate the proper binding condition, which requires that traces be bound. This account may be tenable in the case of (29) because it has been argued that traces of scrambling exhibit strict proper binding effect.[9] On the other hand, a question can be raised for this account of (30) and (31) since there is evidence that traces of NP-movement are not subject to the proper binding condition. If they were, examples of VP-preposing such as those in (34) would be incorrectly excluded.

9 For relevant discussion on this point, see Saito (2003) and the references cited there.

(34) a. *They said the ball might fall into a ditch, and fall into a ditch, it did.*
b. *Mary said she would be praised by the critics, and praised by the critics, she was.*

The structures of the second conjuncts of (34a-b) are shown in (35a-b) respectively.

(35) a. $[_{VP}$ *fall* t_i *into a ditch*] [*it*$_i$ *did* t_{VP}]
b. $[_{VP}$ *praised* t_i *by the critics*] [*she*$_i$ *was* t_{VP}]

Since *fall* is unaccusative, *it* in (35a) originates in the complement position of the verb and moves to TP Spec. Then, the preposed VP contains the trace of this NP-movement. A similar situation obtains in (35b), where a passive VP is preposed.[10]

It is clear at this point that an alternative analysis is required for (30b) and (31b). And Hoshi's analysis of the VP focus construction readily serves this purpose. Note first that the examples in (30) and (31) lack an agentive subject. Since the main verb *su* requires an agent, the *su*'s in these examples cannot be main verbs but must be instances of the expletive *su*. Then, the structure of (30a), for example, is as in (36).

(36)

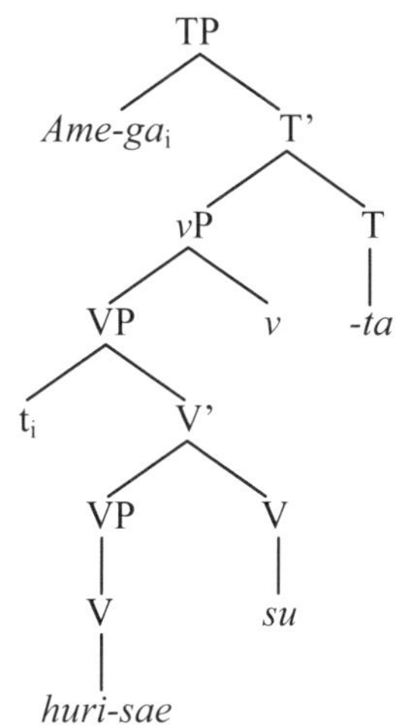

Here, it is crucial that *ame* 'rain' is merged at the spec position of the expletive *su*. Then, *hur* 'fall' must and therefore, can move to the position of *su* in order to discharge its θ-role. As a result, the expletive verb is successfully replaced and there is no violation of Full Interpretation.

If (36) is the structure for (30a), the ungrammaticality of (30b) follows. When the VP headed by *huri-sae* is scrambled, the verb can no longer move to the position of *su* in LF because this would require lowering. Consequently, *ame* 'rain' fails to receive a θ-role and the expletive *su* remains in violation of Full Interpretation. Note that this ac-

10 NP-traces and *wh*-traces exhibit different patterns in this respect. Thus, Maggie Browning (personal communication, 1986) points out that (ib) is hopeless and contrasts sharply with (ia).

(i) a. ?… *ready to marry John, I wonder whether Mary is.*
b. *… *ready to marry* t_i, *I wonder who*$_i$ *Mary is.*

count for (30b) is possible because of Full Interpretation. If an LF representation could contain an expletive verb, the structure in (37) would be possible for (30a).

(37)

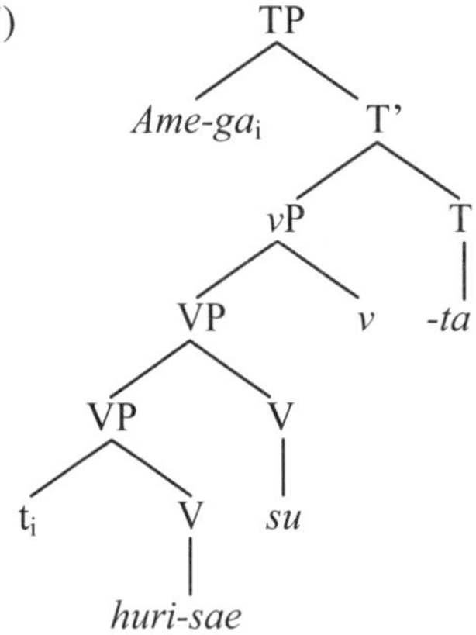

The verb *hur* assigns its θ-role at the base position and hence, does not replace the expletive. But this is, in relevant respects, the structure assumed in Hoji, et al., and as we have seen, it predicts incorrectly that VP-scrambling is possible. Thus, Full Interpretation plays a crucial role in the explanation of (30b), in addition to the Last Resort Principle.

4. Summary and another consequence

In Section 2, I discussed the analysis of the Japanese light verb construction proposed in Saito and Hoshi 2000, and examined the distribution of the expletive verb *su* in this construction. One of the conclusions is that Grimshaw and Mester's (1988) constraint in (15), repeated below in (38), is to be explained not only by the Last Resort Principle but also by Full Interpretation.

(38) At least one internal argument of the θ-role assigning noun must be realized outside the NP it projects.

In Section 3, I updated Hoshi's (1995) analysis of the distribution of the expletive verb in the VP focus construction and presented further evidence for it. The theoretical consequence remains the same: both the Last Resort Principle and Full Interpretation play crucial roles in the analysis. Thus, the discussion in this paper provides strong support for the expletive replacement analysis proposed in Chomsky 1986. As stated at the outset of this paper, the purpose here is not to defend the expletive replacement analysis of the English existential construction against the recent, more refined analysis, say, in Chomsky 1995. It is rather to suggest that there is an insight in the analysis that needs to be reconsidered and reexamined in future research. I also hope that the discussion in this paper will stimulate further research on Japanese expletive verbs to advance the theory.

Before I conclude this paper, I would like to briefly point out a consequence of the proposed analysis for the derivational model of syntax, put forward in Bobaljik 1995 and Nissenbaum 2000, among many others. Based on the idea of cyclic interpretation and cyclic Spell-out, it is proposed in these works that covert movement can be interwoven with overt movement. For example, Bobaljik suggests that covert movement applies basically in the same way as overt movement except that the phonetic features are interpreted at the initial site instead of the landing site. If this conception of covert movement is adopted, then the analysis presented above for (30b), for example, must be slightly modified.

Let us consider (36), i.e., the structure of (30a), again. It was assumed above that overt movement precedes covert movement, and hence, VP-scrambling blocks the LF expletive replacement by the verb *hur* 'fall'. But if covert movement can precede overt movement, another derivation must be considered. That is, the verb can first move to the position of the expletive *su*, leaving behind its phonetic features. Then, the VP can be scrambled to the sentence-initial position. There is no obvious violation of the Last Resort Principle or Full Interpretation with this derivation.

I would like to suggest that the derivation just described is excluded by the proper binding condition.[11] That is, the preposed VP contains the trace left by the covert movement of *hur* 'fall' and this trace violates the condition. This account is consistent with, and hence, provides support for the recent discussion in defense of the proper binding condition in Kuno 2001 and Saito 2003. As noted above, there is evidence that NP-traces are not subject to this condition. At the same time, it is suggested in Lasnik 1999, and Saito and Hoshi 2000, among others, that NP-movement does not leave a trace. The basic idea is that if θ-role assignment can take place in the course of the derivation, then there is no feature of the NP that must be represented at the initial site. This can probably be best illustrated with (36). The verb *hur* 'fall' can first move covertly to the position of the expletive *su*, and assign the theme role to *ame* 'rain' in the Spec position. Then, *ame*, having already been assigned a θ-role, can move to TP Spec overtly. There is evidently no need to postulate a trace in the initial position. And if there are no NP-traces, they could not be subject to the proper binding condition.

On the other hand, *wh*-traces exhibit the proper binding effect as shown in (39).

(39) *[*Which picture of* t_i]$_j$ *does John wonder who*$_i$ *Mary likes* t_j *?*

In this example, *who* is extracted out of *which picture of whom* to the embedded CP Spec and then, the remnant [*which picture of* t] moves to the matrix CP Spec. The latter movement violates Subjacency, but the example is much worse than a normal *wh*-island violation, which suggests that the unbound trace is the main source of ungrammaticality. This is expected because a *wh*-phrase needs to be represented both at the initial site and at the landing site. Its phonetic features and its operator feature are interpreted at the landing site. And the initial site must have those features that make it possible to inter-

---

11 The condition is reformulated as a constraint on the application of Merge in Ausin 1998 and Saito 2003. The proposal in the latter work is that Merge applies only to 'complete constituents', which is defined as in (i).

(i) α is a *complete constituent* $=_{def}$ (1) α is a term, and
(2) if a position within α is a member of a chain γ, then every position of γ is contained within α.

But I will keep referring to the proper binding condition as a condition on traces for ease of exposition.

pret the *wh*-phrase (or its trace) as a variable. Thus, a trace is necessary in order for *wh*-movement to create an operator-variable chain with a variable at its tail.

By the same logic, the covert movement of *hur* 'fall' in (36) must form a chain. The verb must be represented at the landing site so that it can replace the expletive *su*. And it must also be at the initial site because that is where its phonetic features are interpreted. Hence, the movement must leave a copy (or a trace) behind, and we would expect the movement to be constrained by the proper binding condition. Note that the present analysis does not predict that head movement always exhibits the proper binding effect. It does when the moved head must be represented both at the initial site and at the landing site. If this approach is on the right track, the expletive verb replacement offers important data for the investigation of the proper binding effect as well.

References

Ausin, A. 1998. A minimalist approach to the remnant movement constraint. Unpublished manuscript, University of Connecticut.

Bobaljik, J. 1995. *Morphosyntax: The Syntax of Verbal Inflection*. Ph.D. dissertation, MIT.

Chomsky, N. 1986. *Knowledge of Language: Its Nature, Origin, and Use*. New York: Praeger.

Chomsky, N. 1995 *The Minimalist Program*. Cambridge, Mass.: MIT Press.

Grimshaw, J. and A. Mester. 1988. Light verbs and θ-marking. *Linguistic Inquiry* 19, 205–232.

Harada, S.-I. 1973. Counter equi NP deletion. *Annual Bulletin* 7, The Research Institute of Logopedics and Phoniatrics, University of Tokyo, 113–147.

Hoji, H., S. Miyagawa, and H. Tada. 1989. NP-movement in Japanese. Paper presented at WCCFL 8, University of British Columbia.

Hoshi, H. 1995 *Passive, Causative, and Light Verbs: A Study on Theta Role Assignment*. Ph.D. dissertation, University of Connecticut.

Kuno, M. 2001. Why does movement sometimes leave a trace and sometimes not? Unpublished manuscript, University of Tokyo.

Kuroda, S.-Y. 1965 *Generative Grammatical Studies in the Japanese Language*. Ph.D. dissertation, MIT.

Lasnik, H. 1999. Chains of arguments. In: S. D. Epstein and N. Hornstein (eds.), *Working Minimalism*, 189–215. Cambridge, Mass.: MIT Press.

Miyagawa, S. 1989. Light verbs and the ergative hypothesis. *Linguistic Inquiry* 20, 659–668.

Nemoto, N. 1993. *Chains and Case Positions: A Study from Scrambling in Japanese*. Ph.D. dissertation, University of Connecticut.

Nissenbaum, J. 2000. *Investigations of Covert Phrasal Movement*. Ph.D. dissertation, MIT.

Saito, M. 1989. Scrambling as semantically vacuous A'-movement. In: M. Baltin and A. Kroch (eds.), *Alternative Conceptions of Phrase Structure*, 182–200. Chicago: University of Chicago Press.

Saito, M. 2001. Movement and θ-roles: A case study with resultatives. In: Y. Otsu (ed.), *The Proceedings of the Second Tokyo Conference on Psycholinguistics*, 35–60. Tokyo: Hituzi Syobo.

Saito, M. 2003. A derivational approach to the interpretation of scrambling chains. *Lingua* 113, 481–518.

Saito, M. and H. Hoshi. 2000. The Japanese light verb construction and the minimalist program. In: R. Martin, D. Michaels, and J. Uriagereka (eds.), *Step by Step: Essays on Minimalist Syntax in Honor of Howard Lasnik*, 261–295. Cambridge, Mass.: MIT Press.

Sells, P. 1988. More on light verbs and theta-marking. Unpublished manuscript, Stanford University.

Shibatani, M. 1973. Semantics of Japanese causativization. *Foundations of Language* 9, 327–373.

Tada, H. 1993. *A/A-bar Partition in Derivation*. Ph.D. dissertation, MIT.

Terada, M. 1990. *Incorporation and Argument Structure in Japanese*. Ph.D. dissertation, University of Massachusetts, Amherst.

Tsujimura, N. 1990. Ergativity of nouns and case assignment. *Linguistic Inquiry* 21, 277–287.

Agreement and the EPP in Kinyarwanda Applicatives*

Jochen Zeller

1. Introduction

In Government-Binding (GB)-theory and early versions of the Minimalist Program (MP) up to and including Chomsky (1995), it was assumed that agreement relations are established in Spec-Head configurations. For an argument DP to agree with (features of) a head F, the DP hence needed to *move* to [Spec, F], sometimes covertly. This view has changed in Chomsky (2000), however, and it has since been assumed that F can agree with a DP as long as F c-commands the DP and no locality conditions are violated. Movement into specifier positions is hence no longer required for agreement. In Chomsky (2000, 2001), the operation Move is solely triggered by *EPP-features*, which are associated with the functional heads C, T and v.

In his excellent introduction to, and critical discussion of, the MP, Grewendorf (2002) highlights the stipulative character of an account of syntactic movement in terms of the EPP. Formally, the trigger for movement is the EPP-feature: if a syntactic head F is equipped with such a feature (and if the numeration does not include an expletive which could check this feature by being merged into [Spec, F]), then movement has to take place. However, this still leaves open the question of why EPP-features and the associated movement steps are necessary in the first place.

In this paper I propose a possible answer to this question which is based on the idea that not only the functional heads C, T and v, but also the lexical head V, can be equipped with an EPP-feature. The specifier of VP is usually filled with the verb's internal argument, but I suggest that an EPP-feature associated with V may trigger movement of a DP which is not selected by the verb to a second specifier on top of the first. My major claim is that this movement is required to establish agreement between the moved DP and the functional head v, which agrees with the closest DP in its c-command domain. This means that in order for v to agree with a DP located further down in the VP than the verb's argument, this DP must move, but the movement step itself is not triggered by a feature associated with v, but by an EPP-feature associated with the verb.

I substantiate my proposal empirically through a discussion of locative applicative constructions in Kinyarwanda, a Bantu language spoken in Rwanda and its neighbouring countries (Kimenyi 1980, Ngoboka 2005). In section 2, I introduce Kinyarwanda

* I thank Ben Murrell and Dori Posel for their help with this article and Eric Fuß and Katharina Hartmann for their valuable comments and suggestions. A special thanks to Jean Paul Ngoboka for providing the data, including those taken from Ngoboka (2005).

locative applicatives and argue that they are derived by preposition incorporation. Section 3 suggests that the word order in this type of applicative is derived by DP-movement of the applied object to a second [Spec, V], and in section 4 I argue that this movement step is triggered by the EPP-feature of V. Sections 5 and 6 offer a thorough discussion of object asymmetries in Kinyarwanda locative applicatives, which provides the empirical evidence for the theory developed in sections 3 and 4. Section 7 concludes the paper.

2. Locative applicatives and preposition incorporation

Applicative constructions are formed by adding a special morpheme (the applicative marker) to the verb; as a result, a new object DP is added. In Kinyarwanda applicatives, this newly added *applied object* may bear all kinds of thematic roles, depending on the choice of applicative marker and the context (see Kimenyi 1980, Ngoboka 2005). In the applicative in (1b), for example, the applied object is the *instrument*: [1]

(1) a. *Umugabo y-a-tém-ye igití n'úmuhoro.*
man SP-PST-cut-ASP tree with.machete
'The man cut the tree with the machete.'

b. *Umugabo y-a-tém-eesh-eje igití umuhoro.*
man SP-PST-cut-APPL-ASP tree machete
'The man cut the tree with the machete.' (Ngoboka 2005: 109)

In (1a), the verb is transitive and only selects the theme argument *igití*, 'tree', as a DP object; the instrument *umuhoro*, 'machete', is introduced by the preposition *na*, 'with'. In contrast, the verb in (1b) is extended by means of the applicative marker *-eesh-*, and the resulting applicative is a double object construction in which both the theme and the instrument are realised as object DPs.

In this paper, I am concerned with the syntactic properties of *locative applicatives* in Kinyarwanda. Examples are given in (2b) and (3b):

(2) a. *Umufuundi y-o-óme-tse amatáfaári ku rukutá.*
builder SP-PST-stick-ASP bricks on wall
'The builder stuck bricks on the wall.'

b. *Umufuundiy-o-óme-tse-hó urukutá amatáfaári.*
builder SP-PST-stick-ASP-APPL wall bricks
'The builder stuck bricks on the wall.' (Ngoboka 2005: 46)

(3) a. *Umwáana y-a-ménn-ye amáazi mu mwoobo.*
child SP-PST-pour-ASP water in hole
'The child poured water into a hole.'

---

1 Morphemes are glossed as follows: APPL = applicative; ASP = aspect; FV = final vowel; OC = object clitic; PASS = passive; PST = past tense; SP = subject prefix.

b. ? *Umwáana y-a-ménn-ye-<u>mó</u> umwoobo amáazi.*
child SP-PST-pour-ASP-APPL hole water
'The child poured water into a hole.' (Ngoboka 2005: 46)

In the (a)-examples, the goal of the event is realised by a PP headed by a preposition (*ku* and *mu* respectively). In the corresponding applicatives, the applicative markers *-hó*/*-mó* are suffixed to the verb. As a result, the goal argument has become the applied object DP of the derived double object construction.

Applicative constructions in Bantu have received considerable attention in the literature. Researchers such as Bresnan & Moshi (1990), Alsina & Mchombo (1993) or Harford (1993) assume that applicative formation is a morpho-lexical process. According to this view, the applicative morpheme is added in the lexical component of grammar and modifies the argument structure by adding an additional object. In contrast, proponents of a syntactic approach associate the applicative morpheme with an independent syntactic head, although opinions differ concerning the position of the applicative head in the structure. Two prominent syntactic accounts are illustrated in (4):

(4)
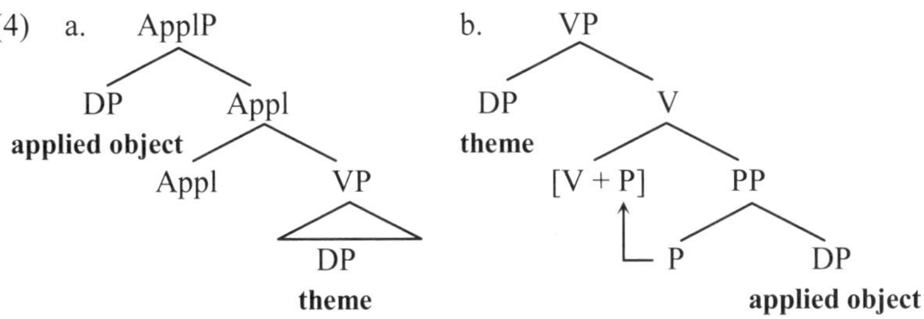

(4a) illustrates a proposal found in Marantz (1993), Pylkkänen (2000), McGinnis (2001) and Anagnostopoulou (2003), among others. Here, the applicative marker (labeled Appl in (4a)) selects the VP as its complement and introduces the applied object in its specifier. The alternative in (4b) shows Baker's (1988, 1992, 1997) *preposition incorporation* analysis of applicatives (also adopted in Nakamura 1997). In Baker's theory, the applicative marker is an affixal preposition and the head of a PP-complement of the verb; due to its affixhood, it undergoes head movement and incorporates into the verb. The applied object is merged as the argument of the applicative marker inside the PP.

I argue in this paper that Kinyarwanda locative applicatives are derived syntactically by preposition incorporation, as shown in (4b). Note that the following examples provide strong evidence against a lexical treatment of locatives:

(5) *Umufuundi y-o-óme-tse urukutá hó amatáfaári.*
builder SP-PST-stick-ASP wall APPL bricks
'The builder stuck bricks on the wall.'

(6) *Umwáana y-a-ménn-ye umwoobo mó amáazi.*
child SP-PST-pour-ASP hole APPL water
'The child poured water into a hole.' (Ngoboka 2005: 46)

(5) and (6) differ from the applicative constructions in (2b) and (3b) above in one small, but important, respect: in contrast to (2b) and (3b), the applicative marker in the examples (5) and (6) is *not attached to the verb*, but appears between the two objects. This situation cannot be explained by lexical theories, in which the applicative marker is treated as a bound morpheme which combines with the verb in the lexicon. However, the data in (5) and (6) are compatible with a syntactic analysis which treats the applicative marker as a syntactically independent, "mobile" element which cliticises either to the verb or to a noun.

Furthermore, I adopt Baker's (1988) analysis represented in (4b) for locative applicatives rather than the alternative syntactic analysis in (4a) for the following reasons. First, the applicative markers *-hó* and *-mó* and the prepositions *ku* and *mu* used in the corresponding dative constructions are phonologically similar – a fact that is consistent with the view that the former elements are clitic-like or affixal variants of the latter. Second, and more importantly, the claim that the locative applicative marker is a syntactic head which introduces the applied object in its specifier does not explain why there are locative applicative constructions *without an applied object*:

(7) *Umufuundi y-o-óme-tse-hó amatáfaári.*
builder SP-PST-stick-ASP-APPL bricks
'The builder stuck bricks there.'

(8) *Umwáana y-a-ménn-ye-mó amáazi.*
child SP-PST-pour-ASP-APPL water
'The child poured water into it/there.' (Ngoboka 2005: 47)

(7) and (8) show that the applied object can be omitted in locative applicatives; when it is absent, the applicative marker is interpreted as a prepositional proform (Kimenyi 1995). This finding does not follow straightforwardly from the analysis underlying the structure in (4a); it would only be compatible with this structure if it is assumed that the specifier associated with a locative Appl-head is optional (or alternatively, that [Spec, Appl] can optionally be filled with a phonetically null applied object). This assumption, however, is problematic, given that overt applied objects are obligatory with all other types of applicatives in Kinyarwanda. Compare the instrumental applicatives in (9) and (1b) above:

(9) **Umugabo y-a-tém-eesh-eje igití.*
man SP-PST-cut-APPL-ASP tree
intended meaning: 'The man cut the tree with something/it.'

If both locative and instrumental applicatives were represented by the structure in (4a), the contrast between (7)/(8) and (9) would require the stipulation that specifiers are optional with locative applicative markers, but obligatory with other Appl-heads. However, if one assumes (as I do) that locative applicatives are represented through a structure like (4b), while an analysis such as (4a) (or even a lexical approach) is adopted for other types of applicatives, then the difference between (7)/(8) and (9) can be derived from the different syntactic representations of these constructions. Whereas the obligatoriness of the applied object in applicatives such as (9) could be seen as a consequence of the obligatory presence of a specifier, the omissibility of the applied object in a locative applicative follows directly from the prepositional status of the applicative marker: complements of locative prepositions are often optional (compare English *He walked out the room* and *He walked out*); in (7) and (8), the prepositional applicative marker is simply used intransitively, and the PP is interpreted as a proform.

Another argument in favour of the analysis in (4b) will be presented in section 6.3, where I show that in certain instances of the locative applicative construction, the applied object is in a position which is c-commanded by the theme. Notice that the structure in (4a) does not include such a position, since here, the applied object is merged into a position from where it c-commands the theme inside the VP. In contrast, the base position of the applied object inside the PP in the structure in (4b) is below that of the theme, and data which show that the applied object can indeed remain in this position therefore provide additional support for the preposition incorporation-analysis. Before I discuss these data, however, I first address the word order facts in Kinyarwanda locatives, which suggest that in examples such as (2b) and (3b), the applied object DP moves out of the PP to a position above the theme.

3. Word order, locality and multiple specifiers

As (10) shows, the applied object obligatorily precedes the theme in locative applicatives:[2]

(10) a. *Umubooyi y-a-ménn-ye-hó umwáana amáazi.* (√ AO > T)
cook SP-PST-pour-ASP-APPL child water
'The cook poured the water on the child.'

b. * *Umubooyi y-a-ménn-ye-hó amáazi umwáana.* (* T > AO)
cook SP-PST-pour-ASP-APPL water child
'The cook poured the water on the child.'

The word order in (10a) does not follow directly from the structure in (4b) and therefore can only be derived by movement. This is in fact the proposal made in Baker (1997), who assumes that the applied object moves from inside the PP-complement of the verb to the specifier of a functional category Asp from where it asymmetrically c-commands the theme in [Spec, V] (see also Nakamura (1997)):

2 In contrast, the word order of other types of applicatives in Kinyarwanda is more flexible, see McGinnis (2001), Ngoboka (2005), Zeller & Ngoboka (2006).

(11)

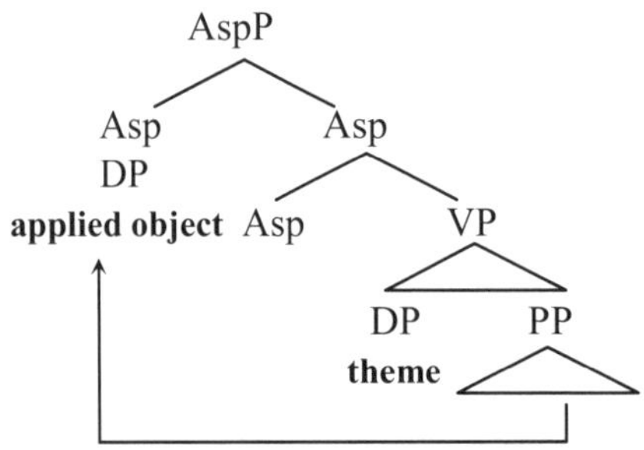

Movement of the applied object is triggerd by case requirements. Baker (1988) argues that the complement of an incorporated preposition can no longer receive oblique case (a proposal which I henceforth adopt). Therefore, the applied object must move to a position where it can receive (or check) structural case, and Baker (1997) takes this position to be [Spec, Asp].

However, this derivation poses a problem with locality. Notice that A-movement of the applied object in (11) crosses the theme DP, which is located in [Spec, V]. The movement step depicted in (11) therefore violates the *Minimal Link Condition* (Chomsky 1995):

(12) *The Minimal Link Condition* (MLC) (Chomsky 1995: 311)
K attracts α only if there is no β, β closer to K than α, such that K attracts β.

The locality concept of 'closeness' in (12) is defined in terms of c-command (see (14) below for a formal definition). Since the theme in [Spec, V] c-commands the applied object inside the PP, it counts as closer to the attracting head Asp than the applied object. Movement of the latter to [Spec, Asp] should therefore be blocked by the MLC.

We seem to be facing a dilemma now. The preposition incorporation-analysis implies that the applied object originates in a position below the theme, while the word order facts force us to assume that the applied object moves to a position from where it c-commands the theme – but any movement which brings the applied object into such a position would have to cross the theme and therefore seems to be ruled out by the MLC.

However, the dilemma is only apparent. As it turns out, there is a movement operation which allows the applied object to cross the theme without violating the MLC: the applied object can move to a *second specifier* of the VP:

(13)

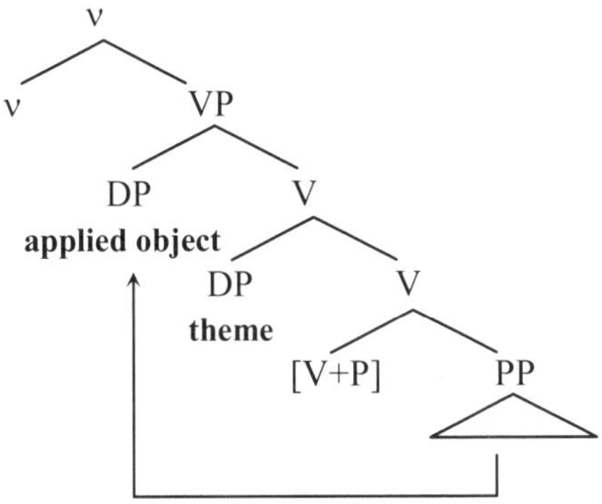

The movement operation in (13) yields the correct surface order applied object > theme. Importantly, it does not violate the MLC: the legitimacy of (13) follows from the characterisation of closeness which is part of Chomsky's (2000: 122) definition of locality:

(14) *Locality*: D(P) is the c-command domain of P, and a matching feature G is closest to P if there is no G' **in D(P)** matching P such that G is in D(G').

In Chomsky (2000, 2001, 2005), the set of uninterpretable φ-features of a functional head is called the *probe* P. This probe has to find a matching *goal* G in its c-command domain (as stated in (14)). Possible goals are the interpretable φ-features of DPs. DPs also have uninterpretable structural case (CASE) features, whose function it is to activate the goal. For a derivation to converge at the interfaces, uninterpretable features must be deleted under agreement. Once a probe has found a matching goal, agreement between F and the DP is established, and the uninterpretable features of F and DP (= F's φ-features and CASE of DP) are erased.

The crucial part of (14) (which I have highlighted in bold) is that for a goal G' to be closer to a probe P than G, G' must be in the c-command domain D(P) of the probe. If we apply this definition of closeness to the movement constraint in (12), then it implies that the theme in (13) does not block movement of the applied object: for β to be closer to K than α, β must be in the c-command domain of K.[3] But heads do not c-command their specifiers. Therefore, if the verb in Kinyarwanda locatives has a feature which attracts the closest DP with a matching feature, it will attract the applied object, since the theme in [Spec, V] is not c-commanded by the verb. Consequently, the applied object can move across the theme to a second [Spec, V], as shown in (13). In contrast, the attracting head in (11) is Asp, which c-commands the theme in [Spec, V]. Therefore, the theme is closer to Asp than the applied object, and movement of the latter across the former would be banned by the MLC.

In Chomsky (2000, 2001), it is argued that DP-movement to specifier positions is triggered exclusively by EPP-features. The derivation in (13) can hence be captured

3 For the sake of simplicity, I treat K in (12) as the attracting head which contains the probe. In Chomsky (1995), K is actually the root node, i.e. the highest projection of the head which contains the attracting feature. Strictly speaking, then, one would have to say that for β to be closer to K than α, β must be in the c-command domain of *the head* (the label) of K. I ignore this complication here.

formally by assuming that the verb in Kinyarwanda locatives may be equipped with an EPP-feature which triggers movement of the applied object to a second specifier of VP. In this respect, V in Kinyarwanda is analogous to v in languages with object shift, where the EPP-feature of v triggers movement of an object DP to a higher [Spec, v] above the selected external argument in the lower [Spec, v] . Similarly, V in Kinyarwanda selects the theme argument in [Spec, V], and if V has an EPP-feature, then the applied object DP moves across the theme and creates a second specifier.[4]

However, as Grewendorf's (2002) remarks remind us, although syntactic movement can be technically implemented through the stipulation of EPP-features, this does not yet explain why movement takes place at all. In the next section, I suggest that, regarding the movement of the applied object in locative applicatives, such an explanation can be found in the way object agreement is established in Kinyarwanda.

4. The EPP and Agree

As noted in the introduction, earlier versions of the MP assume that feature checking under agreement requires a Spec-Head relation between checker and checkee. However, in Chomsky (2000), it is assumed that Spec-Head configurations are no longer required for agreement. Rather, agreement between a probe and a goal is now established by the operation *Agree*, which merely requires the goal to be in the c-command domain of the probe (subject to the locality definition in (14) above). This view of agreement allows to explain agreement between T and nominative subjects in expletive constructions without having to assume covert movement:

(15) *There are several people in the room.*

In (15), the EPP-feature of T is checked by the expletive *there*, which is merged in [Spec, T]. The uninterpretable φ-features of T seek and locate the interpretable φ-features of the subject DP *several people,* which is located inside the VP. Agree can now be established between T and the subject, and the φ-features of T (the probe) and the CASE-feature of the subject DP can be erased in this configuration – covert movement of the subject to [Spec, T] is hence no longer required for feature checking and deletion.

---

4 The idea that V can have an EPP-feature contradicts Chomsky's (2000) assumption that EPP-features are only associated with the so-called "core functional categories" C, T and v. However, I am not aware of any empirical arguments that rule out the possibility of associating EPP-features with lexical categories, particularly given that v, although classified as a functional category, has the selectional properties of a lexical head. Furthermore, as discussed in Radford (2004), the idea of V bearing an EPP-feature may also explain data like the following (see Radford 2004: 359ff.):

(i) *The DA will prove the witness conclusively to have lied.*

(i) is an ECM-construction; the position of the subject of the infinitive in front of the adverb shows that the DP has moved to the matrix clause to check objective case. Since English has no object shift, the DP in (i) must have moved to the highest [Spec, V] of the matrix clause (assuming that the adverb is located in a lower specifier of V), see also Chomsky (2005). Under current assumptions, this movement step must have been triggered by an EPP-feature associated with the matrix verb. (See section 7 for some discussion of the idea that the verb *inherits* the EPP-feature from the functional head v.)

The operation Agree also accounts for object agreement in English-type languages, where the object DP does not undergo object shift but remains in [Spec, V]:

(16)

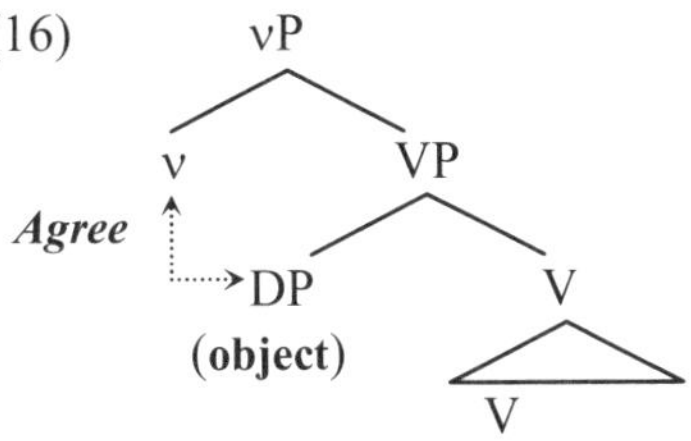

In Chomsky (1995, 2000), it is suggested that the CASE- and φ-features of an object DP are checked against the φ-features of the functional head v under object agreement. This can be achieved on the basis of the structure in (16), where the φ-features of v locate the matching φ-features of the DP in [Spec, V], and the uninterpretable features of v and DP are erased under Agree.

I now propose that object agreement is also the reason for why the applied object moves to a second [Spec, V] in Kinyarwanda locative applicatives. I argue that the applied object moves *in order to reach the Agree-domain of v without violating locality.* Recall that the applied object cannot receive oblique case *in situ* from the incorporated preposition and can only be licensed by structural case. An applied object DP in Kinyarwanda locatives is therefore obligatorily equipped with a CASE-feature, which has to be erased under Agree. But Agree is subject to the locality condition in (14), and v cannot agree with the applied object if the latter remains inside the PP, because the φ-features of the theme DP in [Spec, V] are closer to the probe of v. However, the applied object can agree with v once it is moved to a position above the theme. As was shown in section 3, the only legitimate operation that achieves this result is movement of the applied object to a higher [Spec, V].

This analysis has an interesting consequence. With Baker (1997), it assumes that DP-movement is required in order to establish an agreement relation between the moved DP and a functional head which checks the DP's CASE-feature. Crucially, however, and in contrast to what has been argued by Baker (1997), the resulting agreement relation is *not* established in a Spec-Head configuration between the DP and the attracting head, but via Agree between the DP and a *higher* functional head which c-commands it.

This claim therefore provides one possible answer to the question raised by Grewendorf's (2002) critical comments about the motivation for syntactic movement in the current Minimalist framework. Syntactic movement may still be required in order to establish the right syntactic configuration for feature checking and agreement, but this configuration does not have to be a Spec-Head relation between the attracting feature and the moved constituent. Rather, a head may attract a DP to its specifier in order to allow this DP to check its features against the features of a different head, which is higher in the structure. For such "altruistic attraction" to be possible, the attracting head must itself have a feature which causes the DP to move to its specifier – which is exactly the task performed by an EPP-feature.

5. Object asymmetries in locative applicative constructions

In this and the following section I present a systematic discussion of object asymmetries in Kinyarwanda locative applicatives. The data provide empirical evidence for the theoretical analysis presented in the preceding sections.

Before I turn to the data, let me introduce two additional assumptions which have been made in the literature and which I adopt here. The first concerns object markers in Kinyarwanda, which I take to be incorporated object pronouns rather than agreement markers.[5] Following Baker (1988), I assume that incorporation of its head is one way in which a nominal expression can escape the Case Filter. I therefore assume that if the applied object is a pronominal D which incorporates into the verb by adjoining to V, it neither needs oblique nor structural case. The theme can incorporate into the verb as well, by adjoining to v (to which the verb has moved). Since v and D have φ-features, I assume that if the theme adjoins to v, v's uninterpretable φ-features are checked and deleted.

My second assumption also has to do with case. Since the locative applicative construction is syntactically derived from a construction with only one object DP and a PP (see (4b) above), I assume that, apart from the subject's CASE-feature, only one more CASE-feature is part of the initial numeration of a locative. However, locative applicatives are double object constructions, so if only one object DP can have CASE, the question is how the second DP can be licensed. One possibility is incorporation, as discussed above. Based on Larson's (1988) analysis of double object constructions in English, I assume that, in addition to incorporation, a DP in a locative applicative may also be licensed by getting *inherent case* from the verb. However, this requires that there is a thematic relation between the verb and the DP; therefore, this option is not available for the applied object, which is the argument of the prepositional applicative marker. The only DP which can be licensed by inherent case in locative applicatives is therefore the theme.[6]

Let me now turn to the object asymmetries in Kinyarwanda locative applicatives. In general, it holds that the applied object can adopt what Bresnan & Moshi (1990) call "primary object properties" in the presence of a theme DP, but not vice versa. To put it differently, the two object DPs in Kinyarwanda locatives behave asymmetrically with respect to the following movement operations:

- A-movement in passives
- Incorporation (cliticisation) and Left Dislocation
- A-bar movement in relative clauses

I discuss the relevant data in the following subsections.

---

5 Object markers in Kinyarwanda cannot co-occur with postverbal object DPs, which renders an analysis as agreement markers implausible. See Zeller & Ngoboka (2006) for the relevant data.

6 The fact that the theme can get inherent case from the verb may be a result of P-incorporation: the incorporated preposition transfers its case-assigning properties to the host after incorporation.

5.1 Passivisation

(17) shows that, whereas the applied object can be passivised, the theme cannot:

(17) a. *Umwáana y-a-ménn-w-e-hó amáazi n'úmubooyi.*
child SP-PST-pour-PASS-ASP-APPL water by cook
Lit.: 'The child was poured water on by the cook.'

b. **Amáazi y-a-ménn-w-e-hó umwáana n'úmubooyi.*
water SP-PST-pour-PASS-ASP-APPL child by cook
'The water was poured on the child by the cook.'

The contrast in (17) follows from the analysis outlined in sections 3 and 4. In (17a), the applied object has a CASE-feature, and V has an EPP-feature. The applied object moves to [Spec, V] from where it agrees with v. Assuming that in a passive, v is defective, it follows that the applied object remains active, and can now enter Agree with the uninterpretable φ-featurs of T.[7] Once Agree has erased the probe of T and the CASE-feature of the applied object, the latter can move from the higher [Spec, V] to [Spec, T] to check T's EPP-feature, as in (17a). Alternatively, an expletive *pro* can be inserted into [Spec, T], which checks T's EPP-feature, in which case the applied object can remain in [Spec, V], as shown in (18):

(18) ?*H-aá-menn-w-e-hó umwáana amáazi n'úmubooyi.*
EXPL-PST-pour-PASS-ASP-APPL child water by cook
Lit.: 'It was the child poured water on by the cook.'

Theme passivisation in (17b) is blocked for two reasons. First, due to the fact that the CASE-feature is associated with the applied object, the theme must have inherent case and therefore cannot become the subject of a passive (see Baker 1988; Nakamura 1997). Second, theme passivisation is also blocked by locality (see Ura 1996, McGinnis 2001, Anagnostopoulou 2003): the applied object in the higher [Spec, V] c-commands the theme in the lower [Spec, V] and is hence closer to T (note that I do not assume that two specifiers of the same head are equidistant). Therefore, Agree between T and the theme could not be established even if the theme had a CASE-feature.

Theme passivisation is therefore blocked in (17b) because of the presence of an applied object. This account is confirmed by the possibility of theme passivisation in (19):

(19) *Amáazi y-a-ménn-w-e-mó n'úmubooyi.*
water SP-PST-pour-PASS-ASP-APPL by cook
'The water was poured there by the cook.'

(19) is an applicative in which the applicative marker is used as a prepositional proform (see section 2). Since there is no applied object, the theme is the closest DP to T and can be associated with a CASE-feature. Consequently, theme passivisation is possible.

7 If a functional head is *defective*, the set of its φ-features is incomplete, and the CASE-feature of the DP with which it agrees is not deleted under Agree. The DP therefore remains active and can enter the Agree-operation with another head. This allows for successive cyclic A-movement in Raising and passive constructions, where the same DP checks and deletes the sets of φ-features of a number of functional heads, all but the last one being defective (see Chomsky 2000).

5.2 Incorporation and Left Dislocation

In a locative applicative, only the applied object can be realised as an object marker; in the presence of a full applied object-DP, the theme cannot incorporate. Since left-dislocated topics (in parentheses in (20)) require pronouns in the associated sentence to which they can be anaphorically linked, left dislocation (LD) is also possible only with the applied object:

(20) a. (*Umwáana*) *umubooyi* *y-a-mú-menn-ye-hó* *amáazi.*
child cook SP-PST-OC-pour-ASP-APPL water
'(The child,) the cook poured water on him/her.'

b. *(*Amáazi*) *umubooyi* *y-a-yá-menn-ye-hó* *umwáana.*
water cook SP-PST-OC-pour-ASP-APPL child
'(The water,) the cook poured it on the child.'

In (20a), the applied object has incorporated into the verb as an object marker (in italics) and thereby fulfills the Case Filter; the CASE-feature is associated with the theme. Since there is no applied object DP in need of case-checking, the verb is not associated with an EPP-feature, so there is no movement to a second specifier of VP, and the theme can agree with v, (21). In (20b), however, the applied object has not incorporated into V and therefore must have a CASE-feature, which can only be checked if the applied object moves to [Spec, V] to agree with v. In this configuration, incorporation of the theme is blocked by the MLC, since the applied object in the higher [Spec, V] is now closer to the attracting head v, (22):[8]

(21) [$_{TP}$ *umubooyi* [v [$_{VP}$ *amáazi* [$_{V}$ *-mú-menn-ye-hó* [$_{PP}$ $t_{hó}$ $t_{mú}$]]]]]

Agree *incorporation*

(22) *[$_{TP}$ *umubooyi* [*y-a-yá-menn-ye-hó* [$_{VP}$ *umwáana* [$t_{yá}$ [$_{PP}$ $t_{hó}$ $t_{umwáana}$]]]]]

Agree

no incorporation

---

[8] Lack of space prevents me from discussing in sufficient detail the precise status of incorporation and head adjunction in the MP. Zeller & Ngoboka (2006), following a proposal by Anagnostopoulou (2003), argue that D-incorporation in Kinyarwanda is triggered by gender features of the attracting head and is subject to the MLC. (A gender feature is also associated with the applied object DP in (20b)/(22); it counts as a potential goal and therefore blocks incorporation of the theme.) In contrast, Chomsky (2000: 134) states that head adjunction is "not part of narrow syntax", has no selector and is optional. But, as (20) and other examples below show, head adjunction clearly is influenced by, and has consequences for, syntactic movement operations.

5.3 Relative clause extraction

Finally, (23) shows that only the applied object of a locative applicative can be extracted in a relative clause construction. Relativisation of the theme is not possible:

(23) a. *umwáana umubooyi y-a-∅-ménn-ye-hó amáazi.*
child cook SP-PST-D_{AO}-pour-ASP-APPL water
'the child on whom the cook poured the water'

b. **amáazi umubooyi y-a-∅-ménn-ye-hó umwáana.*
water cook SP-PST-D_{theme}-pour-ASP-APPL child
'the water which the cook poured on the child'

In order to explain the contrast in (23), I adopt the analysis of Kinyarwanda relative clauses put forward in Zeller & Ngoboka (2006), which is based on a proposal by Boeckx (2003). Boeckx (2003) argues that the argument position corresponding to the relativised constituent in a relative clause is occupied by a complex "Big DP" (cf. Uriagereka 1995; Cecchetto 1999, 2000; Belletti 1999) whose head selects the relative operator as its (NP- or DP-) complement. When the relative operator moves to [Spec, C], the D-head of the Big DP is stranded inside the relative clause and now functions as a resumptive pronoun. In Zeller & Ngoboka (2006), we suggest that the stranded head of the Big DP is phonetically null in Kinyarwanda object relative clauses. However, since this D-head acts as a resumptive pronoun, we argue that it must also incorporate, just like other (overt) object pronouns in Kinyarwanda. The relation between the relative operator and the incorporated zero pronoun in relative clauses is hence comparable to the relation between a left-dislocated topic and an incorporated pronoun in LD-constructions.[9]

This analysis implies that the contrast in (23) can be explained along the same lines as (20): In (23a), the zero pronoun which corresponds to the head of the applied object Big DP has incorporated (thus Big DP escapes the Case Filter), and the relative operator has moved to [Spec, C]. The theme checks its CASE-feature against v:

9 Although this parallel suggests that the Big DP-analysis can also be adopted for LD-constructions, a uniform treatment of relative clauses and LD does not explain the crucial difference between these two constructions in Kinyarwanda: while the incorporated pronoun in LD-constructions is overtly realised as an object marker (see (20)), the incorporated D-head in relative clauses in Kinyarwanda is phonetically zero. Since I cannot exclude the possibility that this difference is due to different structural representations of relative clauses and LD-constructions, I remain agnostic about the syntax of LD-constructions in Kinyarwanda in this article (but see Zeller (2005) for an analysis of LD in Zulu, a Southern Bantu language, in terms of the Big DP-proposal).

(24)

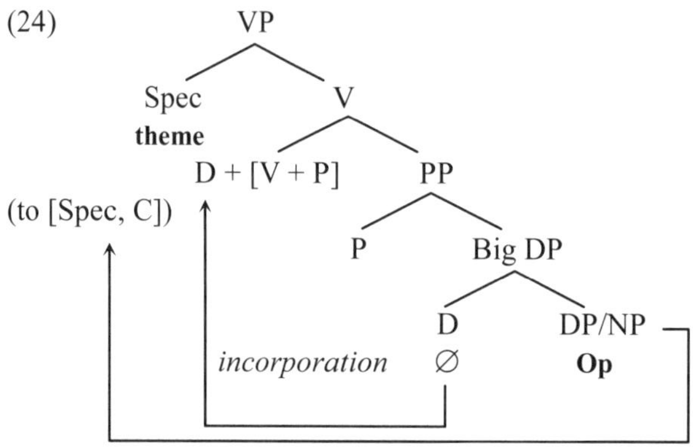

In (23b), however, the applied object bears a CASE-feature and therefore occupies the higher [Spec, V]. Relativisation of the theme is now blocked, because the head of the Big DP corresponding to the theme cannot incorporate into v from the lower [Spec, V], due to the intervening applied object DP.

6. The asymmetries disappear

In this section I provide further evidence for the proposed analysis by looking at contexts in which the theme *can* adopt primary object properties. As I show below, the theme can undergo movement operations under the following conditions:

- The Theme can be passivised, relativised and can incorporate when the applied object is an *incorporated pronoun*
- The Theme can be passivised and can incorporate when the applied object undergoes *relativisation*
- The Theme can be relativised and can incorporate when the applied object is the *subject of a passive*

As I show in the following subsections, the fact that the theme can incorporate, be passivised and relativised whenever the applied object has been subject to any one of these operations follows directly from the analysis that I offered to explain the object asymmetries discussed in section 5.

6.1. Applied object incorporation

As was shown in section 5.2, the applied object can incorporate into the verb, in which case the theme has a CASE-feature and occupies the single specifier of VP (see (21) above). This situation now allows for the theme to be *passivised*:

√ Theme passivisation with incorporated applied object:

(25) (*Umwáana*) *amáazi y-a-mú-menn-w-e-hó n'úmubooyi.*
child water SP-PST-OC-pour-PASS-ASP-APPL by cook
'(The child,) the water was poured on him/her by the cook.'

Since the theme has CASE, and since the applied object does not occupy a higher [Spec, V], the theme can agree with both defective v and T and become the subject of the passive. The grammatical (25) contrasts minimally with the ungrammatical example in (17b), where the applied object is a full DP with CASE, which therefore occupies [Spec, V].

As was shown in (20b) and (22) above, theme *incorporation* is blocked by a full applied object DP in the higher [Spec, V]. However, if the applied object incorporates into V, nothing prevents the theme from incorporating into v, thereby checking the φ-features of v. As a result, both the applied object and the theme may be realised as object markers on the verb in Kinyarwanda; as (26b) and (26c) show, either of the two pronouns can be anaphorically linked to a left-dislocated topic phrase:

√ Theme incorporation with incorporated applied object:

(26) a. *Umubooyi y-a-yá-mu-menn-ye-hó.*
cook SP-PST-OC-OC-pour-ASP-APPL
'The cook poured it on him/her.'

b. *Amáazi, umubooyi y-a-yá-mu-menn-ye-hó.*
water cook SP-PST-OC-OC-pour-ASP-APPL
'The water, the cook poured it on him/her.'

c. *Umwáana, umubooyi y-a-yá-mu-menn-ye-hó.*
child cook SP-PST-OC-OC-pour-ASP-APPL
'The child, the cook poured it on him/her.'

Finally, it is not surprising that incorporation of the applied object also creates a context in which the theme can be *relativised*:

√ Theme relativisation with incorporated applied object:

(27) *amáazi umubooyi y-a-∅-mú-menn-ye-hó*
water cook SP-PST-D_{theme}-OC-pour-ASP-APPL
'the water that the cook poured on him/her'

As was argued in section 5.3, relativisation of the theme requires the zero-head of a Big theme-DP to incorporate. If the applied object is a full DP, it blocks this operation, and theme relativisation is impossible. In (27), however, the applied object is an incorporated pronoun; consequently, the head of the Big theme-DP can incorporate into v, and movement of the relative operator from [Spec, Big D] to [Spec, C] is therefore licensed.

6.2 Applied object relativisation

Relativisation of the applied object implies incorporation of the head of the Big DP corresponding to the applied object; as argued above, Big DP also escapes the Case Filter through this operation. This means that the Big DP does not need to move from the complement position of the applicative marker to [Spec, V], but can remain *in situ*, see (24). We therefore expect that the same operations which were made available for the theme through object marking of the applied object are also available if the applied object is a relative operator. This expectation is borne out:

√ Theme passivisation with relativised applied object:

(28) *umwáana amáazi y-a-∅-ménn-w-e-hó n'úmubooyi*
child water SP-PST-D_{AO}-pour-PASS-ASP-APPL by cook
'the child on whom the water was poured by the cook'

√ Theme incorporation with relativised applied object:

(29) *umwáana umubooyi y-a-yá-∅-menn-ye-hó*
child cook SP-PST-OC-D_{AO}-pour-ASP-APPL
'the child on whom the cook poured it'

Since the Big DP does not need CASE and therefore does not move to [Spec, V], the theme can agree with v and T and move to [Spec, T] in a passive, as in (28), or can incorporate into v as an object marker, (29). (Relativisation of the theme in the context of a relativised applied object is of course excluded on independent grounds.)

6.3 Applied object passivisation

Above I discussed the fact that the theme cannot incorporate into the verb when the applied object is a full DP. Since relativisation also requires incorporation, this option is ruled out for the theme as well. Interestingly, however, if the applied object has become the subject of a passive, both incorporation and relativisation of the theme are possible:

√ Theme incorporation with passivised applied object:

(30) (*Amáazi*), *umwáana y-a-yá-menn-w-e-hó n'úmubooyi.*
water child SP-PST-OC-pour-PASS-ASP-APPL by cook
Lit.: '(The water,) the child was it poured on by the cook.'

√ Theme relativisation with passivised applied object:

(31) *amáazi umwáana y-a-∅-ménn-w-e-hó n'úmubooyi.*
water child SP-PST-D_{Theme}-pour-PASS-ASP-APPL by cook
Lit.: 'the water which the child was poured on by the cook'

As I argued above, theme incorporation into the verb in v is blocked if the applied object-DP is located in the higher [Spec, V], since the applied object is then closer to the attracting head v than the theme. One may therefore suspect that the grammaticality of

(30) and (31) is due to the fact that the applied object is no longer in [Spec, V] in these constructions, but has moved further to [Spec, T]. According to this analysis, only the invisible copy/trace of the applied object in the higher [Spec, V] would intervene between the theme and v in (30) and (31), and since traces do not induce locality violations (Chomsky 2000, 2001), theme movement would be allowed.

However, there is a problem with this approach. If the theme D-head has to wait until the applied object has moved to [Spec, T] before it can move to v, then theme incorporation is countercyclic. In the derivational, bottom-up system of the MP, the theme would have to move and adjoin to v *before* vP merges with T. But then the applied object would still be in [Spec, V], and theme movement would violate the MLC.

More importantly, the following example shows that the applied object does not have to be in [Spec, T] in a passive for theme incorporation to become possible:

(32) *H-aá-ya-menn-w-e-hó* *umwáana* *n'úmubooyi.*
EXPL-PST-OC-pour-PASS-ASP-APPL child by cook
Lit.: 'It (= expl.) was the child poured it (= the water) on by the cook.'

In (32), as in (30), the applied object has been passivised, but a *pro*-expletive has been merged in [Spec, T] to check T's EPP-feature. The applied object *umwáana*, 'child', has remained inside the VP (compare (18) above). The fact that theme incorporation is possible in (32), however, now leads to an important conclusion: in contrast to (18) above, the applied object DP can*not* be in [Spec, V] in (32) – if it was, it would block theme incorporation in the same way that an applied object DP in an active construction blocks theme incorporation (compare (22) in section 5.2). But if the applied object is neither in [Spec, T] nor in [Spec, V], there is only one option left: the applied object DP in (32) must be *in its base position inside the PP*:

(33)

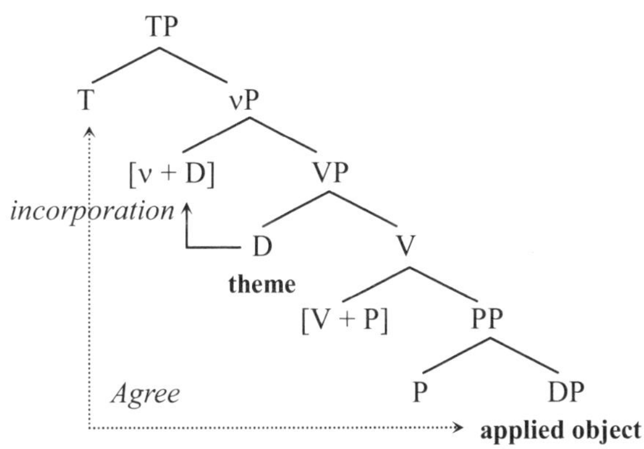

As shown in (33), the operation Agree can be established between the probe of T (its φ-features) and its goal inside the PP (the φ-features of the applied object). (The φ-features of v are checked via incorporation.) In (32), an expletive is then merged in [Spec, T] to check T's EPP-feature. Importantly, however, (33) also serves as the basic structure for the examples in (30) and (31): Once the φ-features of T and the CASE-feature of the applied object have been erased after Agree, the applied object can move to [Spec, T] *directly from its base position inside the PP* to check T's EPP-feature, and we derive

(30) or (31). This means that in these examples, theme incorporation is licensed not because the applied object has vacated the higher [Spec, V], but because it has never been in this position in the first place. Consequently, the above examples are the only instances where the verb in a locative applicative does not have to have an EPP-feature despite the presence of a full applied object DP.

As noted, theme incorporation in (30) and (31) is possible because the applied object has remained inside the PP. However, at the same time, the applied object can remain inside the PP only *because* of theme incorporation – these two aspects of this construction feed each other. Recall the reasons for why the applied object normally has to move to [Spec, V]: if a full theme DP occupies [Spec, V], it blocks agreement between the applied object inside the PP and the functional heads v and T above VP. Therefore, an applied object DP can never remain inside the PP if a theme DP is in [Spec, V], neither in an active construction (recall the ungrammatical word order [theme > applied object], shown in (10b) in section 3) nor in a passive, as shown in (34):

(34) * *H-aá-menn-w-e-hó* *amáazi* *umwáana* *n'úmubooyi.*
EXPL-PST-OC-pour-PASS-ASP-APPL water child by cook
Lit: 'It was the child the water poured on by the cook.'

In (32), however, the theme is no longer in [Spec, V]; as a pronominal D-head, it has moved and adjoined to v. This means that the only element which intervenes between the applied object in the PP and v or T is the copy of the theme-D, but as noted above, copies of moved elements never block feature matching. Furthermore, since a head adjoined to another head does not c-command "outside" (see Chomsky 2000: 117), the D-head adjoined to v in (32) does not block Agree between the applied object and T either (for the same reason, the theme does not block movement of the applied object to [Spec, T] in (30)).[10] Therefore, only if the theme is an object marker can the applied object DP be licensed *in situ*.[11]

To conclude this section, let me finally point out that the example in (32) provides strong evidence for the preposition incorporation-analysis of locative applicatives that I have adopted in this article. In the analyses proposed by e.g. Marantz (1993) or Anagnostopoulou (2003), there is only one syntactic position for the applied object below [Spec, T], namely the specifier position of the applicative marker, in which the applied object originates (see (4a) in section 2). Since the applied object c-commands the theme from this position, the impossibility of theme incorporation in examples such as (20b)/(22) in section 5.2 is explained both by this account and by my analysis: theme incorporation across the applied object violates the MLC – with respect to (20b)/(22), it

---

10 Note that in (31), the Big DP which includes the relative operator is still in [Spec, V] when T agrees with the applied object. This is not a problem, since the relative operator in the complement position of the incorporated D-head does not intervene between T and the applied object in terms of the MLC/Locality, because it does not c-command the applied object.

11 In addition, it is important to note that the applied object is only licensed inside the PP in a passive construction. Theme incorporation in (30)-(32) checks and erases the φ-features of v; therefore, the CASE-feature of the applied object must be checked against the φ-features of T. This is only possible in a passive, since there is no external argument - in an active construction, T's φ-features are checked against those of the thematic subject DP in [Spec, v], and the applied object's CASE-feature would remain unchecked if the applied object stayed inside the PP.

is immaterial whether the applied object is merged or moved into the position from where it c-commands the theme. However, the grammaticality of example (32) is *not* explained by an analysis according to which the applied object originates in a position above the theme, because such an analysis implies that the applied object occupies this position in (32) as well. But then, the applied object would also c-command the theme in this example, and theme incorporation should be blocked by the MLC. The grammaticality of (32) can only be explained by an analysis which assumes that there is also a position for the applied object *below* the theme, which is occupied by the applied object in (32). The existence of such a position is a corollary of the preposition incorporation-theory of applicatives defended here.

7. Conclusion

In Chomsky's (2000, 2001) version of the MP, in which feature checking no longer requires Spec-Head relations, the possibility of DP-movement is captured through the idea that EPP-features of syntactic heads attract DPs to their specifiers. However, this idea does not answer the question of why syntactic movement exists at all. I have presented one possible answer to this question by arguing that an EPP-feature might be needed to trigger a movement operation through which the moved DP circumvents a locality violation. The properties of locative applicatives in Kinyarwanda support an analysis according to which the applied object moves from a position inside the complement of V to a (second) specifier of V in order to be in a position from where it can agree with v and check its CASE-feature. However, this movement step cannot be triggered by v; in order to make it possible, V must have an EPP-feature.

Chomsky (2000, 2001) assumes that movement of a DP to the specifier of a head F with an EPP-feature also requires that F agrees with the DP. For example, before the subject DP moves to [Spec, T] to check T's EPP-feature, Agree has to be established between T and the subject in its base position ([Spec, v]). However, in my analysis, it is v, not the verb, which agrees with the applied object, but it is the verb's EPP-feature which triggers movement to [Spec, V], and Agree between v and the applied object is only established as a result of movement. The question is whether this aspect of my analysis can be reconciled with Chomsky's view on the relation between agreement and the EPP.

A positive answer to this question may be provided by an important idea put forward in Chomsky (2005). There, Chomsky argues that the syntactic categories T and V are not themselves equipped with formal features, but only *inherit* these features from their selecting heads, i.e. C and v (according to Chomsky (2001, 2005), only CP and vP, but not TP and VP, are *phases*, and only phase heads can trigger movement operations). According to this view, it is C, not T, which carries φ-features; T manifests them only as a consequence of being selected by C. Applying this idea to the analysis proposed here, one could assume that the verb is not lexically equipped with an EPP-feature. Rather, this feature, like the φ-features, is associated with v. Since v selects V, it transfers its EPP-feature to the verb; consequently, the applied object is attracted to [Spec, V] from where it enters the Agree-relation with the φ-features of v.[12] According to this view, both Agree and Move are ultimately triggered by features of the functional head v.

It has to be kept in mind that this approach only works if it is indeed the EPP-feature of v, but *not* v's set of φ-features, which is transferred to V in Kinyarwanda. If V inherited v's φ-features as well (as is suggested in Chomsky 2005), then it would be V, not v, which would agree with an object DP in [Spec, V]. Then, however, it would not be clear why the verb in Kinyarwanda can no longer agree with the theme in the lower [Spec, V] once the applied object has moved to the higher specifier. Movement of the applied object causes the theme to lose its "primary object" properties; in order to explain this fact in terms of the MLC, as I have done in this article, it is necessary to assume that the φ-features remain associated with v in Kinyarwanda.

In the light of the latter remark, it would be interesting to explore the possibility that Chomsky's (2005) proposal about the transfer of formal features from the heads C and v to the heads T and V is a parameterised operation. Perhaps languages differ with respect to whether φ-features and EPP-features remain associated with a particular phase head or whether they are transferred to the category selected by the phase head. For example, in languages in which not only the EPP-feature, but also v's φ-features, are transferred to V, the theme may in fact preserve its primary object-properties in applicative constructions, even when the applied object is attracted by V and moves to a second [Spec, V]. It remains to be tested in how far the empirical syntactic properties of applicatives in different languages are compatible with this theoretical approach.

References

Alsina, A. and S. A. Mchombo. 1993. Object asymmetries and the Chichewa applicative construction. In: S. A. Mchombo (ed.), *Theoretical Aspects of Bantu Grammar*, 11–44. Stanford: CSLI Publications.

Anagnostopoulou, E. 2003. *The Syntax of Ditransitives*. Berlin/New York: Mouton de Gruyter.

Baker, M. 1988. *Incorporation: A Theory of Grammatical Function Changing*. Chicago: Chicago University Press.

Baker, M. 1992. Thematic conditions on syntactic structures: evidence from locative applicatives. In: I. M. Roca (ed.), *Thematic Structure: Its Role in Grammar*, 23–46. Berlin: Walter de Gruyter.

Baker, M. 1997. Thematic roles and syntactic structure. In: L. Haegeman (ed.), *Elements of Grammar*, 73-137. Dordrecht: Kluwer.

Belletti, A. 1999. Italian/Romance clitics: structure and derivation. In: H. van Riemsdijk (ed.), *Clitics in the Languages of Europe*, 543–579. Berlin: de Gruyter.

Boeckx, C. 2003. *Islands and Chains*. Amsterdam: John Benjamins.

12 Notice that the idea that T and V inherit features from C and v which then trigger DP-movement to [Spec, V] or [Spec, T] raises problems with the extension condition (cf. Chomsky 1995). For example, in order for V to inherit a formal feature from v, v must first merge with VP, creating vP. However, subsequent movement of a DP to [Spec, V] no longer creates a phrase marker which includes the former vP as a proper part and therefore violates the extension condition. This problem is not addressed in Chomsky (2005).

Bresnan, J. and L. Moshi. 1990. Object asymmetries in comparative Bantu syntax. *Linguistic Inquiry* 21 (2), 147–181.
Cecchetto, C. 1999. A comparative analysis of left and right dislocation in Romance. *Studia Linguistica* 53, 40–67.
Cechetto, C. 2000. Doubling structures and reconstruction. *Probus* 12, 93–126.
Chomsky, N. 1995. *The Minimalist Program*. Cambridge, Mass.: The MIT Press.
Chomsky, N. 2000. Minimalist Inquiries: the framework. In: R. Martin, D. Michaels, and J. Uriagereka (eds.), *Step by Step. Essays on Minimalist Syntax in Honor of Howard Lasnik*, 89-155. Cambridge, Mass.: MIT Press.
Chomsky, N. 2001. Derivation by phase. In: M. Kenstowicz (ed.), *Ken Hale. A Life in Language*, 1-52. Cambridge, Mass.: The MIT Press.
Chomsky, N. 2005. On phases. Manuscript, MIT.
Grewendorf, G. 2002. *Minimalistische Syntax*. Tübingen/Basel: A. Francke.
Harford, C. 1993. The applicative in Chishona and Lexical Mapping Theory. In: S. A. Mchombo (ed.), *Theoretical Aspects of Bantu Grammar*, 93-111. Stanford: CSLI Publications.
Kimenyi, A. 1980. *A Relational Grammar of Kinyarwanda*. Berkeley: University of California Press.
Kimenyi, A. 1995. Kinyarwanda applicatives revisited. Keynote address at the 8th Niger-Congo Syntax-Semantics Workshop, Boston University. Available at http://www.kimenyi.com/kinyarwanda-applicatives-revisited.php.
Larson, R. 1988. On the double object construction. *Linguistic Inquiry* 19, 335–391.
Marantz, A. 1993. Implications of asymmetries in double object constructions. In: S. A. Mchombo (ed.), *Theoretical Aspects of Bantu Grammar*, 113–148. Stanford: CSLI Publications.
McGinnis, M. 2001. Variation in the phase structure of applicatives. In: P. Pica and J. Rooryck (eds.), *Linguistic Variation Yearbook* 1, 101–142.
Nakamura, M. 1997. Object extraction in Bantu applicatives: some implications for minimalism. *Linguistic Inquiry*. 28 (2), 252–278.
Ngoboka, J. P. 2005. A syntactic analysis of Kinyarwanda applicatives. Unpublished MA-thesis, University of KwaZulu-Natal, Durban, South Africa.
Pylkkänen, L. 2000. What applicative heads apply to. In: M. Minnick Fox, A. Williams, and E. Kaiser (eds.), *Proceedings of the 24th Annual Penn Linguistics Colloquium*. UPenn Working Papers in Linguistics. Volume 7.1.
Radford, A. 2004. *Minimalist Syntax*. Cambridge: Cambridge University Press.
Ura, H. 1996. *Multiple Feature-Checking: A Theory of Grammatical Function Splitting*. Doctoral dissertation, MIT, Cambridge, Mass.
Uriagereka, J. 1995. Aspects of the syntax of clitic placement in Western Romance. *Linguistic Inquiry* 26, 79–123.
Zeller, J. 2005. On clitic left dislocation in Zulu. Unpublished manuscript, University of KwaZulu-Natal, Durban, South Africa.
Zeller, J. and J. P. Ngoboka. 2006. Kinyarwanda locative applicatives and the Minimal Link Condition. *Southern African Linguistics and Applied Language Studies* 24 (1), 101–124.

Part III: Grammar

Modularity and the Tic-Toc of Language*

Rainer Dietrich

0. Introduction

If language is a part of our biological endowment, one may ask whether its functioning is subject to the same oscillations of our inner clock as the organs and cells of our body. This question has hardly been considered until now and theoretically there are certainly more options in answering it than just the alternatives "yes" and "no". Assuming that our linguistic capacity is one component of our more general cognitive system, language performance and its circadian variation might well turn out to be in complete synchrony with the variability of our attention and concentration as well. On the other hand, there are also strong theoretical reasons suggesting the existence of a separate daily rhythm of the language 'organ', perhaps even separate rhythms for the various linguistic functions such as the production, comprehension, and articulation of an utterance. In the following, I will briefly summarize the essential arguments for the theory of modularity and in the central section of the paper present some experimental evidence for the existence of an inner clock guiding our linguistic performance, thus strengthening the assumption of an encapsulated modular linguistic faculty.

1. The problem

Fodor's essay on the modularity of mind has successfully reanimated the idea of an integrated language faculty as an encapsulated system of the human mind. In spite of numerous spontaneous critical reactions to the proposal (cf. Swinney 1991, Levy 1996, Karmiloff-Smith 2003, as well as the special issues of *Cognition* 17, 1984 and of *Behavioral and Brain Sciences* 8, 1985), the concept has had a wide appeal to the psycholinguistic community, spreading through the labs and into the journals as a handy formula for talking about language in the same way as about vision, auditory perception and syndromes of cognitive disorders. There are, however, some "buts" connected to it. Visual perception has figured as one of the prime candidates for a modularized system and a prototype input system independent of central control, cf. Marr (1982). After approximately a decade of investigating gaze movements, the role of central control in

* This work was partially supported by a grant from the Gottlieb Daimler- and Karl Benz-Foundation to the research project ClockWORK. I thank Guido Kiecker, Katja Kühn, Maren Peters, Kathrin Sommer, and Katja Suckow for the tremendous help in running the experiments, Dieter Kunz and Joachim Kramer for discussions of the clinical matters, Wolf Lesener and Wolfgang Krössler for their help in the statistical analyses, Sue Olsen for the red pencil, and, especially, Till Roenneberg for his continuous chronobiological coaching of every move we did in his field and an anonymous reviewer for having scrutinized the originally submitted version of this text.

visual perception has become increasingly clearer and vision is no longer conceived as a passive input system but, rather, as an active intake procedure guided by shifts of attention; cf. Findley & Gilchrist (2003). Thesen et al. (2004) report on neurological evidence indicating that the sensory systems interact in a synaesthetical fashion with one another which is not compatible with the idea of an informational encapsulation.

The idea of language being implemented in a fixed neural architecture is facing challenging evidence as well. Neurological symptoms do not clearly dissociate different types of aphasia and their match with patterns of linguistics disorders is less than satisfactory; see Caplan (1992) and – more recently – Poeppel and Hickok (2004). While Caplan has argued for a microscopic reanalysis of every single dysfunction using the performance of an unimpaired speaker as a checklist, Poeppel & Hickok adhere to a global perspective and suggest "to refocus some attention on the broader organization" (p.2). This recommendation takes us back to where we started. Fodor's view of language being essentially a modular system is not incompatible with the discovery of an extremely complex internal architecture. The more complex the path of an external signal from the sensory transducers up to its final destination at the conceptual level turns out to be, the more plausible it is to assume that the entire system's processing is mandatory, fast, domain specific etc. But how can the modularity of a whole system be tested experimentally?

2. To everything there is a time

Front workers in psycholinguistic labs are familiar with the everyday experience that subjects' performance in language experiments varies over the course of the day. This is attributed to attention or fatigue without further ado since the relevant data reveal differences but not absolute values. In the present context, however, this observation might be interesting. Thus, Fodor's definition of a modular cognitive system lists nine properties. While six of them describe characteristics of the systems of information processing, properties 7 to 9 are concerned with the modules' biological hardware. "Roughly, modular cognitive systems are (…) hardwired, autonomous and not assembled" (Fodor 1982: 37). Let us assume that the ups and downs of reaction times and error proportions have the same biological source as feeling more or less hungry, more or less sleepy, and more or less susceptible to sexual stimulation. The source is not only homeostasis, a metabolic reaction to insulin or adrenalin values. It is also a reflex of a general phenomenon, ubiquitous in the world of living organisms, namely the temporal oscillation of the so-called inner clock; cf. the comprehensive introduction by Foster & Kreitzman (2004). Oscillation can be slow or fast and, correspondingly, there are long and short rhythms of life in nature. The rhythm that dictates most drastically the pace of human life is, probably, the circadian one. Its sinus wave of roughly 24 hours determines the phases of wake and sleep, the up and down of body temperature, the production of melatonin and cholesterol, the activity of the liver and other organs and – perhaps – of the language system as well; see fig. 1 below.

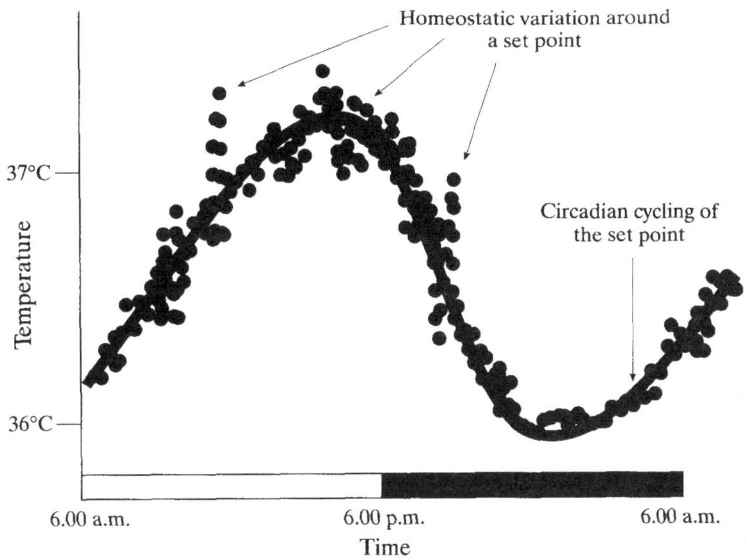

Fig. 1: Human body temperature: Circadian cycling (line) and homeostatic fluctuation (dots); Foster & Kreitzman (2004: 54; Fig. 4.1)[1]

In short, the inner clock is a central oscillator which directly or indirectly beats the time for every cell in the human organism and, hence, for all human activities over the course of a day. It is implemented in a population of some 20,000 neurons situated in the brain above the chiasmatic crossing of the optic nerves. That is why it was dubbed SCN (= suprachiasmatic nucleus). SCN's oscillatory activity is set by the organism's perception of light and darkness via special receptors situated in the retina. SCN also provides signals to entrain peripheral oscillators as, for instance, those of the heart, the lung etc.; see fig. 2.

1 With kind permission of Yale University Press.

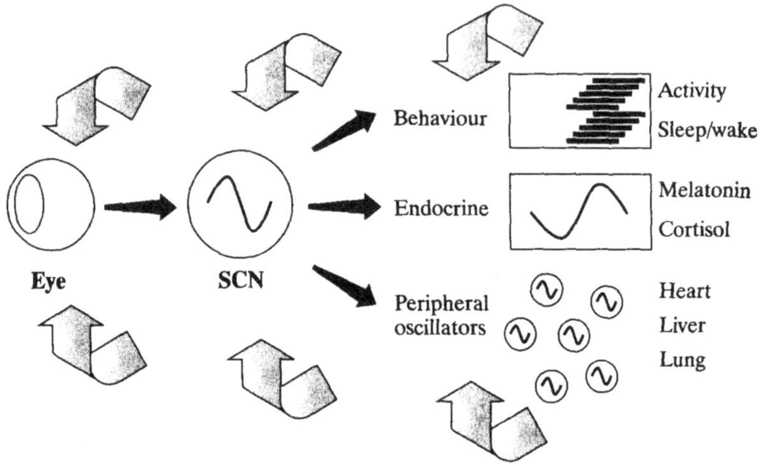

Fig. 2: Schematic view of the circadian system in mammals. Foster & Kreitzman (2004: 78; Fig. 5.5)[2]

Upon this background, it is not implausible to ask whether the inner clock might play a role in the performance of the language organ as well.

3. Empirical evidence

Considering on the one hand the assumption by Fodor and others of a neurological basis of language and the ubiquitous presence of a circadian rhythm in the human organism on the other, we can surmise that the human central nervous system is provided with time signals as is every other cell system of the body. Being an autonomous modular system, the language organ might very well feature a peripheral oscillator of its own whose existence in turn would bear testimony to the modularity thesis.

3.1 Survey of previous studies

The assumption of an individual time marker in the language module is at once plausible and bold. Its plausibility is founded in personal experience and introspection. "Our cognitive abilities change rhythmically over a 24-hour period. Tooth pain is lowest after lunch, proof reading and swimming are best performed in the evening;" (Foster & Kreitzman 2004: 11). The idea is bold in light of the many caveats surrounding neurological models of the linguistic map of the human cortex. Nevertheless, there are some results from experimental approaches indicating that a clock may mark time in the language organ. So far, all of them are behavioural studies.

2 With kind permission of Yale University Press.

The susceptibility of the auditory transducer of linguistic signals to circadian rhythm was already investigated by Lingenhöhl as early as 1980. The project's objective was to determine whether the time of day and two kinds of drugs, both of them sedative, affect the performance of auditory perception in a different manner. Fifty adult subjects of German nationality listened to dichotic presentations of ten sets of phonetically controlled numerals of two and four syllables and twenty sets of words at different levels of loudness at two different times of day, namely 7 a.m. and 7 p.m. The subjects were asked to listen attentively and to repeat what they had gathered from the input. The mean decibel value of a 50 % proportion of correct responses for all subjects was calculated under three experimental conditions, "no drugs" (I), 30 subjects, "50 mg valium" (II), 10 subjects, and "1 quantity of Adalin, a sleeping pill" (III), 10 subjects. The results of the no drugs-sample are shown in tab. 1.

performance / stimulus material / time of day	right ear: mean loudness (dB)	left ear: mean loudness (dB)
50 % correct / numerals / 7 a.m.	23,8 / +- 0,5	21,0/ +- 0,5
50 % correct / numerals / 7 p.m.	23,4/ +- 0,6	21,5 /+- 0,7
50 % correct / words / 7 a.m.	33,8/ +- 1,1	33,1/ +- 1,3
50 % correct / words / 7 p.m.	33,9/ +-1,3	33,8 /+- 0,7
100 % correct / numerals / 7 a.m.	36,7/ +-0,4	37,2 /+- 0,8
100 % correct / numerals / 7 p.m.	35,4 /+-0,6	34,8/+- 0,8
100 % correct / words / 7 a.m.	64,0 /+- 0,7	62,9/+- 2,3
100 % correct / words / 7 p.m.	63,8 +- 0,6	63,2/+- 0,6

Tab. 1: Performance in auditory perception (numerals and words) over time of day (Lingenhöhl 1980)

No significant circadian effect on the auditory perception of words was observed. There was, however, a significant effect on the type of stimulus. At lower loudness, numerals are better understood than words.

Looking deeper into language processing, Reinberg et al. (1988) investigated the susceptibility of linguistically impaired vs. linguistically unimpaired school children to circadian rhythm. Their performance in prelexical access to syllabic and sentence comprehension was tested in two separate experiments. Sixteen linguistically unimpaired school children (male and female from 7 to 9 years of age) and 16 matched children with language disorders (half of them in therapeutic treatment) listened to ten nonwords of four or five syllables which they were asked to repeat immediately. The tests were run at four different times of day, 9 a.m., 11 a.m., 3.30 p.m., and 7.30 p.m. Reaction times and errors were measured. Leaving the linguistically impaired children aside, both the reaction time (rt) and the percentage of errors showed significant circadian effects: rt, $p < 0.0001$, errors, $p < 0.01$. Their maximum performance was at 7.30 p.m.; their maximum body temperature peaked at 3.30 p.m.

The sentence comprehension data as well displayed a significant circadian variation. The maximum performance of the normal children was, however, at 9 o'clock in the morning. Due to the sparseness of information on the testing procedure, the findings of this study are difficult to assess.

3.2 Our own experiments

To test the two central hypotheses, the circadian rhythm-hypothesis and the modularity hypothesis, we conducted an experiment comparable to the procedure used by Reinberg et al. (1988) in our lab with unimpaired adult subjects. The first hypothesis is the more general of the two and follows directly from the chronobiological model of mammals' circadian rhythm. This model assumes a regular 24-hour sequence of phases in every human. Since human mental subsystems such as perception, memory, attention and problem solving must be assumed to have a physiological basis in the brain, it is also plausible to expect them to be subject to some chronological oscillatory force.

The circadian rhythm-hypothesis: The human brain features a daily pattern of sequential phases varying from more to less capacity in sentence comprehension.

The second hypothesis is more specific than the first one. It is based on modularity theory, more precisely on the autonomy assumption of this theory. The autonomy assumption suggests that the processing of a modular cognitive system is essentially independent of central control. Being organizationally encapsulated, the language module as well can be expected to feature a circadian rhythm of its own, different from that of, e.g., working memory.

The modularity hypothesis: The linguistic system has a circadian rhythm of its own, different from those of, e.g., reckoning, general attention, visual perception etc.

It becomes immediately obvious that the modularity hypothesis is methodologically flawed because there is no method by which the prediction can be falsified conclusively. We will come back to this difficulty in the discussion.

Design and methods
The design and the methods of the experiment more or less paralleled the procedure of a so- called constant routine (CR). In a professional CR, a small group of subjects stay in the lab for 48 hours or even more of testing under continuous medical control, lying in a bed without sleeping and only being given water and small portions of light food. In advance, their life style is synchronized over seven days in a preparatory procedure to include normal wake and sleep-intervals, no drugs, normal food etc.

Our experiment was a shortened approximation to a constant routine procedure. Four unimpaired, male subjects, 24 to 34 years of age, were continuously tested every two hours over a period of fourteen hours from 9 a.m. to 11 p.m. For measuring the fundamental circadian phases, body temperature was taken every two hours, as well. In advance, they were synchronized over a period of one week in which they were asked to avoid drugs, cigarettes, and coffee, to sleep regularly according to their normal needs and to take notes of their daily sleeping times in a sleep protocol. For the purpose of a later reshifting of divergent individual chronotype dependent sets of data, the Munich Chronotype Questionnaire (MCTQ) was also administered to them during the preparatory week.

Material

To test the variation of general attention, we used a standardized version of the d2 paper and pencil letter cancellation test; see Brickenkamp & Zillmer (1998). This test measures the processing speed and the quality of the performance. The subjects have to learn a small set of symbols, the letter 'd' with two strokes above or beneath it or one above and one beneath. For the test procedure, the subject is presented a test sheet with lines and rows as symbols, targets and distracters, and is asked to spot and to cancel the targets as fast and correctly as possible. There is a time-out per line set of 20 sec as a standard. Figure 2 shows an example of a test sheet.

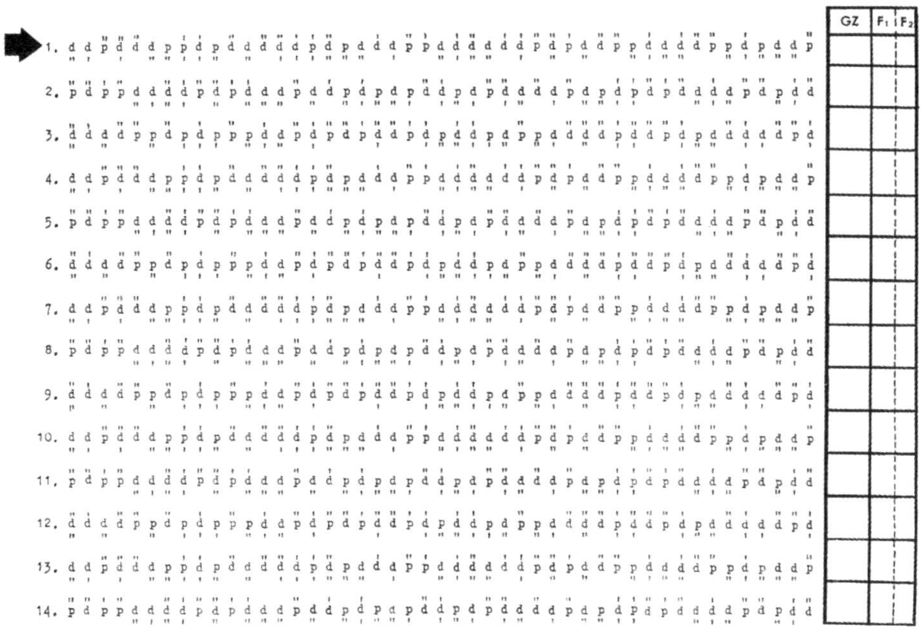

Fig. 3: Letter cancellation test of attention; example of a test sheet.

The performance in reading comprehension was measured by a selfpaced reading task of sentence comprehension, cf. Schriefers, Friederici & Kühn (1995). The subjects read 64 complex German sentences and 64 fillers. The experimental items contained a relative clause in a predicative nominative NP with a relative pronoun ambiguous between a subject and object reading in the initial position of the relative clause, cf. the examples under (1) and (2).

(1) Das sind die Banken, *die*$_{\text{[akk. plural]}}$ die Diebin betrogen *hat*. ('These are the bank houses which the thief cheated.' → 'The bank houses were cheated by the thief')

(2) Das sind die Banken, *die* $_{\text{[nom. plural]}}$ die Diebin betrogen *haben*. ('These are the bank houses which cheated the thief.' → 'The thief was cheated.')

Each experimental sentence was followed by a yes/no probe question displayed on the screen, i.e., “Wurde die Diebin betrogen?” (‘Has the thief been cheated?’). To answer, the subject had to hit the y- or n-key of the keyboard as quickly as possible. Hitting the spacebar started the next trial. Each word’s reading time, answer latency, and number of errors were measured.

Results
Figure 4 shows the variation in the mean temperature of four subjects over the fourteen hours of the experiment.

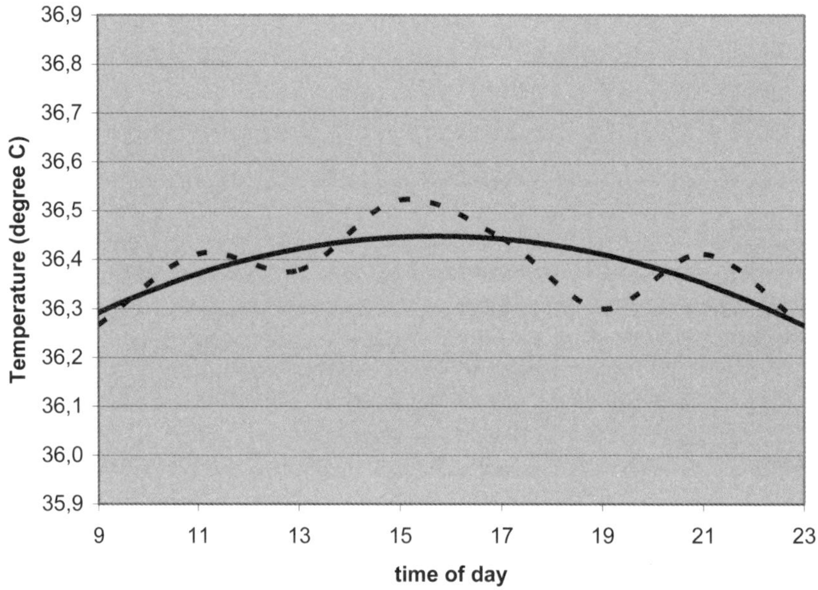

Fig. 4: Mean temperature of four subjects from 9 a.m. to 11 p.m. not corrected for chronotype variation (plus trend line).

To all appearances, the curve matches perfectly with the corresponding time segment of the prototype curve of human circadian rhythm as displayed in Fig.1 above.

The attention measures could not be evaluated. The subjects worked through the test much faster than average and made almost no errors, neither missing targets nor false alarm errors. Obviously, the test’s standard time limit per line of 20 sec was much too long and the test, therefore, yielded a massive ceiling effect. For the comparison of linguistic performance with general attention, we therefore had to substitute the attention measures with the results of another experiment with comparable subjects, cf. the next section.

The comprehension test provided three sets of data, the reading times of the critical utterance segments, latencies of question answering and number of errors. There was considerable individual variation in all data and, therefore, the resulting means do not allow for more than a descriptive statistical evaluation of the individual subjects. Fig. 5 shows the times of day and the mean reading times for the auxiliary (subject no. 3).

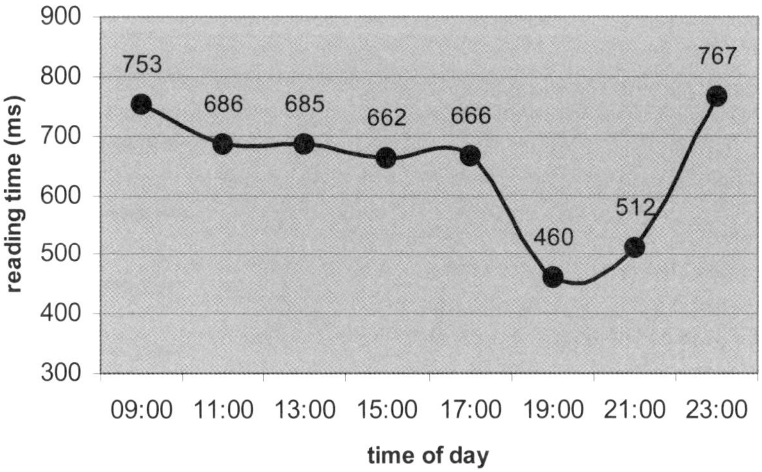

Fig. 5: Mean reading times of the disambiguating auxiliary; 64 experimental trials at eight times of day; subject no. 3.

This picture suggests a clear negative correlation of parsing performance in sentence comprehension and body temperature, cf. fig.4. Notice, however, that the parsing process is not the whole story of sentence comprehension as measured by the answering latencies.

Fig. 6 shows the mean relative differentials from the total rt-mean of the subjects' answering latencies.

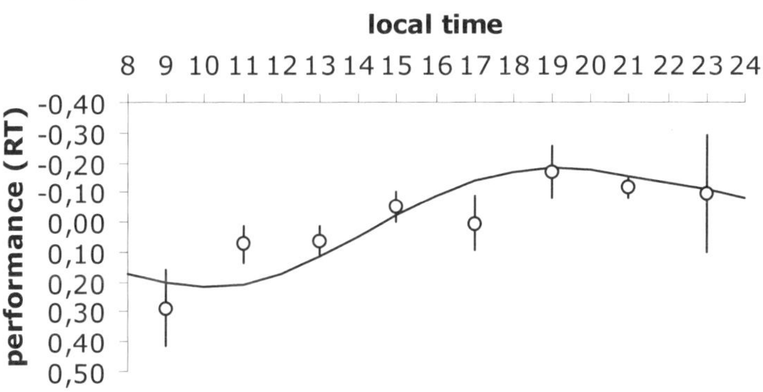

Fig. 6: Mean performance in sentence comprehension of four subjects at eight times of day. Notice that the short, i.e. the better, latencies are graphically displayed above the 0-mark.

This becomes much more visible when we look at the more detailed temperature data in figure 7 below.

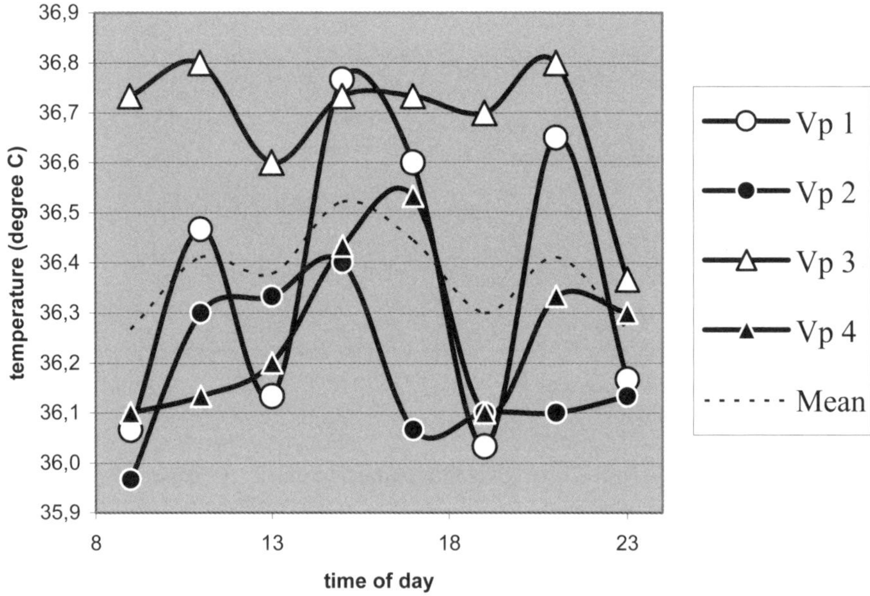

Fig. 7: Body temperature (raw data) of four subjects plus mean temperature at four times of day during the experiment.

As these data show, there is a clear upper peak around 3 p.m. and a bottom peak around 7 p.m. This corresponds to two peaks in the question answering data; see figure 8. The t-test[3] analysis yields a significant effect of time of day upon the sentence comprehension performance of the four subjects.

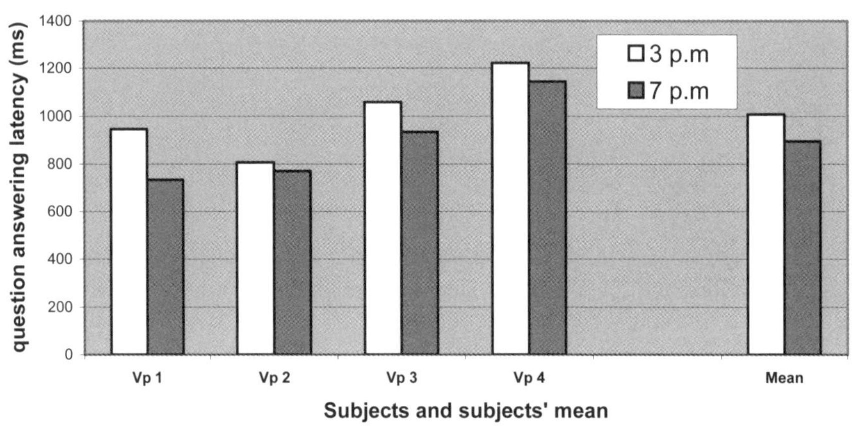

Fig. 8: Question answering latencies at two times of day. The difference is significant; t-Test: $p = 0.0004$.

3 The t-test analyzes the differences between group means of data in respect of significance.

What can be concluded from these data? First of all, the results do not represent hard data and, hence, do not allow a decisive evaluation of the circadian rhythm hypothesis. Nevertheless, I still think that there is no reason to bluntly reject the assumption that the language faculty is subject to a circadian rhythm. Both the on-line parsing process and the question answering latency curve do suggest a more-than-chance probability for the existence of at least one circadian peak in language comprehension, a maximum performance at around seven o'clock in the evening. What makes the latency data especially remarkable is their complementary distribution with respect to the temperature curve featuring a peak at about 3 p.m. and a clear decline at 7 p.m., i.e. at the time of the maximum performance in reading comprehension.

As for the modularity hypothesis, the results of the general attention measures cannot be used for a comparison with the linguistic data due to methodological flaws. Nevertheless, there are data from another chronobiological experiment that do allow for at least a rough estimation. Stolz' (1986) lab examined the effect of the time of day on several cognitive and psychomotor systems of eight male German adult subjects between 20 and 34 years of age. Applying a couple of different experimental methods, the project aimed at gaining insight into the circadian rhythm effects on visual perception (visually evoked potentials; VEPs), psychomotor mechanisms (maximal strength of grip force, finger tapping), and attention (letter cancellation test). The independent variable was the circadian rhythm (body temperature, potassium level, and urine volume). The subjects were synchronized over a period of five days before the constant routine was conducted and were then tested in two series of 30 hours each, taking measures in 3-h intervals. The results showed significant circadian effects on all dependent variables. The relevant fields for testing the modularity hypothesis are performance in visual perception and general attention. The VEPs (N 80, P 100) show the shortest latencies around 5 p.m. and are slowest at 1 a.m. and 5 a.m., cf. figure 9.

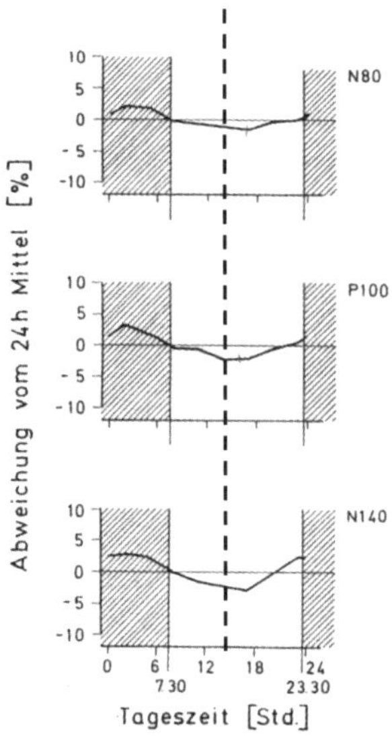

Fig. 9: Latency differentials of visually evoked potentials at different times of day; Stolz (1986); the 5 a.m.-line was added by R.D.

Performance in general attention, measured by means of the same d2-test we had used, is best at 5 p.m. and features its minimal performance peak around 2 at night. When we compare this with the findings on the circadian distribution of language performance levels there is no doubt that the modularity hypothesis is not falsified. Language processing has a circadian rhythm and its phases differ clearly from those of general attention and from the inner clock of visual perception.

4. Discussion

The objective of this paper was to contribute to the modularity issue from a chronobiological perspective. After a period of rather unanimous agreement, more recent chronometrical and neurophysiological data have begun to weaken the strengths of the modularity claim. The findings of the chronobiological experiments reported here, though limited in number, indicate that the language organ/linguistic faculty is subject to a specific circadian rhythm which manifests itself in a low performance phase in the early afternoon and with a maximum phase in the evening. Certainly, some considerable flaws in the experimental evidence have to be admitted. The basis of enduring evidence is small. Not all of the experiments satisfy the methodological standards set for constant routines without exceptions. Nevertheless it has still been possible to present results from different independent experiments with different tasks and different types of subjects which all point in the same direction: On the basis of a precisely formulated theoretical argument, some empirical evidence has been brought to light in favour of the assumption that there is an inner clock marking time for the language faculty.

References

Brickenkamp, R. & E. Zillmer. 1998. *The d2 Test of Attention.* Göttingen: Hogrefe & Huber.

Caplan, D. 1992. *Language: Structure,Processes, and Disorders.* Cambridge, Mass.: MIT Press.

Findley, J. M. & I. D. Gilchrist. 2003. *Active Vision.* Oxford: Oxford University Press.

Fodor, J. A. 1983. *The Modularity of Mind. An Essay on Faculty Psychology.* Cambridge, Mass.: MIT Press.

Foster, R. & L. Kreitzman. 2004. *The Rhythms of Life.* London: Yale University Press.

Lingenhöhl, H. 1980. *Einfluss von Tageszeit und Sedierung auf die Ergebnisse im Sprachaudiogramm.* Doctoral thesis, University of Tübingen.

Karmilloff-Smith, A. 2003. *Beyond Modularity: A Developmental Perspective on Cognitive Science.* Cambridge, Mass.: MIT Press.

Levy, Y. 1996. Modularity of language reconsidered. *Brain and Language* 55, 240–263.

Marr, D. 1982. *Vision. A Computational Investigation into Human Representation and Processing of Visual Information.* San Francisco: Freeman and Co.

Poeppel, D. & G. Hickok. 2004. Toward a new functional anatomy of language. *Cognition* 92, 1–12.

Reinberg, A., C. Ugolini, Y. Motohashi, C. Dravigny, A. Bicakova-Rocher, and F. Levi. 1988. Diurnal rhythms in performance tests of school children with and without language disorders. *Chronobiol. Int.* 5(3), 291–299.

Schriefers, H., A. D. Friederici, and K. Kühn. 1995. The processing of locally ambiguous relative clauses in German. *Journal of Memory and Language* 34, 499–520.

Stolz, G. 1986. *Die circadiane Rhythmik visuell evozierter Potentiale.* Doctoral thesis, University of Ulm.

Swinney, D. A. 1991. The resolution of indeterminacy during language comprehension: perspectives on modularity in lexical, structural and pragmatic process. In: Simpson, G. B. (ed.), *Word and Sentence*, 367–385. Amsterdam: Elsevier Publishers B.V.

Thesen, T., J. F. Vibell, G. A. Calvert, and R. A. Österbauer. 2004. Neuroimaging of multisensory processing in vision, audition, touch and olfaction. *Cognitive Processes* 5, 84–93.

A Skeptical Note on the Syntax of Speech Acts and Point of View

Hans-Martin Gärtner and Markus Steinbach

so bin ich nun auf dem harten Felsen angelangt,
und mein Spaten biegt sich zurück
Ludwig Wittgenstein

1. Introduction

Speas and Tenny (2003) (henceforth S&T) have put forward a forceful and thought-provoking argument that the relation between syntax and pragmatics can be modeled more insightfully if syntax takes over a larger share of the burden. In particular, S&T suggest that well-established tools from formal syntax rather than philosophico-conceptual *a priori* considerations or logico-semantic discourse representations provide the correct constraints for grammatical encodings and their empirical consequences. The domains of application are grammaticalized illocutionary force, i.e. sentence mood, and point of view related phenomena. Combining recent work on functional projections in the C-domain (Cinque 1999; Rizzi 1997) and on the mapping between lexicon and argument structure (Hale and Keyser 2002), S&T hypothesize that grammatically relevant properties for the mentioned domains are encoded in or deducible from the following syntactic configurations.

(1)

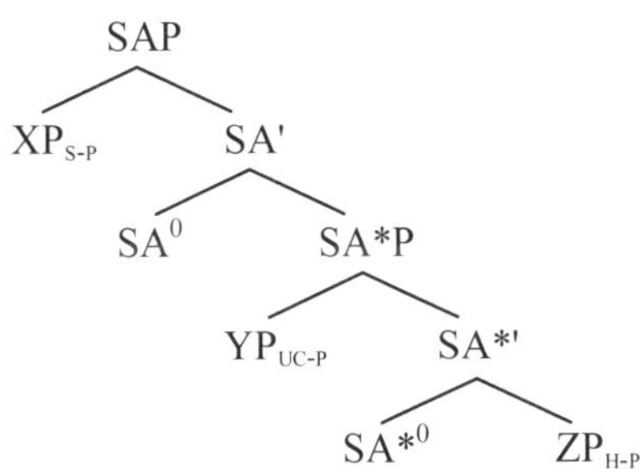

(2)

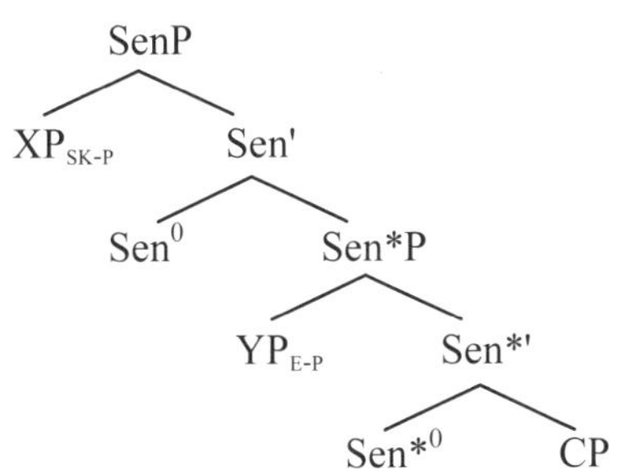

(1) represents the formal encoding of sentence mood in terms of a layered "speech act phrase" ($SA^{(*)}P$) configurationally defining the "P(ragmatic)-roles" SPEAKER (S-P), UTTERANCE CONTENT (UC-P), and HEARER (H-P).[1] (2) represents the formal basis for point of view phenomena in terms of a layered "sentience phrase" ($Sen^{(*)}P$) configurationally defining the P-roles SEAT OF KNOWLEDGE (SK-P) and EVIDENCE (E-P).[2] SenP is plugged into SAP as specifier of SA*P and thus, acquiring the P-role UC-P, encodes the utterance content. Sen*P further embeds CP, which stands as placeholder for the remainder of the sentence.

Crucially, S&T claim that (1) forms the basis for explaining why grammars usually encode no more than four basic sentence moods, namely, *declarative*, *interrogative*, *imperative*, and *subjunctive*.[3] Varying the finiteness of $YP_{UC\text{-}P}$ in (1), S&T define declarative [+finite] and subjunctive [–finite] mood.[4] The other two types are transformationally derived from (1) by an application of a counterpart of dative shift (cf. Larson 1988), as shown in (3).

(3)

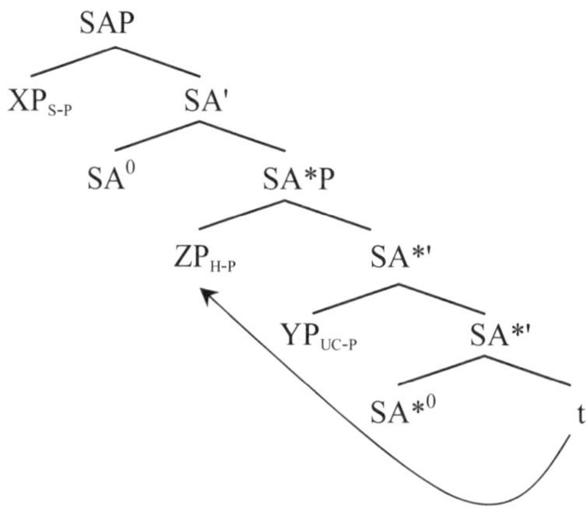

Again, the finiteness of $YP_{UC\text{-}P}$ makes a difference, defining an interrogative if [+finite] and an imperative if [–finite] is chosen. Independent formal assumptions ensure that this exhausts the possibilities. Likewise (1) and (2) together define the maximal range of P-roles, guaranteeing that no more than three "sentient" P-roles, i.e. S-P, H-P, and SK-P can be grammatically active. S&T then go on to provide empirical evidence for their sys-

1 There is an intentional analogy to the configurational definition of Θ-roles AGENT (A-Θ), THEME (TH-Θ), and GOAL (G-Θ) in approaches like Hale and Keyser (2002).

2 With some hesitation, S&T take SenP to be a unification of the more familiar "Evaluation Phrase" (EvalP ≈ SenP) and "Evidential Phrase" (EvidP ≈ Sen*P), for which one may consult Cinque (1999).

3 Sadock and Zwicky (1985) find only three major ("frequent") types, excluding S&T's "subjunctive". We follow S&T's terminology, noting that a proper distinction between formal sentence types and "functional" sentence moods is called for (cf. for example Grewendorf and Zaefferer 1991). It is also interesting to note that Grewendorf (2002: 66ff.) relabels Rizzi's influential "ForceP", the counterpart of "SAP", as "TypeP". This shift toward a more neutral formal characterization of sentences is extremely useful, especially so when it comes to subordination.

4 The standard use of (root) "subjunctives" is to express wishes (S&T: 318). Thus "optative" may be an alternative label for this sentence mood. S&T note the tendency of subjunctives to occur in embedded structures.

tem, which they claim is predictable in terms of formal notions like c-command, control via binding, and locality.

The aim of our discussion is to show that S&T's approach is inadequate in substantial respects. In particular we will show in section 2 that S&T's inventory of sentence moods is a product of arbitrary stipulation rather than formal syntactic deduction. In section 3 we will further demonstrate that S&T's empirical arguments in favor of their system cannot be upheld in the light of both technical problems and empirical counter-evidence. We take this as an interesting and important result in so far as it seriously challenges the kind of syntactic reductionism S&T are pursuing. It is our impression that more satisfactory revisions of S&T's system will only be forthcoming if one allows for substantial and explicit assumptions from an outside theory of language and discourse to interact with syntax.

2. Deriving the inventory of sentence moods and the size of SAP and SenP

S&T take "inspiration" for postulating structures (1)/(2) from Hale and Keyser (2002) (henceforth H&K), who set out to configurationally define the bounds on possible lexical projections and thus indirectly define the notion "possible word".[5] However, (1) and (2) correspond to only one out of four admissible basic configurations allowed by H&K, namely, the maximal one. Structures (4a)-(4c), although illustrated in S&T (p. 320), are discarded as candidates without discussion.

(4)

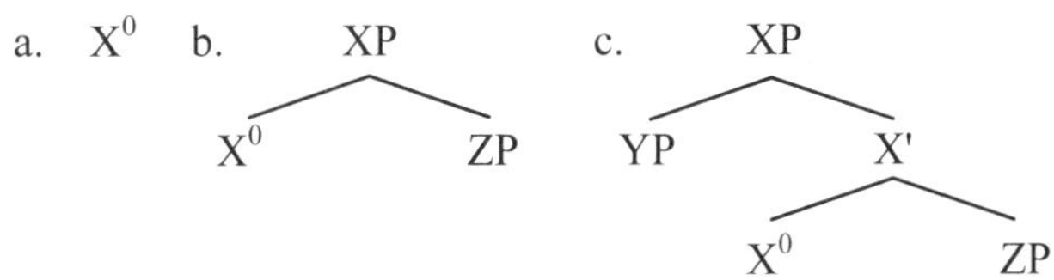

In order to motivate such an omission, it would, as a first step, have to be argued why languages could not dispose of four different basic SAP-domains. Otherwise, (4a)-(4c) could be taken to define between three and five additional basic sentence moods, bringing the total inventory up to between seven and nine, contrary to common assumptions.[6] As a second step, it would have to be argued why (1) is the correct choice for SAP rather than any of (4a)-(4c). It is unclear whether it would suffice to claim that $SA^{(*)0}$ is supposed to encode an abstract predicate of communiction, there being a whole range of different such predicates with different argument structures.[7] Such a claim would not be

5 For a critique of H&K's approach, see Fodor and Lepore (1999).

6 We assume that finiteness of at least ZP, i.e. the complement of X^0 could be the source for variation as before. Sadock and Zwicky's (1985) subtypes and minor types might potentially be mapped onto such additional structures. See also Truckenbrodt (2004), who argues for a "hearer-less" subtype of interrogatives in German.

7 For examples, see McCawley (1977) and Ballmer and Brennenstuhl (1981). From this perspective it would actually be plausible to argue for an extended SAP-structure introducing an additional "aboutness

based on purely formal grounds anyway. Likewise, recourse to the theory of functional projections would at most rule out (4a), if we take it as given that functional heads require at least one argument. Clearly, the obvious point that (1) meets the "virtual conceptual necessities" of any standard theory of communication (cf. Shannon and Weaver 1949) by containing representations of a sender ($XP_{S\text{-}P}$), a receiver ($ZP_{H\text{-}P}$), and a message ($YP_{UC\text{-}P}$), would lie outside formal syntax, which according to S&T provides a superior explanatory perspective nevertheless.

Sidelining this fundamental question, S&T instead touch on the issue of checking resources and recursivity as imposing further restrictions. It is conceded right away that "[i]n most of Hale and Keyser's work, it is not clear how recursion is to be ruled out" (p. 339, fn. 5). In fact, in dealing with double object configurations, H&K (p. 163) appeal to a "recursive (b)-type structure embedded in an (a)-type structure", where (b)-type corresponds to (4c). This shows that no principled ban on recursion can be derived from H&K's theory. Neither does such a ban seem to be part of current approaches to functional projections, where e.g. Rizzi (1997), driven by empirical considerations, makes use of * (the Kleene star operator) to allow for an unbounded number of TopPs in the C-domain.

S&T's more substantial attempt at limiting the size of SAP/SenP involves the idea that "the head can move only once" (p. 319), which itself would follow from the "stipulation [. . .] that each head can only have one feature to check (or, all features must be checked in the same position)" (p. 339, fn. 7). Of course, the *one-move*-constraint has a trivial reading once pied piping is disregarded. Thus, classical V^0-to-I^0-to-C^0-movement by adjunction involves only one instance of V^0-movement, given that $[_{I^0}\ V^0\ I^0\ ]$-to-C^0 counts as I^0-movement that pied pipes V^0. But, of course, ruling out pied piping in general does not seem to be desirable. Thus, imposing a *one-feature-per-head*-constraint on checking may be a more promising strategy. Assume SA^{*0} has an attractable feature, $^{\wedge}\!f$, and SA^0 an attracting one, $^{\vee}\!f$. If these cancel after SA^{*0}-to-SA^0 in (1), the size of the structure is fixed.[8] However, the projections in (1)/(2) have to connect to the sentence they are heading. In particular, given that Sen*P is modeled after Cinque's (1999) "Evidential Phrase", which can be realized by overt morphology in some languages, Sen^{*0} would have to check such features against (a category enclosing) V^0 eventually.[9] Thus, in addition to a $^{\wedge}g$, canceling against some $^{\vee}g$ of Sen^0, Sen^{*0} needs some $^{\vee}h$ in order to license evidential morphology in combination with a substantial bearer. Consequently, there is no deeper (formal) reason why SA^0 could not have some $^{\wedge}i$ in addition to its $^{\vee}\!f$, i.e. no reason why SAP could not be extended further up, licensing additional (more complex) sentence moods.[10] The theoretical point about these considerations is that

argument": "speaker communicates to hearer *about* x that . . . " Of course, this function is often coded sentence-internally in terms of TopP.

8 We disregard the possibility of allowing multiple specifiers as argued for by Chomsky (1995).

9 It is an open question to what extent S&T's approach is intended to capture Cinque's (1999) idea that the functional projections in question should host semantically compatible adverbs in their specifiers.

10 Note that Ross (1970:224) modeled the speech act domain as a full-fledged (performative) superordinate clause, i.e. as a structure of category S. Translated into current views about clause structure one would minimally expect IP or (AgrP and) TP to embed SAP. In such a system, SA^0 would clearly need an additional $^{\wedge}i$ in order to be licensed by I^0. Use of TP would further evoke the difficult issue of how to interpret $SA^{(*)0}$ temporally.

generative syntax in general and head movement in particular are formally designed to enable recursion. Thus, local bits of non-recursive syntax have to be motivated by independent principles.

Another arbitrary feature of S&T's system is the exclusive appeal to dative-shift, i.e. 3-2 promotion, as generating additional sentence moods. Passive- or middle-like 2-1 promotions, which can be found in H&K's toolkit, are not considered. Certainly, S&T could not claim that 3-2 diatheses are less marked and therefore more suitable for defining unmarked sentence moods. Instead, 2-1 promotions would seem to lead to unwelcome empirical predictions having to do with binding and locality.[11] The crucial point is that a multiplication of possible sentence moods by application of 2-1 promotions is ruled out by mere stipulation. The underlying formal syntax does not of itself warrant such an exclusion.

3. Explaining the behavior of indexicals and logophor(ic reflexive)s in terms of c-command and locality

Let us move on to the independent empirical evidence S&T put forward in favor of their syntax-based approach. The main claim is that c-command and locality are the proper tools for treating (the interpretation of) among other things indexicals, logophors and logophoric reflexives. The aim of our discussion is to demonstrate that S&T's system leads into a tangle of puzzles and inconsistencies. We take this as a serious challenge, our intuition being that the problems are less a matter of purely technical failure but of a misconception about the division of labor between core syntax and more peripheral systems.

One purported advantage of bringing locality of binding into the picture has to do with Slave (Athapaskan) 1st and 2nd person pronouns in complements to "direct discourse verbs" (Rice 1986). The highly intricate facts warrant the following generalization (S&T: 325).

(5) a. Slave first person pronominal is bound by the most local SPEAKER.
    b. Slave second person pronominal is bound by the most local HEARER.

S&T follow Tsoulas and Kural (1999) in taking indexical pronouns to be operator-bound variables. They further assume that bindings are computed over structures involving representations of the P-roles S-P and H-P as well as their Θ-theoretic counterparts, all of which function as variable-binders.[12] Thus, as shown schematically in (6) (S&T:325), the interpretation of *you* changes if there is an overt c-commanding expression bearing H-Θ in the sentence (coreference is indicated by cosuperscription).

---

[11] See section 3. As can be inferred from our discussion there, making YP the most prominent argument in SAP would lead to numerous cases of binding/control failure and thus, potential uninterpretability.

[12] Trying to determine the exact nature of this structural level would take our discussion too far afield.

(6) a. [$_{SAP}$ $^{i}XP_{S\text{-}P}$ $^{j}ZP_{H\text{-}P}$ [$_{CP}$ ^{k}*Simon*$_{S\text{-}\Theta}$ *told* ^{n}*him*$_{H\text{-}\Theta}$ [$_{CP}$ ^{n}*you not visit* ^{k}*me*]]]
b. [$_{SAP}$ $^{i}XP_{S\text{-}P}$ $^{j}ZP_{H\text{-}P}$ [$_{CP}$ ^{k}*Simon*$_{S\text{-}\Theta}$ *said* [$_{CP}$ ^{j}*you not visit* ^{k}*me*]]]

While (5) makes correct predictions for (6), it clearly fails to rule out cases like (7), i.e. cases where the bound indexical is outside of the reported clause.[13]

(7) a. *[$_{SAP}$ $^{i}XP_{S\text{-}P}$ $^{j}ZP_{H\text{-}P}$ [$_{CP}$ ^{k}*Simon*$_{S\text{-}\Theta}$ *told* ^{k}*myself*$_{H\text{-}\Theta}$ [$_{CP}$ *(that) it is raining*]]]
b. *[$_{SAP}$ $^{i}XP_{S\text{-}P}$ $^{j}ZP_{H\text{-}P}$ [$_{CP}$ ^{k}*Simon*$_{S\text{-}\Theta}$ *said* [$_{CP}$ *when* ^{k}*I was asked*] [$_{CP}$ *(that) it is raining*]]]

What S&T's approach misses is that it is the fact that reports are sufficiently like speech acts which allows shifting of indexicals. (7) would be correctly ruled out right away if an embedded SAP-layer were employed for the complement of "direct discourse verbs". The S-P and H-P roles of this SAP would have to be sensitive to higher S-Θ and H-Θ if present, but the binding of 1st and 2nd person indexicals would have to be the exclusive job of (bearers of) S-P and H-P. Yet, S&T (p. 338) explicitly opt against embedding SAP in an attempt at stearing clear of the infinite regress dilemma lurking behind the traditional performative analysis.[14]

Another technical deficiency concerns c-command, which is a prerequisite for syntactic operator-variable binding. A swift glance at (1) reveals that $ZP_{H\text{-}P}$ does not c-command the remainder of the sentence, i.e. $YP_{UC\text{-}P}$, in declaratives and subjunctives. Thus, indexical *you* should go unbound and therefore be uninterpretable in structures bearing these sentence moods, contrary to fact of course. S&T's representations in (6) are misleading in this respect. Now, heeding the operator-like nature of the binders one could, as a remedy, postulate an additional QR-step adjoining $ZP_{H\text{-}P}$ to a sufficiently high projection. This, however, jeopardizes what motivated the distinction between (1) and (3), since, according to S&T, declaratives and interrogatives differ essentially according to whether $XP_{S\text{-}P}$ or $ZP_{H\text{-}P}$ are coindexed with $XP_{SK\text{-}P}$ in Spec,SenP (p. 334f.). S&T "take this coindexing to be a sort of control, which requires that the controller c-command the controlee" (p. 335). Now, it may be reasonable to assume that such a kind of "SK-control" requires the controller to be in an A-position, while QR targets A-bar-positions and thus leads to A-bar-binding. This would sufficiently distinguish SK-control from "indexical binding". We then have to interpret the P-role positions in (1)-(3) as A-positions, something made plausible by the analogy to H&K's lexical projections.

The QR-solution just sketched has further disadvantages though in that it leads to a weak crossover configuration and should be ruled out independently.[15] A properly QR-ed $ZP_{H\text{-}P}$ would bind its trace, i.e. a variable, in the complement position of SA*P and could thus not bind another (indexical) variable inside YP at the same time.

Likewise, SK-control, which, as we have seen, S&T take to presuppose c-command, leads into substantial difficulties. Consider the licensing of a logophoric SELF-anaphor in (8) (cf. Ross 1970: 228).

---

13 Although we do not have the necessary data from Slave, we conjecture that the bindings and interpretations in question are not well-formed in this (or any other, except, perhaps, a sign) language.

14 In this respect S&T diverges from Speas (2004). For an early critique of the performative analysis, as developed by Ross (1970), see Grewendorf (1972).

15 See Heim and Kratzer (1998: chapter 10) for ample discussion.

(8) [$_{\text{SAP}}$ ^{i}XP$_{\text{S-P}}$ [$_{\text{SenP}}$ ^{i}XP$_{\text{SK-P}}$ [$_{\text{CP}}$ *This paper was written by Ann and* ^{i}*myself*]]]

Coindexation of the (bearer of the role) "seat of knowledge" with XP$_{\text{S-P}}$ ("speaker") is S&T's encoding of the logophoric role SOURCE, taken to be a primitive in Sells (1987). Given that XP$_{\text{S-P}}$ is the highest specifier of the sentence, c-command is guaranteed. However, it has repeatedly been pointed out in the literature (e.g. Reinhart and Reuland 1991: 317; Sells 1987: 451) that antecedents need not c-command logophors. Take for example the licensing of Japanese *zibun* in (9) from Sells (1987: 454).

(9) *Taroo wa* ^{i}*Takasi kara* [*Yosiko ga* ^{i}*zibun o nikundeiru to*] *kiita.*
Taroo TOP Takasi from Yosiko SUBJ self OBJ be.hating COMP heard
'Taroo heard from ^{i}Takasi that Yosiko hated ^{i}him.'

Here, *Takasi* is the source of information. However, irrespective of whether this would provide that DP with something like SK-Θ to be coindexed with and control *zibun* directly or with S-Θ to be coindexed with and control an intermediate XP$_{\text{SK-P}}$ in (some) Spec,SenP, *Takasi* is not in a position to c-command its controlee given that it is inside a PP, [$_{\text{PP}}$ *Takasi kara*]. Again, a purely formal syntactic perspective seems to be misguided.[16]

Apart from this technical issue, there are substantial empirical problems, too. Thus, consider the contrast in (10) (S&T: 335).

(10) a. *Honestly, Mary knew the victim.*
b. *Honestly, who knew the victim?*

In a declarative, *honestly* is speaker-oriented, while it is hearer-oriented in an interrogative. According to S&T this follows if the interpretation of *honestly* is dependent on SK-control, which itself underlies a closeness principle. XP$_{\text{S-P}}$ is the closest controller of XP$_{\text{SK-P}}$ in declaratives, (1), trivially, while ZP$_{\text{H-P}}$ is closest in interrogatives, (3).[17] However, in imperatives, which are also defined by configuration (3), speaker-orientation of speech-act adverbials is permitted, contrary to S&T's predictions, i.e. in violation of the closeness principle. This is shown in (11).[18]

(11) *Seriously, don't waste your time with polemics!*

---

16 We are well aware of the possibility to reanalyze PPs in ways that they become "transparent" for c-command by DP (cf. Cinque 2004). The point is that such costly reanalysis, even if it can be made theoretically watertight without voiding c-command as a formal tool, is superfluous once one allows an interpretational notion of source of information into the system as has been done by Sells (1987). The same issue of c-command failure arises (e.g. in Icelandic) for genitive subjects of DP ("John's opinion") and probably combinations like ("We gathered *from John's report* that ...").

17 S&T systematically confuse the terms "closest" and "highest" in their crucial section 3. This may have to do with their earlier assumption that English indexicals, as opposed to their Slave counterparts, are bound by the highest potential binder (p. 326). We have shown that indexical binding and SK-control have to be kept apart.

18 Other examples, like those in (i), are admittedly slightly marked.

(i) *Honestly / Frankly, leave me alone!*

It is clear, however, that the speech act adverbials must be construed as speaker-oriented not hearer-oriented. Thanks to Chris Potts for providing native speaker judgments. The argument can be reproduced for German.

Also, the evaluator responsible for the interpretation of adverbs like *unfortunately* in (12) is claimed to vary with sentence mood (S&T: 335).

(12) a. *Mary unfortunately knew the victim.*
 b. *Who unfortunately knew the victim?*

Yet, (13) indicates that speaker-orientation must be an option in interrogatives too. Again this violates the closeness principle.

(13) a. *Why did John unfortunately leave? # Something I personally find extremely fortunate.*
 b. *Who knew the victim that unfortunately died?*

If *unfortunately* in (13a) were hearer-oriented, the continuation made by one and the same speaker should not sound contradictory. Equally, the acceptability of (14) would seem to come as a surprise, given that $ZP_{H\text{-}P}$ as a closer controller should block speaker-orientation of *myself* in interrogatives.[19]

(14) *Why shouldn't this paper be written by Ann and myself?*

Again, our suspicion is confirmed that purely formal syntactic mechanisms, locality in the case at hand, do not provide any systematic account of the facts.

4. Conclusion

In this paper, we have argued against S&T's thought-provoking attempt at reducing an explanation of the inventory of sentence moods and the licensing of indexicals, logophor(ic reflexive)s and other point of view related phenomena to formal syntax. In particular, we showed that S&T's method for keeping the inventory of (major) sentence moods small boils down to arbitrary stipulation rather than formal syntactic deduction. This involves (i) arbitrary in-/exclusion of deep structures underlying $SA^{(*)}P$, (ii) arbitrary in-/exclusion of diatheses allowed to operate on $SA^{(*)}P$, and (iii) arbitrary constraints on recursivity and feature checking (resources).

We also showed the inadequacy of S&T's employment of formal syntactic tools (c-command/binding/control/closeness) in dealing with empirical phenomena. Thus, c-command is not systematically required for a proper account of logophoric reflexives. Also, requiring c-command for an operator-binding account of indexical pronouns leads to a dilemma. Either 2nd person pronouns like *you* go unbound in declaratives and subjunctives due to c-command failure. Or, if one allows an additional movement step for the binding constituents another dilemma arises. If the movement step is an instance of A-movement, S&T's sentence mood distinctions threaten to break down. If it is an instance of A-bar-movement, a violation of the weak crossover constraint follows.

[19] Thanks to Philippa Cook, Laura Downing, and Benjamin Shaer for native speaker judgments. Note that S&T could not assume that (14) is fine because the actual speech act involved may be more of a suggestion than of an information-seeking question. Their approach is explicity designed to capture abstract sentence moods instead of full-fledged speech act representations (p. 317).

Equally, closeness does not play any systematic role in deriving the facts contrary to S&T's assumptions. Thus, (i) accounting for Slave (Athapaskan) indexical shift in the complement of "direct discourse verbs" is in need of additional localization, which, however, conflicts with S&T's ban on embedding SAP. (ii) The behavior of speech-act adverbs in imperatives and the behavior of evaluative adverbs and logophoric reflexives in interrogatives is not dependent on the closest binder, i.e. it is not hearer-oriented.

As already indicated, we take the deficiencies of S&T's system as an interesting and important result in so far as they seriously challenge the kind of syntactic reductionism S&T are pursuing. It is our impression that more satisfactory revisions of S&T's system will only be forthcoming if one allows for substantial and explicit assumptions from an outside theory of language and discourse to interact with syntax.

References

Ballmer, T. and W. Brennenstuhl. 1981. *Speech Act Classification*. Heidelberg: Springer.

Chomsky, N. 1995. *The Minimalist Program*. Cambridge MA: MIT Press.

Cinque, G. 1999. *Adverbs and Functional Heads*. Oxford: Oxford University Press.

Cinque, G. 2004. Issues in adverbial syntax. *Lingua* 114, 683–710.

Fodor, J. and E. Lepore. 1999. Impossible words? *Linguistic Inquiry* 30, 445–453.

Grewendorf, G. 1972. Sprache ohne Kontext. Zur Kritik der performativen Analyse. In: Wunderlich, D. (ed.), *Linguistische Pragmatik*, 144–182. Frankfurt/M.: Athenäum.

Grewendorf, G. 2002. *Minimalistische Syntax*. Tübingen: A. Francke.

Grewendorf, G. and D. Zaefferer, 1991. Theorien der Satzmodi. In: A. v. Stechow and D. Wunderlich (eds.), *Semantik. Ein Handbuch zeitgenössischer Forschung*, 270–286. Berlin: de Gruyter.

Hale, K. and S. Keyser. 2002. *Prolegomenon to a Theory of Argument Structure*. Cambridge MA: MIT Press.

Heim, I. and A. Kratzer. 1998. *Semantics in Generative Grammar*. Oxford: Blackwell.

Larson, R. 1988. On the double object construction. *Linguistic Inquiry* 19, 335–391.

McCawley, J. 1977. Remarks on the lexicography of performative verbs. In: A.R. Rogers, Robert Wall, and J.P. Murphy (eds.), *Proceedings of the Texas Conference on Performatives, Presuppositions, and Implicatures*, 13–25. Washington D.C.: Center for Applied Linguistics.

Reinhart, T. and E. Reuland. 1991. Anaphors and logophors: an argument structure perspective. In: J. Koster and E. Reuland (eds.), *Long-Distance Anaphora*, 283–321. Cambridge: CUP.

Rice, K. 1986. Some remarks on direct and indirect speech in Slave (Northern Athapaskan). In: F. Coulmas (ed.), *Direct and Indirect Speech*, 47–76. Berlin: Mouton de Gruyter.

Rizzi, L. 1997. The fine structure of the left periphery. In: L. Haegeman (ed.), *Elements of Grammar*, 281–337. Dordrecht: Kluwer.

Ross, J. R. 1970. On declarative sentences. In: R. Jacobs and P. Rosenbaum (eds.), *Readings in English Transformational Grammar*, 222–272. Waltham MA: Ginn.

Sadock, J. and A. Zwicky. 1985. Speech act distinctions in syntax. In: T. Shopen (ed.), *Language Typology and Syntactic Description I: Clause Structure*, 155–196. Cambridge: CUP.

Sells, P. 1987. Aspects of logophoricity. *Linguistic Inquiry* 18, 445–479.

Shannon, C. and W. Weaver. 1949. *The Mathematical Theory of Communication*. Urbana: University of Illinois Press.

Speas, M. 2004. Evidentiality, logophoricity, and the syntactic representation of pragmatic features. *Lingua* 114, 255–276.

Speas, M. and C. Tenny. 2003. Configurational properties of point of view roles. In: A. Di Sciullo (ed.), *Asymmetry in Grammar*, 315–344. Amsterdam: John Benjamins.

Truckenbrodt, H. 2004. Zur Strukturbedeutung von Interrogativsätzen. *Linguistische Berichte* 199, 313-350.

Tsoulas, G. and M. Kural. 1999. Indexical pronouns as bound variables. In: S. Bird, A. Carnie, J. Haugen, and P. Norquest (eds.), *Proceedings of the 18th West Coast Conference on Formal Linguistics*, 545–557. Somerville MA: Cascadilla Press.

The First Traces of Language. On Darwin's Package Hypothesis[1]

Georg Meggle

0. Introduction

Are there any archaeo-anthropological problems to which a philosopher (of the analytical school) could help find the solution? Yes, there are. The problem of the origin of human language is one example. Maybe an inquiry into this topic will also contribute to the answer of the question whether the reason for the Neanderthal's lack of success in evolutionary history was his linguistic inferiority.[2]

My contribution to the very old problem of the origin of language is on the one hand trivially obvious; on the other hand, it entails far reaching consequences. To anticipate my result: In the context of this specific problem and research, I recommend that archaeologists should end their debate. No more discussion! That would be the best solution.

In *The Origin of Humankind* (1994) Richard Leakey reports on the latest empirical data in research on the evolution of the human linguistic ability. He does so in a form that is clear and comprehensible, also for non-experts. I use his book as a guideline in orientating myself in this non-philosophical field of expertise. Frankly, my knowledge of the state of the art in the research on human evolution hardly exceeds the contents of this book. (And, as you will see, knowing this much is quite enough to get solved our or his problem, respectively.)

I will, for a start, take on Leakey's role as research-reviewer – only later (in § II), I will return to my acquired role as philosopher.

1. Darwin's Package Hypothesis

1.1 Speaking in the voice of Leakey: The title of my book is, of course, meant to remind the reader of Darwin's *On the Origin of Species*. Not without deeper reason: Darwin's position concerning the origins of mankind is the starting point of the whole debate. In the following, I will address the course this debate has taken and the set of problems yet unresolved. My starting point is, more precisely, the DPH (Darwin's Package Hypothesis):

---

1 This paper is the English (and in its contents slightly changed) version of an (up to now unpublished) lecture given at the Archaeological Symposium in honour of Sabine Rieckhoff, University of Leipzig, Nov 06, 2004.

2 This is the reformulation of the old problem of the origin of human language by Paul Mellers in Nature, Vol. 432 (Nov 25, 2004).

DPH: The evolution of genuine human characteristics did not happen in an isolated way (one by one), but en bloc – in a package. This package consists of:

- upright walk
- fabrication of stone weapons
- intensifying social interaction/communication
- special cognitive qualities
- brain volume/ brain structure
- speech organs
- language

1.2 The DPH is an empirical hypothesis. Thus, there must be a way to test it empirically. Different disciplines are involved in this process of verification. On the one hand, concerned with the human being as a natural form of life, there is *biology* (especially biological anthropology) with its subsections:

- anatomy
- molecular biology
- neurology

On the other hand, concerning human beings as social beings, there is *archaeology* with its task to find and to interpret the works humans have created, in particular:

- technical works (tools)
- cultural works (especially artistic ones)

1.3 The DPH is already refuted. Even the first two characteristics in this package, upright walk and the production of stone weapons, are – from an evolutionary point of view – not linked to one another. Molecular research suggests that we have been bipeds for about 7 million years; the archaeological evidence, however, supports the thesis that we have been producing stone weapons only from about 2 million years ago, that is a good 5 million years later. *Quod erat demonstrandum.* Upright walk does definitely not belong into Darwin's package.

1.4 My question (note: Leakey's question) is now: What about the rest of the package? My position: For the rest of the characteristics the evidence is inconsistent, too. But in this case not only for the respective different characteristics, but even, much worse, for one and the same characteristic feature. I even declare the worst case: The evidence is contradictory even with regard to the most important of the features, namely the final one – linguistic ability. When did we acquire this ability?

In contrast bones and stones, the words and thoughts of the first humans cannot be dug up anymore, after millions of years. This means we have circumstantial evidence at best. Again, we have this evidence from both biology and from archaeology.

Both disciplines assume (like all other disciplines, too) that *language* has something to do with *cognitive skills*. In this respect, biologically oriented anthropologists primarily start with the notion that, firstly, language and thought have something to do with the *brain*, and, secondly, that the ability to speak is connected with our *speech apparatus*. Archaeologists, however, primarily assume that our cognitive skills are manifest in the things we do and make, especially in the things we generate, in short, in our *products*.

1.5 Since when do we have *linguistic ability*? As mentioned before, bio-anthropologists and archaeologists – at least some of them – make completely different suggestions. The anthropo-biologists' suggestion is that we have a rather long tradition of linguistic ability; at least since we have made the transformation from *Homo habilis* to *Homo erectus* – that is, roughly 2 million years ago. Archaeologists, however, particularly those with an understanding of the arts, say that we have had the ability to speak only for a very short time-span, since the Neo-Paleolithic revolution some 30.000 to 35.000 years ago. If the above mentioned period of 2 million years would be equated to one month, then, from this evening's point of view, the respective span of time (suggested by archaeologists) would be merely some hours; it would be as if linguistic ability had just happened today, just this morning.

1.6 Let me clarify these different points in a graph. In my book there are, on page 136, three small charts. For clarity's sake, let me present them one after the other.

First the two biological graphs:

(Bio I)

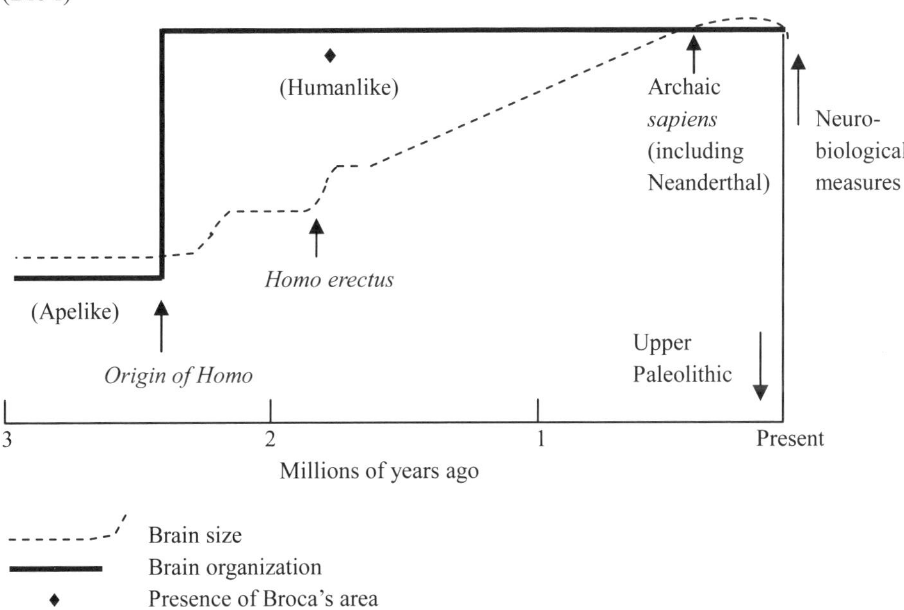

This *anatomy figure* (Bio I) shows the development of *volume and structure of our brain*. The *dotted line* refers to the volume; the *regular line* to the structure. It is true that according to figure Bio I [i.e., the regular line] we are "humans" since we are bipeds (this means, as mentioned, since about 7 million years). But the human *homo*, as this figure indicates, only enters the stage a little less than 2 ½ million years ago.

Homo's brain volume is 50 % larger than that of his predecessor. And with him the brain structure has altered, too. Since then, by the way, there has hardly been any other change: Our brain structure today is still the same as it was back then. Further, you can see that since our hominization our present brain volume has enlarged three times. The brain volume of the archaic *Homo sapiens* was even a little larger than ours today, though our current brain wirings are the furthest developed. Naturally, this "shrinking of the brain" in *Homo sapiens* is a better solution for our mothers and for our children with respect to the birth procedure and accordingly lowers the birth mortality rate.

Importantly, the Broca centre (i.e. the centre which is decisive for linguistic ability), too, has existed shortly before the appearance of *Homo erectus*, i.e., for around 2 million years. Now, is there anybody who can tell me why this centre of linguistic ability should have already existed even when there was no use for linguistic ability?

1.7 Now to the second chart (Bio II)[3]

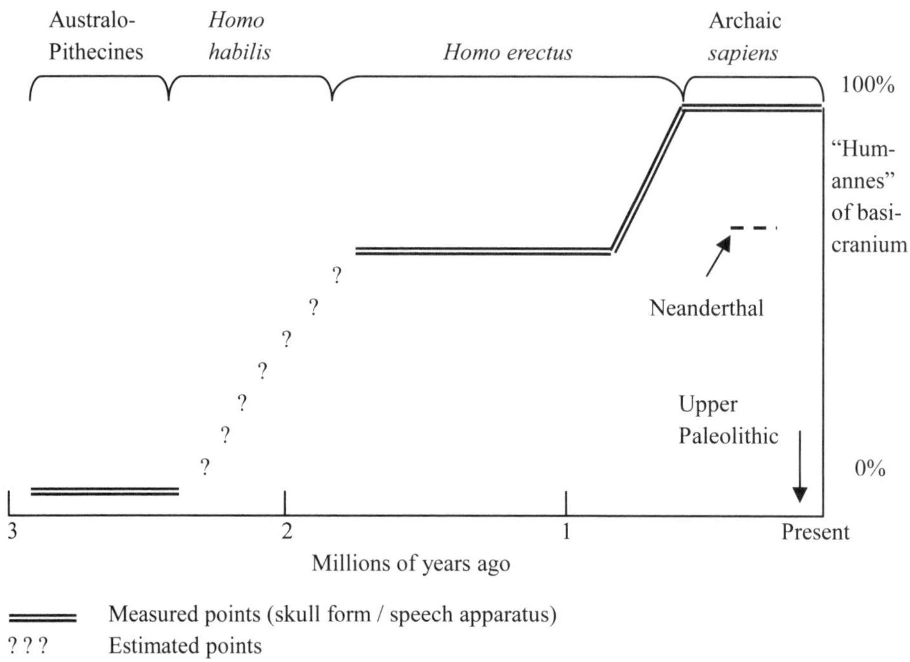

This chart's *double line* indicates how closely the *skull form* of our ancestors resembles ours today. This is important to know as the skull form correlates with the functioning of our organs of speech (tongue, larynx, etc.). According to this chart, we people of today are the absolute measure: 100 %. When we move backwards, i.e. from the "present" to the left, then we see that even our direct predecessor, the archaic *sapiens*, has had the same skull that we have – thus, regarding speech ability, he, too, would have

3 In this figure and in the next one I changed for the sake of their intended combination (in the figure in § 2.0 below) the presentation of the lines, but not their routes, of course.

been able to use a cellphone. *Homo erectus*, on the other hand, had, compared to the pre-homo period, only covered half of the skull's development distance. Maybe my colleague Meggle will come back to the famous Neanderthal paradox [indeed: see below 3.7]. According to this paradox, the Neanderthal's linguistic ability was much worse than the one of our predecessor, the archaic *sapiens* who came into being more than half a million years before him.

1.8 Let me combine the results of the two bio-figures: The beginning of our linguistic ability corresponds with the appearance of *Homo erectus* (1.7 million years ago) and from then on (along with the improvement of the brain) this ability keeps developing, while our mouths only loosen up within our archaic *sapiens* phase.

So much for the biological-anthropological data.

1.9 The archaeological figure speaks a completely different language.

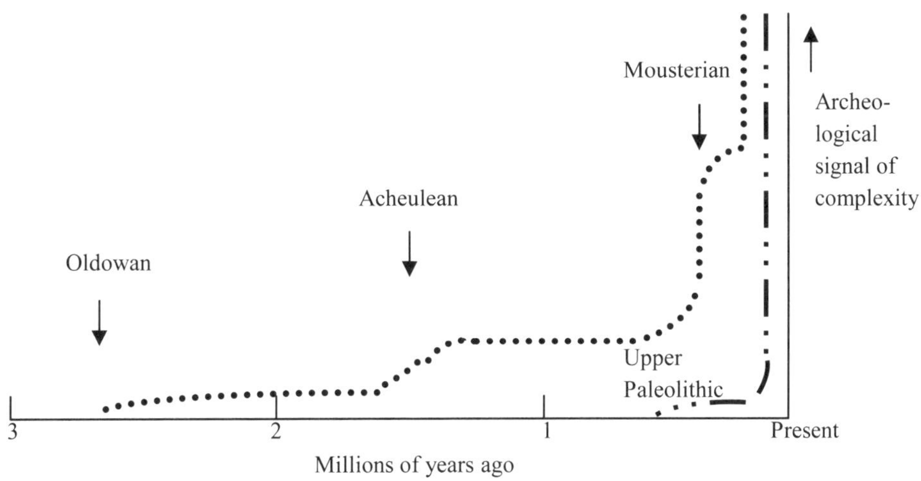

•••••• Number and standardization of stone tool types
— · · — Production of "arts" objects

This figure marks the *development of the craftsmanship of our stone works* with a *bold-dotted-line* and with — · · — our *development to become real artists*. Humanity did not have to wait for the Renaissance to see its artistic facilities being fully fledged; in form of the unbelievably beautiful and skilfully made cave drawings this had happened as early as in the Neo-Paleolithic revolution, i.e. about 30.000 years earlier. And this revolution was preceded by a significant progress of our abilities in the production of stone tools.

We have been using stones as tools for more than 2½ million years; at first – as the Oldowan culture indicates – only as rough stone chips. These came about more or less by accident, when two stones were struck against one other. A million years later, we begin to consciously work on the stones in such a way that they become stone wedges. Then, our repertoire increases again only after another one million years. Yet, it does so rapidly. We turn to the production of many more types of tools than before; and we do not need millions of years for the improvement of these types, only thousands of years.

1.10 Therefore: *If* language has something to do with our mental abilities and if these abilities, as archaeologists suppose, manifest themselves both in our good and in our bad products (tools and weapons), *then* the evidence shows that the actual jump in our language development reaches back about a maximum of 200.000 years; a minimum of 35.000 years is certain, as our cave paintings prove. The remaining controversial point is whether art presupposes language or whether, as some people believe, the referential function of language could only be introduced via pictures, via pictographic images, indicating that it is language which presupposes art.

2. Is Darwin right?

2.0 So much from Richard Leakey. Now back to me [GM]. Let me first summarize the above conclusion by combining the results of all three figures in one, a procedure – I know, I know – being not completely accurate, but still very revealing. At least the time axis remains the same in the new figure.

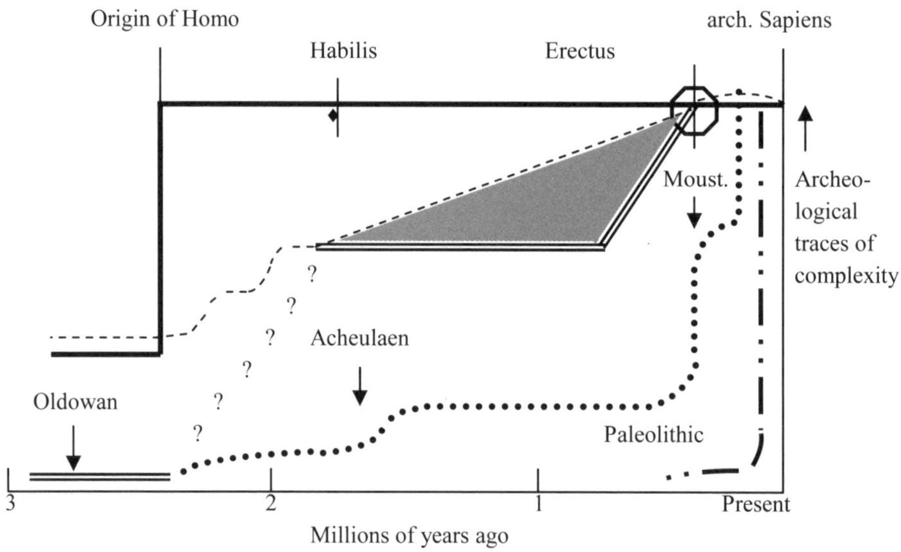

- - - - - - - -/	Brain size
———	Brain organization
♦	Presence of Broca's area
═══	skull form / speech apparatus – measured points
? ? ?	Estimated points of skull form
·······	Number and standardization of stone tool types
— · · —	Production of "arts" objects
	BTP[4]

Does this threefold result, as described to us by my colleague Leakey, really contradict Darwin's package hypothesis? Not at all.

2.1 *Point 1:* There are stunning correspondences between the results. Some of the registered developmental jumps correlate. We find the most obvious example in the beginning: the origin of our Homo can be defined (equivalently) in four ways, namely by

(a) the beginning of stone tool manufacturing (bold dotted line);
(b) the big jump – see the regular line – in brain structure;
(c) the 50 % jump in the enlargement of the brain volume (weakly dotted line);
(d) the beginning of the development of the skull form / towards speech (double line)

I take this to be an incredibly strong point *in favour of* the Darwin package (minus the feature of us being bipeds)! Surely, Darwin's package hypothesis has more weight the more we move back in time towards our origins. After all, this hypothesis was meant to explain these very beginnings. Or, in other words: the later the phases we study in our evolution the more the developments of our formerly converging characteristics might diverge.

And finally: With regards to the structure of our brain, the first big developmental jump was the *decisive* one. Since then, this structure has essentially remained the same until today. Therefore, we can from now on neglect the brain structure (regular line). The remaining question is only to what extent the DPH is valid for the rest of the development package.

2.2 *Point 2:* The next developmental jump also happens more or less simultaneously on all three levels, with a give or take of some (negligible) thousands of years in evolution. The mutation from *Homo habilis* to *Homo erectus* can be defined in manifold, equivalent ways, too, namely via

(a) the transition to planned stone production (e.g. blueprint: stone wedge);
(b) the fact that half of the developmental distance towards our speech apparatus is covered;
(c) the doubling of our brain volume in comparison to apes.

---

4 Dubbing this point as "BTP/Brain Threshold Point" is an invention of my own.

2.3 *Point 3:* The transition from *Homo erectus* to the *Archaic sapiens* manifests itself in at least two ways: in the step from the Acheulean culture to the Mousterian culture (i.e. from stone wedges to systematic stone manufacturing); and this very step is roughly simultaneous with the full development of our speech apparatus. And maybe with the associated point on the brain volume line we reach what one could call the critical brain threshold point (short: the BTP). The mass of the respective brain, at any rate, is already quite close to ours.

I would bet that this was the real conversation starter. To speak with Leakey, one more time – we owe a lot to our *Homo erectus*. Homo erectus invents the use of fire; with his stone wedges he introduces hunting; along with hunting fresh meat becomes part of our diet; he does not run slower than our best Olympic runners; and he was the first to leave Africa and explore other continents.

2.4 The only divergences I can see lie within the phase of 1 million years of the *Homo erectus*.

One of the divergences is situated at the end of the phase, with the appearance of the *Archaic sapiens*. The brain volume line lacks the respective jump in relation to both the dotted bold line and the double line. But I have already revealed my explanation for this: BTP, or, *qualitative* threshold points can do without *quantitative* jumps.

The second divergence is located within the Erectus phase. The brain volume increases, but the speech apparatus develops unrelated from it for a full span of 500.000 years. How can this be explained? And, most of all: Is this fact still compatible with Darwin's package hypothesis?

2.5 It surely is. I have nearly automatically two explanations to offer. The first is an obvious one and the same as above: BTP. Only when this point is reached, the evolution of the skull can catch up. Remember: A pair of scales can be built in such a way that it only tips over when the container on one of the sides is filled to the last drop. Note: Threshold value functions do by no means exclude correlations with other developmental lines.

But my second explanation is by far my favourite:

2.6 I like it better because it takes into account a deeper level of explanation. It also illuminates the first explanation further. Moreover, it originates directly from my own main field of research – the General Theory of Communication. But, slowly, step by step:

The debate surrounding the verification of Darwin's package hypothesis, as presented so far, entails many suppositions. One of them is, to repeat: The DPH only makes sense, *if* the cognitive abilities which are involved in our linguistic abilities also manifest themselves in our non-linguistic works. This leads to a long overdue question: Is this *if-clause* really valid? Do the cognitive skills which are necessary for the fabrication of stone tools factually presuppose the existence of language?

2.7 We can limit ourselves to the only problematic DPH phase, that is, limit ourselves to considerations concerning the Acheulean products. These are the products of our dear

Homo erectus. Do the Acheulean stone wedges presuppose the existence of language in their producers? This is a question tailored for a philosopher. It is exactly this question Christian Strahm, my befriended emeritus in archaeology, posed to me on August 16, 2004. Is it really possible that he is the only archaeologist who has ever turned to a philosopher with this question? A pity, if this was the case. For without this shyness towards philosophers many sleepless nights and many months, if not years, of unnecessary work might have been spared to some archaeologists.

3. Language, speech, and nonverbal communication

3.1 Do the Acheulean stone wedges presuppose language? It is obvious how a good (analytical) philosopher has to reply, simply by answering with another question: *"Language" in which sense?*

I was amazed to find that Christian and the few other archaeologists whom I managed to meet until now at first apparently did not seem to understand this last question. In the meantime, I discovered the reasons for this: They are representative for almost the entirety of the community of archaeologists who are concerned not with the development of our communicative ability in general, but only with the development of our *strict linguistic* ability in particular – and thus they (wrongly) equate *language* with *spoken language*. This, by the way, is also what our bio-anthropologist colleague Richard Leakey did wrong.

The supposition is completely unsubstantiated. There are, for example, (different variants of) deaf-mute languages. And those are, concerning every question of our daily life, no less informative than our various verbalizations. And there are many different forms of strictly embodied ways of communication.

3.2 Verbal (spoken or written, respectively) languages, on the one hand, versus nonverbal languages, on the other – this difference is closely related to (but, of course, not to be identified with) two fundamentally distinct conceptions of language: language as a *system*, on the one hand, and language as a special mode of *action,* on the other.[5]

When we differentiate, as is common, between syntax, semantics and pragmatics, then this distinction can also be expressed in the following way: in the system-concept of language the reference point of all language-theoretical reflections is the syntax (grammar), whereas in the action-theoretical approach it is, by definition, pragmatics (in the wide sense, i.e., for communicative purposes not yet presupposing the existence of a language to be used for these purposes).

The grammar of a language system divides the infinitely large amount of complex signs which are constructible out of the respective basic signs into well-formed versus non-well--formed entities. Linguistic meanings in this framework are nothing but entities that are assigned to the respective signs (basic signs as well as complex signs).

---

5 For this point, see esp. G. Grewendorf, *Sprache als Organ, Sprache als Lebensform*, Frankfurt, 1995.

These entities can either be from the outside world or from the inside world (concepts, mental representations or ideas, let's say).

Language as *action*, on the other hand, primarily orientates itself on something that can be accomplished *non-verbally*, too: Communication. Our gestures and other 'signs' get their (group-related common communicative) meaning only through conventional use, guided by rules, in the respective populations and contexts of utilisation.

According to this pragmatic approach, what we *mean* – in the double sense of what we try to communicate and what we intend (mean to do) – primarily shows up in the things we do. To have language at one's disposal means to dispose of certain group-specific conventional forms of communication. Conceptual distinctions manifest themselves in a corresponding differentiated *knowing how*.

3.3 It is obvious which of these two conceptions has been serving as an orientation for the prevailing language paradigm of prehistoric archaeology, so far; its results have entered Leakey's history of traces. And it should be equally obvious what the shift of perspective will bring about when rethinking the research paradoxes which result from the system-concept of language. To put it in a short formula:

Linguistic ability is not the same as *speech ability*; linguistic ability (in a wide sense) is group-specific ability of communication – by any means; whereas speech ability relates to the possession of an equipment and apparatus (organs) for the production of distinctive sound sequences.

Leakey's arguments contra Darwin's Package Hypothesis (DPH) result from the (false) equation of linguistic ability and speech ability; Leakey's arguments against the DPH are therefore insubstantial.

3.4 The fact that certain objections against Darwin's package are not cogent, does, of course, not imply that this package is already proven right. It remains to be shown that the new concept of language does much more justice to Darwin's hypothesis. This is the case.

It is Leakey's achievement (by summing up the state of the art of research) to have drawn our attention to a visible explanatory gap concerning the alleged correlation of brain and language developments of the human being. This gap matches exactly the triangle that results from figures I-II-III above. In excerpt:

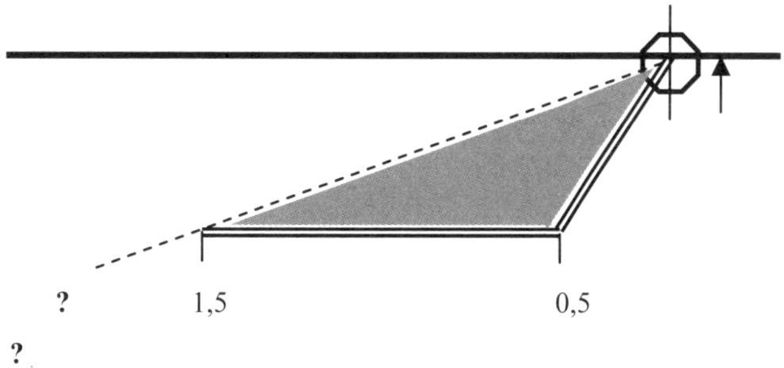

The corners of the triangle are: a) the 'point' about 1½ million years back in time (coinciding with the beginnings of the Acheulean culture) when the development of our brain volume (thin dotted line) uncouples from the development of our mouth tools (double line); b) the point about 1 million years later (bottom right) from where this development of mouth and jaw made a major step towards our tools of speech of today; and c) BTP (brain threshold point).

3.5 If there was no such triangle there would not be any problem – even for Leakey - with Darwin's hypothesis. How can a newly orientated archaeology (regarding language theory) cope with this problem? Very easily.

The so far unanswered question is: (*) what actually did our ancestors do for a million years with their perpetually developing brain and the improving cognitive capacities that went along with it? Or, as an evolutionary equivalent: (**) why did the brain of Homo erectus develop into the one of the Archaic Homo sapiens – while the development of our speech equipment was quasi at a standstill during all that time?

3.6 You already know the answers. The answer to question (*) is: They communicated, what else? And this in an increasingly improved way – in the sense of more diverse, more refined, more cooperative etc. ways of communicating. This is the only way to explain the transition from the Oldowan to the Acheulean culture (from mere stone chips to targeted forming of stone wedges). And the answer to (**), again the evolutionary equivalent: The targeted production of stone wedges made a much more differentiated *know-how in production* necessary, beginning with the teaching and learning structures of this production; and this improvement of production technology would be unthinkable without a corresponding *communicative know-how*.

The 1 million year triangle – an objection *against* Darwin's package? Not at all. It is rather a very strong argument *in favour of* this package hypothesis.

3.7 In order to not disappoint Leakey's expectation (see above 1.7) about tackling the issue of the Neanderthal paradox, here is a brief thought: Why aren't the brains of the Neanderthals *less developed* than ours (our ancestors')? (See again the Bio-II figure on page 4.) Because the Neanderthal *wasn't less communicative* than we were. This already suffices to dissolve the alleged paradox.

Why, then, did he die out? This is an additional problem; we can only make an educated guess: *If* his extinction has something to do with his less developed speech equipment, at all, then his departure was probably due to the fact that a larger verbal linguistic communication competence bears a huge advantage in comparison to non-verbal gesturing communication: It functions, E.g., much better by night – even in absolute darkness. Our *Archaic Sapiens* predecessor would have been stupid if he had not utilized this advantage during nocturnal raids and hold-ups.

3.8 Leakey himself actually has a communicative know-how explanation ready at hand. Why did he ignore it? Because it is so simple and self-evident. Blinded by the language system concept and its fixation on *speech* instead of *communication* Leakey and most of his colleagues have just not perceived this most simple of all explanations or did not want to.

Although we can never know for certain what daily life was like in the earliest times of Homo erectus, we can use the rich archaeological evidence of site 50 and our imagination, to re-create such a scene, 1.5 million years ago. (p. 73 f.) –

and then Leakey describes how such a day may have looked like 1.5 million years ago:

The males hunt, the females gather: it's a system that works spectacularly well for our group and for as long as anyone can remember. … females, naked apart from an animal skin … [use] digging sticks, which allow … to unearth deeply buried, succulent tubers [etc. etc.] The youngsters do what all human youngsters do: they play games of pretending, games that foreshadow their adult lives. [One of the female kids] persuades one of the females to show her, again, how to make stone tools. Patiently, the woman brings two lava cobbles together with a swift, sharp blow. A perfect flake flies off. With studied determination, the girl tries to do the same, but without success. The woman takes hold of the girl's hands and guides them through the required action, in slow motion. … Making sharp flakes is harder than it looks, and the skill is taught mainly through demonstration, not verbal instruction. (p. 74 f.)

This *example from everyday life* intuitively illustrates in a beautiful way how the acquisition of technical skills goes hand in hand with mimetic (i.e. communicative) know-how. But for the *theory* of language development (as represented by Leakey, op.cit.) this story apparently does not play any role.

3.9 Leakey himself must at this point have completely forgotten about Darwin's package. How else could it be, as tool production and language development in his view have entirely uncoupled as soon as "in the beginnings of Homo erectus" (loc. cit.), so that he writes about a Homo erectus group and their actions on a faraway day in our prehistory: "They are chattering loudly, some of their exchanges obvious social repartee, some the discussion of today's plans". (p. 74) Chattering loud voices, infused with jesting words – even without developed organs of speech? Seriously?

It was probably a distinction like this (between implicit knowledge and explicit theory) that Wittgenstein had in mind when he occasionally demanded: "Do not think, look!". This is exactly my recommendation to those anthropologists/archaeologists that deal with the evolution of language.

3.10 There is a very simple explanation for the communicative ignorance in the anthropological view: Non-verbal communication (as mentioned above) does not leave immediate traces; at least none that could be reconstructed with methods of anatomy. This is different with speech. This is why we have developed our speech equipment:

Pharynx, vocal chords, laryngeal shift, cellphone etc. Thus, the methods of bio-anthropology are excused concerning their ignorance on the issue. Yet, this excuse is not valid for social anthropology; and yet, social anthropology also subscribed to the language sign system approach for a long time – and thus attempted to communicate with our next of kin with the help of simple material linguistic signs (e.g. cards with different colours and forms).

But this is already old news. The vast scope of social anthropological progress after this necessary shift towards the language-action-paradigm is evident. Evidence can be seen for example in Leipzig, right next door to the Institute of Philosophy. Just visit 'Pongo Land' or the webpages of the *Max Planck Institute for Evolutionary Anthropology* which is cooperating with this research centre: www.eva.mpg.de.[6]

References

Darwin, C. 1859. *On the Origin of Species*. London: John Murray.

Leakey, R. 1994. *The Origin of Humankind*. London: Weidenfeld and Nicholson.

Meggle, G. 1997[2]. *Grundbegriffe der Kommunikation*. Berlin & New York: de Gruyter.

Meggle, G. *Handlungstheoretische Semantik* via www.uni-leipzig.de/~philos/meggle/index.php. Link *Schriftenauswahl Online*, entry (HTS).

Meggle, G. 2003. Kommunikatives Verstehen – Die Grundzüge. In: H. Richter and H.W. Schmitz (eds.), *Kommunikation – ein Schlüsselbegriff der Humanwissenschaften?* 341–352. Münster: Nodus-Publikation. Also via: www.uni-leipzig.de/~philos/meggle/index.php. *Schriftenverzeichnis Online*, entry (2003c).

Grewendorf, G. 1995. *Sprache als Organ, Sprache als Lebensform.* Frankfurt: Suhrkamp.

Strahm, C. 1998. Die kulturelle Evolution des Menschen. *Freiburger Universitätsblätter* 139, 91–109.

Tomasello, M. 1999. *The Cultural Origins of Human Cognition*. Cambridge, Mass. and London: Harvard University Press.

6 For eva's most relevant publication in this respect see M. Tomasello, *The Cultural Origins of Human Cognition*, 1999.

Comprehensibility in Forensic Linguistics – New Perspectives for Frame Semantics*

Monika Rathert

Auch juristische Probleme entstehen, wenn die Sprache feiert.[1]
Günther Grewendorf

1. Introduction

During the last two decades, Günther Grewendorf has worked intensively in the field of forensic linguistics. In a number of inspiring papers and talks (Grewendorf 1985, 1990a, 1990b, 1991, 1992a, 1992b, 1997, 2000a, 2000b, 2000c, 2001, 2002), he pointed out that forensic linguistics needs the insights and methods developed within modern theoretical linguistics. Jurists need linguists, and the more theoretically informed the linguists are, the better the overall interdisciplinary collaboration will be.

I believe Günther is perfectly right. Moreover, I claim that theoretical linguistics will get vital feedback from forensic linguistics. In this paper, I demonstrate this claim with respect to the notion "comprehensibility". It has often been said that legal texts are incomprehensible for the layman. This global judgment has been made concrete in various ways, psycholinguistic experiments have been set up to examine the actual degree of comprehension. Drawing on these experiments, attempts have been made to optimize legal texts for the layman. However, the results of this research are unsatisfactory in several respects. First, there are different sorts of laymen, some understand more (e.g. a law student), some less. Text optimization, especially in the context of the *plain language movement*, usually aims at meeting the needs of a layman understanding next to nothing. Second, there is no principled way to predict comprehension (for whatever kind of layman). Instead, comprehension is just measured empirically. We demand more from a linguistic theory. Third and maybe most important, the results of this research are not applicable in everyday life. A lawyer who wants to check a letter to his client in

* I would like to thank my colleagues from the *Computational Linguistics & Phonetics* Department in Saarbrücken, Aljoscha Burchardt, Anette Frank, and Stephan Walter, for fruitful discussions about Frame Semantics. Günther Grewendorf was the first to draw my attention to the topic of forensic linguistics. This paper would not exist without the wonderful cooperation I experienced from him. I would like to thank him very much. I am looking forward to our two books on the topic we are going to write and edit. Part of this research was presented at the workshops "Language of the law" (June 2005, Saarbrücken) and "The Language and the Law Conference" (May 2006, Düsseldorf). Thanks to the audience for comments and discussions. Thanks to Jan Engberg and an anonymous reviewer for valuable comments. Thanks to Patrick Brandt and Eric Fuß for their help with the manuscript, their support, and – last but not least – for their patience. Thanks to Annette Crowther for correcting my English.

1 Grewendorf (1992b:39).

terms of comprehensibility does not have anything at hand to do this. He still has to use his intuition.

The paper is structured as follows. Section 2 presents linguistic approaches to comprehensibility in general. Section 3 is devoted to comprehensibility within forensic linguistics. The theoretical and practical shortcomings of this research will be shown. Section 4 presents a new approach to comprehensibility within forensic linguistics. *Frame Semantics* as a shallow semantic approach will be used to predict the comprehensibility problems of a legal text.

2. Approaches to comprehensibility in theoretical linguistics

There are essentially three different approaches to comprehensibility. The oldest is within readability research, where readability is expressed in linear equations (section 2.1). The Hamburg School of the 70s assumes several dimensions of comprehensibility (section 2.2): *simplicity, structure, brevity, stimulation.* Finally, the propositional approach by Kintsch and his collaborators in the late 70s is the first to take psychological processing as an important aspect of comprehensibility into account (section 2.3).

2.1 Readability yardsticks

There is a long tradition on readability research in linguistics, dating back to the 30s of the past century. Readability is seen as a textual feature that has nothing to do with the addressee. Roughly, complex words and complex sentences make texts less readable.

Technically, readability is treated with regression analysis. The dependent variable y is called *readability ease*, the independent variables are x_1 (*number of syllables*), x_2 (*number of words*), and x_3 (*number of sentences*). With multiple linear regression, it is assumed that y can be explained by a linear combination of x_1, x_2, and x_3. The function is a linear (straight-line) equation.

There are more than 50 formulas for English, the most famous one by Flesch (1948:225). Flesch's *reading ease* may vary between 0 and 100, the higher it is, the more readable a text is. This is the formula and an example:

(1) RE (reading ease) = 206.835 – 0.846*wl – 1.015*sl
wl = word length, syllables per 100 words
sl = sentence length in words

(2) example from Flesch (1948:225):
The man with the pain in the nonexistent hand was an indigent, and Rovenstine was working before a large gallery of student anasthetists and visitors when he exorcised the ghosts that were paining him.
values for the independent variables: sl = 34, wl = 162
RE = 206.835 – 0.846*162 – 1.015*34 = 35.3

For German, the formula must be changed because the value for *wl* is usually higher. Amstad (1978:80) develops a German version of Flesch's formula:

(3) RE (reading ease) = 180 – sl – 58.5*wl
wl = syllables per word

Such a simple approach to comprehensibility seems very attractive, because it can easily be applied in automated processing. But we encounter serious problems. First, the ideal of linguistic simplicity (short sentences, simple words) which is part of the analysis may be apt for newspaper texts, but can certainly not be seen as an ideal for legal texts. Newspaper texts address a broad audience, whereas legal texts (like all specialized texts) address a small audience of highly educated professionals. Rigorously simplifying legal texts means artificially reducing the technical vocabulary that is needed – most of it cannot be translated into plain language without a loss in precision. As for the layman, it will not suffice to simplify the words and shorten the sentences. If the law and its language as such are hard to understand, it will not help to simplify the linguistic surface. Second, the formulas do not take the internal structuring of texts into account. A badly structured, nearly senseless text will get the same RE-value as a well structured text if the sentences are of equal length and the words are equally complex on average. Psycholinguistic experiments validating the formulas are actually evidence for a different comprehensibility than the one expressed in the formulas. Thus, it is not correct to establish correlations.

2.2 The Hamburg School

The Hamburg School of the 70s (Langer et al. (1974)) uses a semantic differential to determine factors of comprehensibility. About 20 adjectives referring to style and structure (like *coherent, circuitous, catchy* etc.) are offered to ten experts as a means to judge many texts of different kinds. The judgments are to be given in a scalar form, ranging from +2 to –2. In a second step, correlations between the scalar judgments are calculated and finally brought together in a multi-factorial analysis of comprehensibility. The Hamburg School establishes four dimensions of comprehensibility: *simplicity, structure, brevity, stimulation*. They are usually displayed as graphs, like in (4):

(4) dimension: *simplicity*

Simplicity	+2 +1 0 –1 –2	*Intricacy*
simple exposition short, simple sentences common words terminology explained concrete vivid		complicated exposition long sentences, embeddings uncommon words terminology unexplained abstract not vivid

Subjects mark the numerical values (from +2 to –2) of all features from all four dimensions. A text achieves optimum comprehensibility where the following four conditions apply:

(5) a. the text has an average simplicity-factor of 0.5 to +2 (i.e., the text is rather simple)
b. the text has an average structure-factor of 0.5 to +2 (i.e., the text is well ordered)
c. the text has an average brevity-factor of –0.5 to +1.5 (i.e., the text is neither too short nor too long)
d. the text has an average stimulation-factor harmonizing with the factor of structure (i.e., if a text is almost unstructured, it should contain few elements of stimulation; if a text is well structured, it should contain much stimulation)

The problem with this approach is its impressionism. Taking the features for *simplicity* in (4) as an example, most of them are subjective and intuitive. Moreover, features referring to content (like *terminology explained*) occur alongside with those referring to structure (*simple exposition*) or style (*vivid*). The comprehensibility of a text might be reflected in the features of the four dimensions, but it is not clear, how. Again, as with the readability yardsticks in the last section, the ideal of linguistic simplicity is certainly not an ideal for legal texts. There does not seem to be a theory behind the empirical attesting of the four dimensions. It is unclear why we have to have four (and not more or less) dimensions. It is not worked out in which ways the four dimensions interact (with the exception of *stimulation* and *structure* – and also for this case we do not know whether the correlation of the values for *stimulation* and *structure* is more or less important than a high value for *structure* as such).

Amstad (1978:85f.) tested his German version of Flesch's readability formula against the texts of Langer et al. (1974). Surprisingly, he makes the same predictions as for comprehensibility. Thus, the more elaborate system of Langer et al. (1974) does not seem to deliver better results.

2.3 The propositional approach

A more linguistically informed approach to comprehensibility is proposed in Kintsch and Dijk (1978), Kintsch and Vipond (1979), and Kintsch (1988). The motivation for this approach was twofold. First, the process of learning and understanding texts was poorly understood at this time. Second, the semantic content of texts could not be described in a more than impressionistic manner.

To start with the second point, the propositional approach tries to apply a simplified version of formal semantics to ordinary texts. In the style of Fillmore (1968), propositions are defined as n-tuples of *concepts*, with a concept consisting of a *predicate* P and *arguments* $A_1, \ldots, A_n$. Here is an example:

(6) *John loves Mary.* → (LOVE, John, Mary) — proposition (p)/ concept

predicate (LOVE) — arguments (John, Mary)

Propositions may be *predicative*, *modifying* or *connective*. With a *predicative* proposition, the verb simply has the role of the predicate, (6) serving as an example. *Modifying* propositions are those where arguments are modified, *connective* propositions express relations between propositions:

(7) modifying propositions
All PhD students sleep badly. →
p1 (SLEEP, PhD students)
p2 (quantification: PhD students, all)
p3 (attribute of, p1, badly)

(8) connective propositions
Schools are important for children not to stupefy. →
p1 (IMPORTANT, schools)
p2 (STUPEFY, children)
p3 (negation, p2)
p4 (in order to, p3, p1)

The propositional approach is not only interested in the semantics of individual sentences like those in (6)-(8), but is heading for semantic representations of texts. As a text consists of propositions, coherent texts must somehow consist of a net of propositions. This net is called *coherence graph*. Cf. the following example:

(9) *The Greeks loved beautiful pieces of art. When the Romans overwhelmed the Greeks, they imitated the Greeks. By doing so, they learnt how to produce beautiful pieces of art.*
p1 (LOVE, Greeks, art)
p2 (beautiful, art)
p3 (when, p4, p5)
p4 (OVERWHELM, Romans, Greeks)
p5 (IMITATE, Romans, Greeks)
p6 (LEARN, Romans, p8)
p7 (consequence, p5, p6)
p8 (PRODUCE, Romans, p2)

(10)

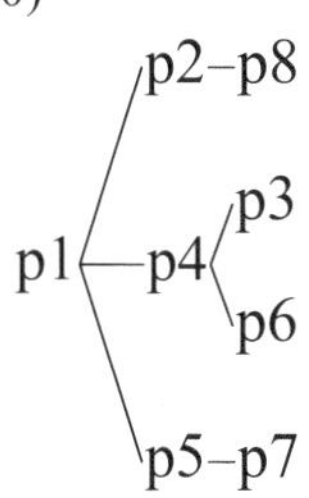

All propositions of a text must be represented within a single coherence graph. The propositional approach claims that texts are either fully understood or not understood at all. Texts cannot be understood “more or less”. The reader infers propositions (or relations between propositions) only if the single coherence graph cannot be constructed.

The propositional approach also contributes to the psychological aspects of text processing. The claim is that text processing is a recursive process involving the following steps:

(11) a. A chunk (= set of propositions) is taken into the working memory
b. A coherence graph is constructed from the chunk (also in the working memory)
c. Some part of the coherence graph is taken over to the short-term memory
d. This part is finally taken over to the long-term memory

The most important part of the recursion is (9-c), because subjects make a selection at this point. The part of the coherence graph which is taken over to the short-term memory is the so called *leading edge* (the left lower edge), printed in bold in the following example:

(12)

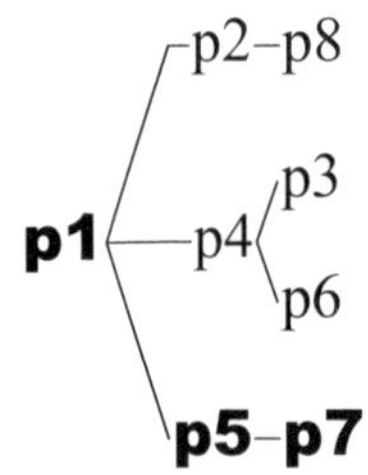

There are three possibilities to handle incoherence within the working memory. First, a copy of some suitable proposition can be taken from the long-term memory, this is called *reinstatement*. Second, all or at least some graphs within the long-term memory can undergo *reorganization*. Third, subjects may repair incoherences with *inferences* drawn from the text in a free manner, or drawn from the knowledge within the long-term memory.

The propositional approach to comprehensibility surely is an innovation. However, the assumed semantics is very abstract; there is no connection to syntax. The semantics as such is relatively primitive, compared to the standard of formal semantics e.g. in Heim and Kratzer (1998). The mode of description in this propositional approach is semi-formal. As arguments of predicates we have mere words or whole propositions. Quantification, modalization, tense, polarity and similar notions cannot be formalized here. Grabowski (1991) already explains why sentences like *Geißler called the Greens slobs* cannot be treated with this semantics – the correct propositional representation demands too much, namely that the pragmatics of the sentence is understood, it must be clear that the representation differs from *Geißler called Mrs Wright Jenny*. Well-known example sentences like *The morning star is the evening star* (where *morning star* and *evening star* have the same meaning but a different sense) or John Perry's example *Heimson believes that he is Hume* (where *he* cannot be just anaphoric, Heimson is not as crazy to believe that Heimson is Hume; on the other hand, Heimson does not only believe that Hume is Hume) cannot be treated. – But there are also problems with the processing part of the approach. It is not clear why subjects should only start drawing inferences when incoherence comes into play. One would rather expect some inferring simultaneously to normal processing. Also, what happens to the rest of the *coherence graphs*, after the *leading edge* has been cut out? It is neither very plausible that the whole rest is forgotten, nor should it be taken for granted that all *leading edges* are transferred into long-time memory. What is forgotten or memorized depends on the subject's interest and knowledge. Finally, this system surely cannot be automatized.

3. Comprehensibility in forensic linguistics

There is a long tradition of complaints about the incomprehensibility of legal texts. Frederick the Great already said about the Prussian law "Es ist aber sehr Dikke und gesetze müssen kurz und nicht Weitläufig seindt (*It is very bulky, but laws should be short and not extensive*)."[2] He demanded that the "möglichste Deutlichkeit und Simplicite eines natürlichen und ungezwungenen Erzählerstyls" (*utmost clarity and simplicity of a natural and effortless narrative style*) should be attained and that the following must be avoided: "affectirte Wendungen, neu gemachte Worte ... imgleichen Termini technici, wenn sie nicht allgemein bekannter Verständlichkeit seien" (*affected style, neologisms... technical terms if they are not generally known*).[3]

The most frequent complaints raised in the German context are the following[4]:

(13)
a. nominal style (nominalizations, nominal compounds)
b. impersonal style (passive)
c. abbreviations
d. long sentences with many embeddings
e. official style ("Kanzleistil", *stylum curiae*)
f. outdated terms (e.g. *Kraftdroschke* meaning 'car')
g. terms with a meaning differing from the usual one (e.g. *Sache* has a terminological meaning in the legal context which it does not have in everyday's speech)
h. vague terms (e.g. "geringe Menge", "niedrige Beweggründe")

All issues in (13) refer to style, and most of them could be amended. But comprehensibility problems in the legal context go beyond style; they have a very serious background. Incomprehensible laws threaten the separation of powers, cf. the following quote from Hassemer (1992:73):[5]

---

2 Cited according to Grewendorf (1992b:37).
3 Cited according to Lorenz (1991:295f.).
4 Cf. Hoffmann (1992), Lerch (2004), Pfeiffer et al. (1987), and Jaspersen (1998).
5 Cf. also Grewendorf (1992b:35).

Im gewaltenteilenden Staat hat die Legislative gegenüber den beiden anderen Gewalten *de iure* eine Leit- und Begrenzungsfunktion, welche sie über das Instrument der Gesetze ausübt. Im Bereich der Rechtsprechung nennt man das "Bindung an das Gesetz" und assoziiert damit, dass der Richter das Gesetz nur zu vollziehen habe, ihm inhaltlich nichts hinzufügen dürfe, ihm gänzlich unterworfen sein solle. [...] Gelingt es dem Gesetzgeber [...] nicht, sich so präzise auszudrücken, dass der semantische Spielraum der von ihm verwendeten Begriffe klar ist, so setzt er die Judikative und die Exekutive auf den Platz, der vom Prinzip der Gewaltenteilung ihm vorbehalten ist. Die Teilung der Gewalten funktioniert nicht mehr, weil die Trennung ihrer jeweiligen legitimen Bedürfnisse verwischt ist. Die eine bestellt das Feld der anderen, schlimmer noch: Man weiß nicht genau, wo die Grenzsteine stehen. Diese Grenzsteine sind zu einem guten Teil sprachliche Gebilde.

Within a state that established the separation of powers, the legislature has de jure a leading function against the other powers. This leading power is executed with the instrument of the law. In adjudication this is called "being bound by the law". The association with this is that the judge may only execute the law; he is not allowed to add his own thoughts to it. In case the legislature is unable to produce precise laws, the judiciary and the executive must take the power of the legislature. The separation of powers no longer exists in this situation. Nobody knows where the responsibilities of the different powers start and end because the boundaries between them fade. These boundaries are to a large degree linguistic in nature.

Thus, the aim of comprehensibility seems to be intimately linked to the enlightened type of state in which powers are separated. However, Herberger (1983:20) has shown that there is no correlation between a certain type of state and the aim of comprehensibility: rather, he found that comprehensibility has been regarded as valuable as such, independent of the form of government.

The struggle for comprehensibility has been made explicit in a number of ways. In Germany, the *Gemeinsame Geschäftsordnung der Bundesministerien zur Gesetzgebung* states that "Gesetze müssen sprachlich einwandfrei und sollen so weit wie möglich für jedermann verständlich gefasst sein" (*laws must be linguistically correct and should be intelligible to everyone*).[6] The famous principle of civil freedom *nullum crimen, nulla poena sine lege* (Art. 103 II GG, §1 StGB) breaks down into several principles, among them the *lex certa* principle ("Bestimmtheitsgebot"): the legislative powers must formulate norms as precise as possible.

However, comprehensibility or incomprehensibility does not exist as such. It depends on the addressee and his/her knowledge and interests. Objective features of texts (like sentence length, frequency of nominalizations, depth of embeddings etc) are no predictors of comprehensibility.[7]

Empirical studies about comprehensibility in the field of forensic linguistics will often surprise the reader by their methodology. A representative and recent example is Hansen-Schirra and Neumann (2004). In this paper, judgments from the German Federal Constitutional Court (*Bundesverfassungsgericht*) are compared with the corresponding press releases in terms of comprehensibility. The empirical base of the paper is not very large – the corpus of judgments contains 14,000 words (this corresponds to about 25 pages of text), press releases are 15,000 words, and a third corpus called *reference corpus* contains 15,000 words. The press releases were written by a single judge –

6 GGO II § 35 Abs. 1, cited from Hoffmann (1992).

7 See Chall (1958), Weber and Noschka-Roos (1988), Hoffmann (1992).

thus, the investigation of the press releases really is an investigation of the language of a single person, a lawyer having the job to write press releases. This weakens all conclusions considerably. The content of the reference corpus is not clear, it is only said that it contains different sorts of texts. The function of the reference corpus seems to be that of a linguistic standard, it is at least always compared to the other two corpora. The first step in the analysis is to choose (morpho)syntactic and lexical features that may hinder comprehensibility (Hansen-Schirra and Neumann (2004:171)). The authors choose a subset of the well-known features in (13), without giving any justification for their selection. For each feature, the authors compare its frequency in the judgment corpus with the frequency in the press releases corpus and with the frequency in the reference corpus. Exact numbers are given for all features, percentages etc. But the theory behind the numbers is at best hazardous. Two examples may illustrate this:

Example 1:

Lange Sätze. Längere Sätze dürften im Gegensatz zu kurzen Sätzen längere Verarbeitungszeiten beim Lesen erfordern. Somit erschweren sie insbesondere in Kombination mit den unten diskutierten komplexen Nominalphrasen das Verständnis der Texte. (Hansen-Schirra and Neumann 2004:172)

Long sentences. Longer sentences will take more processing time than short sentences. Therefore, they will decrease comprehensibility, especially if combined with complex NPs.

The first statement is just trivial – of course it takes longer to read a longer sentence. The second statement is an unchecked conjecture about the comprehensibility of "long" (how long is long?) sentences. The authors did not check the real understanding with any of the standard psycholinguistic tests (like cloze procedure, eye tracking, multiple choice questions etc.). They did not check one of the three corpora; they merely report beliefs about comprehensibility. Another dangerous point about the second statement is the combinatorics. If "long" sentences are hard to comprehend, this should be checked empirically as a parameter, leaving all other parameters like the mentioned "complexity of noun phrases" constant. There is a tremendous amount of literature on psycholinguistic experiments (see the overviews in Christmann (2004) or Schendera (2002)), and it is hard to understand why such simple issues like the separation of parameters are not known in applied linguistics.

Example 2:

Formelhafte Wendungen. […] Zunächst ist anzunehmen, dass in den Urteilen durch die wiederholte Verwendung von formelhaften Ausdrücken der Gesamtwortschatz weniger umfangreich ist als im Referenzkorpus und somit das Type-Token-Verhältnis niedriger ist. Die Urteile weisen mit 38,70% in der Tat den am wenigsten umfangreichsten, das Referenzkorpus mit einem Wert von 53,38% den umfangreichsten Wortschatz auf. Die Pressemitteilungen liegen mit […] 43,02% zwischen den anderen Werten […]. Dies deutet darauf hin, dass die Autorinnen der Pressemitteilungen einen breiteren Wortschatz verwendet haben als die Autorinnen der Entscheidungen, was die Texte weniger repetitiv und somit ansprechender […] macht. […] Diese Verbesserung der Lesbarkeit sollte auch positive Auswirkungen auf die Verständlichkeit der Texte haben. (Hansen-Schirra and Neumann 2004: 173-181)

Formulaic terms. Due to the frequent repetition of formulaic terms, it is probable that judgments have a less comprehensive vocabulary and a lower type-token-ratio than the reference corpus has. And indeed, the judgments have the least comprehensive vocabulary (38.7%) whereas the reference corpus has the richest vocabulary (53.38%). The press releases are in between (43.02%). This means that the authors of the press releases have a greater vocabulary than the authors of the judgments, and this in turn makes the press releases less repetitive and thus more vivid. Enhancing readability in this way should have a positive effect on comprehensibility.

First, it becomes obvious that the reference corpus has no function at all in this paper. The goal of the paper is to show that the press releases are easier to comprehend than the judgments (Hansen-Schirra and Neumann 2004: 168). If a "better" type-token-ratio (better for whom?) alone suffices to attest more stimulation (in the terms of Langer et al. 1974), and more stimulation means enhanced comprehensibility, then the goal of the paper is achieved. The comparison with the reference corpus is not necessary to prove the point. Second, nobody was asked if he/ she judged the press releases more vivid and if this in turn enhances his/ her comprehension. Third and again, the paper neglects that different addressees have different needs.

Many papers in the field are like Hansen-Schirra and Neumann (2004), where a small textual basis combines with a lack of empirical validation. The notion of comprehensibility remains a theory-free and empiricism-free notion.

But not all papers are like this. A notable exception is Dietrich and Kühn (2000). This study clarifies how much and which parts of the content of a legal text are understood by expert and non-expert readers. The legal text was about conditions of insurance policies. The second experiment shows that both types of subjects (expert and non-expert readers) have substantial comprehensibility problems. This is not due to the lack of acquaintance with legal terms, but due to the lack of expert knowledge, knowledge about insurance law.

Given all this, what is needed in forensic linguistics' comprehensibility research is the following:

(14) a. *A method to model the meaning (the semantics) of a text*
We need to have a model for the maximally understandable content of a text. The propositional approach by Kintsch and Dijk (1978), Kintsch and Vipond (1979), and Kintsch (1988) must be pursued and improved.

b. *Text meaning representations must be automated*
For practical reasons, we need a software tool, as lawyers and laymen shall also use comprehensibility tests.

c. *Inferences and interactions with connected pieces of knowledge must be shown*
The findings of Dietrich and Kühn (2000) should be taken very seriously. If expert knowledge is indispensable, the model must indicate it.

d. *The model must be empirically validated*
It is necessary that psycholinguistic tests determining individual comprehension shape the model.

4. New perspectives for *Frame Semantics*

There does not seem to be much interdisciplinary collaboration between computational linguists and jurists, the DFG-project CORTE[8] being an exception. However, it seems to me that many of the desiderata in (14) could receive inspiring suggestions from the part of computational linguistics.

Within computational linguistics, "shallow" modelling of meaning in terms of predicate-argument structures has lately regained interest. For example, the Berkeley *FrameNet* project (Fillmore et al. (2003)) is developing a semantic lexicon for a core set of English, based on Fillmore's *Frame Semantics* (Fillmore (1976)). Frame semantics models the semantics of predicates in terms of *frames*. A frame describes a conceptual structure or a prototypical situation together with a set of involved semantic roles or *frame elements*. Frame elements thus identify the participants of the situation. Taking the example of VERDICT, the frame VERDICT has the roles CASE, CHARGES, DEFENDANT, JUDGE, and FINDING:

(15) [The jury]$_{\text{JUDGE}}$ convicted [him]$_{\text{DEFENDANT}}$ [on the counts of theft]$_{\text{CHARGES}}$

Elements like verbs, nouns, or adjectives evoke certain frames with their lexical semantics. As *FrameNet* (currently) focuses on these three open categories, the meaning representation is necessarily partial.

How a combination of *FrameNet's* frame semantics with a deep syntactic analysis can be used as a shallow meaning representation of texts has been discussed in Burchardt et al. (2005). Here, frames are not only evoked by the lexicon, but also in the course of inferential processes, or via meaning postulates. To show how such an analysis is represented, the essentials from the case study in Burchardt et al. (2005) are reproduced here. This is the text:

(16) *In the first trial in the world in connection with the terrorist attacks of 11 September 2001, the Higher Regional Court of Hamburg has passed down the maximum sentence. Mounir al Motassadeq will spend 15 years in prison. The 28-year-old Moroccan was found guilty as an accessory to murder in more than 3000 cases.*

A partial representation with local roles will look like this:

---

8 *Computerlinguistische Methoden für die Rechtsterminologie*, Manfred Pinkal, Stephan Walter (http://www.coli.uni-saarland.de/projects/corte/). The goal of CORTE is to provide knowledge-based technological support for work in the legal domain using techniques from computational linguistics based on a sound theoretical understanding of the creation, semantics and use of legal terminology. The aims of the project are: developing a linguistic model of definitions in legal text, building computational linguistic tools for the automatic extraction of such definitions, and finally exploring methods for the exploitation of the extracts in terminologies for the legal domain.

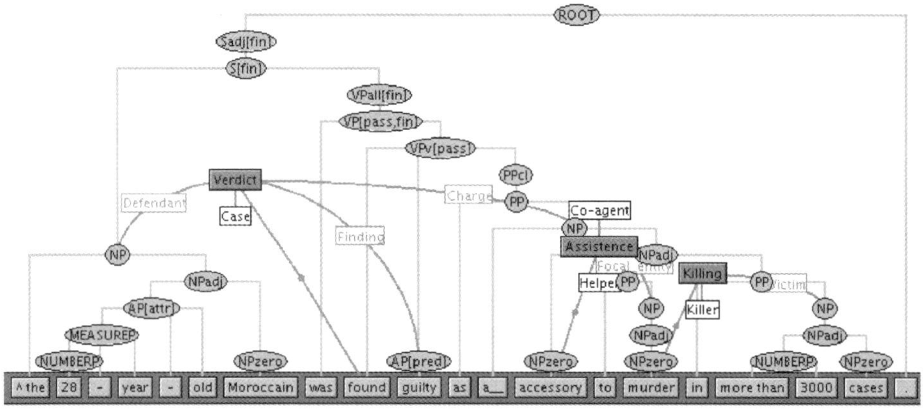

The last sentence from (16) is printed at the bottom. An LFG parser produces the syntactic structure for the sentence, the frames and their roles associated with it are provided automatically, see Erk and Pado (2006). In the diagram above, frames are in grey rectangles (like VERDICT at the left, ASSISTENCE and KILLING at the right), roles are white rectangles with a grey margin (like DEFENDANT, CASE, FINDING, CHARGE as roles of the frame VERDICT). Erk and Pado's *Shalmaneser* is used in the RTE-system of Burchardt and Frank for robust, broad-coverage semantic annotation (Burchardt and Frank 2006).

The frame VERDICT (see left hand side) has the filler-NP *The 28-year-old Moroccan* for its role DEFENDANT, whereas its role CASE has no filler. What does this give us? I believe it is the first usable tool with which a new model of comprehensibility can be developed. In the spirit of the propositional approach by Kintsch and Dijk (1978), Kintsch and Vipond (1979), and Kintsch (1988), we have a model for the maximally understandable content of a text. It is not yet usable for laymen, but it is an automated software tool that can be made more user-friendly.

The denser the frame net is (with super- and sub-frames) the better a text should be understood. It would be interesting to check whether relations between frames as defined in *FrameNet* can be attested empirically in comprehensibility tests. Defined prominent relations are *inheritance* and *subframe*. If a frame *a* inherits from frame *b*, then all frame elements of *b* are also available for *a*. The frame ARREST inherits PATIENT and AGENT from the frame INTENTIONALLY_AFFECT. The *subframe* relation is involved in larger frames, often called 'scenario frames', which represent complex events being composed of temporally ordered sub-events. As an example, CRIMINAL_PROCESS contains the subframes ARREST, TRIAL, SENTENCING (among others), having a fixed causal and temporal order with respect to each other. Frame elements are usually also inherited to the subframes. To stay with the example subframes ARREST and TRIAL, the same person who is arrested will also be the 'patient' in the trial.

Frames that cannot be integrated should cause comprehensibility problems. In the diagram above, the role CASE of the frame VERDICT cannot be filled. This should decrease comprehensibility. On the other hand, syntactic material that fills no frame should also decrease comprehensibility, because it adds material that does not contribute to conceptual structure. Of course, this does not apply to nodes whose daughters are fillers: in the diagram above, the node S[fin] is no filler, but its daughter NP bears the role DEFENDANT.

Inferences, i.e., inferred relations between frames, can also be shown. One would expect a difference in comprehensibility between overt and inferred relation, and also differences between sub- and super-frames. Referring to the desiderata in (14), (14-d) must be fulfilled: *the model must be empirically validated.*

However, we have a model at hand here. This is much more than most studies on comprehensibility come up with.

5. Conclusion

Coming back to Günther's Wittgenstein variation *Auch juristische Probleme entstehen, wenn die Sprache feiert* – problems of a language celebrating itself are the well-known problems of comprehensibility the forensic linguist and the jurist have to deal with.

The frame annotations leading to partial text meaning representations as described in Burchardt et al. (2005) aim at applications in practical NLP tasks, such as Information Extraction or Question Answering. No one from *Frame Semantics* has ever looked for an application like forensic linguistics, and forensic linguists have not yet started using the methods from computational linguistics. This is a concrete point where interdisciplinary collaboration should start.

In the introduction, I claimed that theoretical linguistics will get vital feedback from forensic linguistics. The productive feedback for *FrameNet* and *Frame Semantics* in this case is being exposed to a question always asked in forensic comprehensibility research: what is it that we understand, on word and on sentence level? To put it more concrete, the declared aim of *FrameNet* and *Frame Semantics* is to build up lexical databases, to do lexicography. A newer aim is to annotate continuous texts, as a demonstration of how this tool can contribute to text understanding.[9] However, as this research is empirically based on written corpora only, it is not clear what the psycholinguistic correlates of the frames (and their hierarchical and inferred relations) are. In other words: *FrameNet* and *Frame Semantics* should expose themselves to an empirical validation of their powerful and promising model.

9 See the Berkeley site for current work carried out: http://framenet.icsi.berkeley.edu/.

References

Amstad, T. 1978. *Wie verständlich sind unsere Zeitungen?* Doctoral dissertation: Universität Zürich.

Burchardt, A. and A. Frank. 2006. Approximating textual entailment with LFG and Frame Net Frames. To appear in: *Proceedings of the 2nd PASCAL Recognizing Textual Entailment Workshop*, 92–97. Venice.

Burchardt, A., A. Frank, and M. Pinkal. 2005. Building text meaning representations from contextually related frames. *Proceedings of IWCS* 6, Tilburg.

Chall, J.S. 1958. *Readability: an Appraisal of Research and Application.* Columbus: Ohio State University.

Christmann, U. 2004. Verstehens- und Verständlichkeitsmessung. Methodische Ansätze in der Anwendungsforschung. In: K. Lerch (ed.), *Sprache des Rechts. Band 1: Recht Verstehen. Verständlichkeit, Missverständlichkeit und Unverständlichkeit von Recht*, 33–62. Berlin, New York: Walter de Gruyter.

Dietrich, R. and K. Kühn. 2000. Transparent oder verständlich oder wie was verstanden wird - Eine empirische Untersuchung zum Verstehen eines juristischen Textes. *Zeitschrift für Literaturwissenschaft und Linguistik* 118: 67–95.

Erk, K. and S. Pado. 2006. *Shalmaneser* – a toolchain for shallow semantic parsing. To appear in *Proceedings of LREC-06*. Genoa.

Fillmore, C. J. 1968. The case for case. In: E. Bach and R.T. Harms (eds.), *Universals of Linguistic Theory*, 1–88. New York: Holt, Rinehart amd Winston.

Fillmore, C. J. 1976. Frame semantics and the nature of language. *Annals of the New York Academy of Sciences: Conference on the Origin and Development of Language and Speech* 280: 20–32.

Fillmore, C. J., C. R. Johnson, and M. R. L. Petruck. 2003. Background to FrameNet. *International Journal of Lexicography* 16: 235–250.

Flesch, R. 1948. A new readability yardstick. *Journal of Applied Psychology* 32: 221–233.

Grabowski, J. 1991. *Der Propositionale Ansatz der Textverständlichkeit: Kohärenz, Interessantheit und Behalten*. Münster: Aschendorff.

Grewendorf, G. 1985. Vertrauen ins Mißtrauen. Linguistische Analysen zur Argumentation des Bundesverfassungsgerichts anläßlich des Urteils zur vorzeitigen Auflösung des 9. Deutschen Bundestages. In: J. Dyck (ed.), *Rhetorik. Ein Internationales Jahrbuch*, 101–115. Tübingen: Niemeyer.

Grewendorf, G. 1990a. XY… Ungelöst. Zu Theorie und Praxis forensischer Linguistik. In: H. Kniffka (ed.), *Texte zu Theorie und Praxis Forensischer Linguistik*, 247–288. Tübingen: Niemeyer.

Grewendorf, G. 1990b. Theoretische Probleme der Forensischen Linguistik. Guest lecture, University of Bonn, 23.10.1990.

Grewendorf, G. 1991. Rechtskultur als Sprachkultur. Theoretische Aspekte der forensischen Sprachanalyse. Guest lecture, RTWH Aachen, 20.12.1991.

Grewendorf, G. (ed.). 1992a. *Rechtskultur als Sprachkultur. Zur Forensischen Funktion der Sprachanalyse*. Frankfurt am Main: Suhrkamp.

Grewendorf, G. 1992b. Rechtskultur als Sprachkultur. Der sprachanalytische Sachverstand im Recht. In: G. Grewendorf (ed.), *Rechtskultur als Sprachkultur. Zur Forensischen Funktion der Sprachanalyse*, 11–41. Frankfurt am Main: Suhrkamp.

Grewendorf, G. 1997. Sprache und Recht. Guest lecture, Akademie für gesprochenes Wort, Stuttgart, 17. 01. 1997.

Grewendorf, G. 2000a. Die sprachliche Pflege des Rechts. Linguistische Aspekte der Rechtsprüfung. *Zeitschrift für Literaturwissenschaft und Linguistik* 118: 96–114.

Grewendorf, G. 2000b. Gibt es syntaktische Spuren? In: C. Baldauf (ed.), *2. Symposion Autorenerkennung. Tagungsband des Bundeskriminalamts*, 83–96. Wiesbaden: BKA.

Grewendorf, G. 2000c. Gibt es syntaktische Spuren? Paper presented at the *Symposium "Autorenerkennung"*, Bundeskriminalamt Wiesbaden, 3.–5.4.2000.

Grewendorf, G. 2001. Kein Patent auf Leben! Zur Pragmatik des Patentrechts. Paper presented at the conference "Sprache des Rechts", Berlin-Brandenburgische Akademie der Wissenschaften, Blankensee bei Berlin, 2.–4.4.2001.

Grewendorf, G. 2002. "Kein Patent auf Leben!" Zur Pragmatik des Patentrechts. *Zeitschrift für Literaturwissenschaft und Linguistik* 128: 8–33.

Hansen-Schirra, S. and S. Neumann. 2004. Linguistische Verständlichmachung in der juristischen Realität. In: K. Lerch (ed.), *Die Sprache des Rechts. Band 1: Recht Verstehen. Verständlichkeit, Missverständlichkeit und Unverständlichkeit von Recht*, 167–184. Berlin: Walter de Gruyter.

Hassemer, W. 1992. Richtiges Recht durch richtiges Sprechen? Zum Analogieverbot im Strafrecht. In: G. Grewendorf (ed.), *Rechtskultur als Sprachkultur. Zur Forensischen Funktion der Sprachanalyse*, 71–92. Frankfurt am Main: Suhrkamp.

Heim, I. and A. Kratzer. 1998. *Semantics in Generative Grammar*. Oxford: Blackwell.

Herberger, M. 1983. Unverständlichkeit des Rechts. Anmerkungen zur historischen Entwicklung des Problems und des Problembewußtseins. In: R. Wassermann and J. Petersen (eds.), *Recht und Sprache*, 19–40. Heidelberg: C. F. Müller.

Hoffmann, L. 1992. Wie verständlich können Gesetze sein? In: G. Grewendorf (ed.), *Rechtskultur als Sprachkultur. Zur Forensischen Funktion der Sprachanalyse*, 122–154. Frankfurt am Main: Suhrkamp.

Jaspersen, A. 1998. *Über die mangelnde Verständlichkeit des Rechts für den Laien.* Doctoral dissertation, Universät Bonn.

Kintsch, W. and T. van Dijk. 1978. Toward a model of text comprehension and production. *Psychological Review* 85: 363–394.

Kintsch, W. and D. Vipond. 1979. Reading comprehension and readability in educational practice and psychological theory. In: L. Nilsson (ed.), *Perspectives on Memory Research*, 329–365. Hillsdale: Erlbaum.

Kintsch, W. 1988. The role of knowledge in discourse comprehension: a construction-integration model. *Psychological review* 95: 163-182.

Langer, I., F. von Schulz, and R. Tausch. 1974. *Verständlichkeit in Schule, Verwaltung, Politik und Wissenschaft*. München, Basel: Ernst Reinhardt.

Lerch, K. (ed.). 2004. *Sprache des Rechts. Band 1: Recht Verstehen. Verständlichkeit, Missverständlichkeit und Unverständlichkeit von Recht.* Berlin: Walter de Gruyter.

Lorenz, R. 1991. Recht - Sprache - Begriff. In: J. Eckert and H. Hattenhauer (eds.), *Sprache - Recht - Geschichte. Rechtshistorisches Kolloquium 5.–9. 6. 1990*, 295–311. Heidelberg: C. F. Müller.

Pfeiffer, O. E., E. Strouhal, and R. Wodak. 1987. *Recht auf Sprache. Verstehen und Verständlichkeit von Gesetzen*. Wien: Orac.

Schendera, C. F. G. 2002. Der Nutzen wissenschaftlicher Verfahren zur Verbesserung der Verständlichkeit von normativen Texten: Überblick, ausgewählte Beispiele, Diskussion. *Zeitschrift für Literaturwissenschaft und Linguistik* 128: 98–113.

Weber, T. and A. Noschka-Roos. 1988. *Texte im Technischen Museum. Textformulierung und Gestaltung, Verständlichkeit, Testmöglichkeiten*. Berlin: Materialien aus dem Institut für Museumskunde, Staatliche Museen Preußischer Kulturbesitz Berlin.

Not *Only I*: Notes on the Syntax of Focus Binding

Tom Roeper

1. Introduction

Heim (1991) has brought attention to the role of indexicals in sentences like (1):

(1) *Only I got a question that I could answer.*

Partee (1989) noted a similar effect for sentences like (2):

(2) *I'm the only one around here who will admit that I might be wrong.*

In (2), the indexical binding has to proceed via a pronominal *the only one* that leads into a *wh*-relative clause. Kratzer (1998, 2006) has explored a variety of morphological variations on how these indexicals interact and exhibit agreement. In informal notes (Roeper 1999), I observed that proper nouns can also participate in this form of binding:

(3) *Only Mary still looks like Mary in these old pictures.*

In what follows we will describe, in syntactic terms, a variety of contexts where this connection both succeeds and fails, without specifying exactly how semantic representations might be involved, but rather observing from the other side of the interface what syntactic dimensions are engaged.

1.1 Semantics

First, it should be observed that whatever semantics are used here (structured meanings and alternative semantics have been explored), the effects must capture cases like (4):

(4) *Even I got a question that I couldn't answer.*

Here the same binding occurs, but the determination of the relevant set entails whatever value scale is implied by the word *even*.

1.2 Delicate judgments and a deeper method

Second, we would like to freely observe that, while the core intuitions about proper nouns are "delicate", in current parlance, they move in a common direction. Thus while one may not always be confident in some judgments, one can be confident in the method of searching through variation to see if it takes a common direction. In particular, many grammaticality judgments become clearer once a larger array of relevant facts are brought into focus. For instance, VP-ellipsis judgments usually become more liberal as the sentence becomes more complex, cf. (5).

(5) *A lot of this material can be presented in an informal and accessible fashion, and often I do.*
(Hardt (to appear), quotation from an interview with Noam Chomsky)

In comparison, a shorter version seems decidedly worse:

(6) *? *A lot of this material can be presented, and often I do.*

The *only Mary*-cases like (3) by contrast become systematically worse, not better, in complex environments, as we shall show. Therefore, the collective data points to an elaboration of what constitutes an "intervenor", what makes a sentence complex, in the sense that has become common in describing syntactic barriers.

2. The breadth of the phenomenon

What constituent categories allow set-induction? We argue that all categories do. Consider now the variety of proper nouns which, in a suitable context, allow set-induction:

(7) Proper names:
a. *In these dim photos, only Mary still looks like Mary.*

The primary reading here is that therefore, Bill does not look like Bill anymore in the mentioned photos – nobody except Mary looks like themselves anymore in them. Likewise we have:

(8) a. *Only Mary can act like Mary in the culture of Greenwich village* (*noone else can act like themselves there*).
b. *Only Mary can act like Mary in a foreign culture* (*the rest of us can't really be ourselves*).
c. *Only Fred still talks like Fred in front of royalty* (*everyone else puts on a phony accent*).

(9) Other proper nouns:
a. *Only Mt Hood still looks like Mt Hood even in winter* (*no other mountain looks like itself in winter*).
b. *Only Christmas is always like Christmas in every culture* (*no other festivity is always the same in every culture*).
c. *Only my birthday makes me feel like my birthday in my old hometown* (*Christmas does not feel like Christmas in my old hometown*).

(10) Indefinites:
a. *Only a very odd person still looks like a very odd person after using modern make-up techniques* (*ordinary people don't look ordinary*).
b. *Only a very loud person sounds like a loud person in the old radio tapes* (*other accents don't come through as themselves*).
c. *Only a very very tall person looks like a very very tall person from a great distance* (*small people don't look like small people, they just look normal*).

(11) Definites:
In these old pictures, only the living room still looks like the living room (*no other room looks like itself*).

Other kinds of indexicals, which carry a point of view (henceforth: POV), also seem to allow this behavior:

(12) Point of view (POV) indexicals:
a. *Only now can one appreciate now* (*then one could not appreciate then* or *then one could not appreciate now*).
b. *Only here can one praise here* (*there one cannot praise there*).

It seems to be a truly free phenomenon, applicable to VP gerunds as well:

(13) a. *Only ballet dancing looks like ballet dancing in a still photo* (*breakdance doesn't look like breakdance, squaredance doesn't look like squaredance... in a still photo*).
b. *Only being nice is appreciated as being nice* (*acting any other way is misinterpreted*).

If these examples are sustainable, then it is clear that the element which can "associate with focus" is not limited to a set of lexical types (like pronouns) or syntactic types, or specificity types. To buttress out this view, we undertook an informal experiment in a linguistics class, and asked undergraduates to complete sentence (14).

(14) *Only Mary looks like Mary, no one else....*

14 out of 18 undergraduates gave as an answer: *looks like themselves* rather than *looks like Mary*. This showed that the interpretation is readily available and not an obscure alternative.

3. Divergence

Where do these structures begin to diverge? Notice that we get a sharp degradation when we substitute *person* for the pronoun in the Partee example:

(15) ?? *I am the only person around here who will admit that I might be wrong.*

The pragmatics of the intended meaning is so strong that no other meaning is plausible; why should a person other than me be caused to admit something about me. Yet, the grammaticality seems to decline, which suggests structural factors interfering with our intuitive sense of well-formedness. Notably we find that where the non-set reading is plausible it will take over easily:

(16) *I am the only person around here who thinks that I might be wrong.*

This has the reading that other people might think that I am wrong far more naturally than the reading that they might be wrong. While *only I* can form a chain with *the only one who*, it seems that proper nouns do not so readily accept that connection:

(17) ?? *In this picture, Mary is the only one who looks like Mary, no one else does.*

Anything other than *one* leads to much stronger ungrammaticality:

(18) a. ?* *Only Mt Hood is a mountain that looks like Mt Hood.*
b. ?* *In these old photos, only Ana is wearing a dress that makes Ana look good (everyone else...)*

It is not clear that an equivalent to *one* for location exists, so that *place* does not allow us to build a chain between the two instances of the proper name:

(19) **Only Mt Hood is a place that retains the majesty of Mt Hood in all weather.*

These examples suggest that indexicals more readily agree with a pronoun *one* than with proper names, which is not surprising since they are also pronouns. It implies that more than co-indexing is involved, namely a whole array of features, including [pronominal] for all types of constituents and perhaps all types of binding.[1]

4. Association with focus

Now we can begin to ask what the best explanatory tools will be. Rooth (1992) describes the phenomenon of association with focus where non-identical elements are linked, cf. (20):

(20) a. *Only John wears TWO hats every day.*

This is commonly described as a leftward movement at LF of the focal element, very roughly as (20'), where both *someone* and *x* have the status of traces/copied elements.

(20') [*Only* John, TWO [someone *is wearing* x-number *hats*]]

Notably this association with focus can occur across typical barriers:

(20) b. *Only John knows someone who wears* TWO *hats every day.*

It is generally assumed that the apparatus for focus binding is not within a movement theory if it does not obey obvious barriers. We will simply sketch an approach to the possibility of extraction from islands that continues to assume the presence of real movement. There have been many claims that subjacency is a weak effect and therefore

1 Kratzer (2006) pursues the features behind agreement in these contexts in a paper "Minimal Pronouns" presented in Saarbrücken, which follows Kratzer (1998).

it is not very unusual to seek a method to circumvent islands rather than abandon movement. We will discuss alternatives briefly below.

First, it is known that some quantifiers, like *each*, are also able to scope out of islands [e.g. *there is some fact that bothers each man*]. Terada (2003) has made the interesting argument that *each* has both an anaphoric feature and a distributive feature to account for the fact that it is linked to the subject anaphorically but confers distributivity on an adjective (cf. 21a,c) in a Spec-Head Agreement relation (sketched here):

(21) a. *The boys arrived* $[_{spec}$ *each* $[_{head}$ *alone*]].
[+anaph] [+anaph, +dist] [+dist]
b. * *The boys arrived each together.*
c. *The boys arrived each together with his mother.*

In (21b) distributivity fails, but is restored in (21c). We can carry the idea over to the *only* phenomenon. Let us assume, unlike for *wh*-words that are attracted to an invisible [+wh] feature, a link to a visible *only* creates an anaphoric connection. However, compositional interpretation of *TWO* requires movement of a focal feature to an LF adjacent position. We will not explore this idea further, but suggest that it may be an avenue through which one can explain the evasion of islandhood. Thus we will assume both that leftward movement occurs and that an anaphoric binding relation is present. We can capture that additional feature for all our data if we assume a rule that establishes binding under phonetic and semantic identity:

Binding: 1. Co-index identical words in a c-command domain as anaphorically bound.
2. At LF, move focal material to a position accessible to the *only*-operator (via feature-checking mechanism).
Only I … I
[+Foc] ← [+uFoc]
3. Treat identical foci as bound variables subject to joint set-induction from *only*.
Only Ix … Ix
Only → induces contrast set for all x

Now we need to define the movement path in such a way as to permit intervention.

4.1 C-command

It is clear that c-command is a requirement for the set-induction cases (as pointed out to me by R. Bhatt (pc)):

(23) *Only my daughter helped me, Jane's daughter didn't…*

For (23), one might marginally obtain set-induction in the first clause, but it does not continue into the elided VP: Jane's daughter didn't help me, not Jane. This is predicted if c-command is a requirement on the binding relation. Given c-command, if we assume that leftward movement underlies the bound focus in all instances of binding, then the

LF analysis of focus movement for these phrases suggests that we should expect barrier effects for movement.

4.2 Perspectivals

External perspective adverbs block long-distance movement:

(24) **How did John say* t_{how} *that he unfortunately could win the game* t_{how}.

Adverbs like *unfortunately*, *luckily* or *certainly* all imply an outside perspective, which Speas (1999) calls the "Seat of Knowledge". Potts (2005) discusses expressives like *damn* or *sure* and related terms which he argues can be interpreted at the root of the clause, but require what he calls a multi-dimensional semantic representation. This translates into the claim that the expressive features percolate to the CP of the clause. They also appear to shift POV to a speaker. We find that these outside perspectives weakly block set-induction (cf. (25)), and that expressives also block it (cf. (26)).

(25) a. ?? *Only I unfortunately got a question that I could answer.*
b. ?? *Only Mary unfortunately got a question that Mary could answer.*

Expressives also block:

(26) a. ?? *Only Mary sure looks like Mary in these old pictures.*
b. ??? *Only I sure got a question that I could answer.*

Again the reading remains the same, but the judgment of grammaticality seems to decline. Ultimately, we need to explain exactly how the movement chain experiences each kind of interruption.

4.3 Point of view (POV)

It is commonly assumed that external perspective adverbs reflect a speaker POV, but a more careful look indicates that they imply a general POV. Sentences like *Unfortunately, Bush is president* are assertions that it is unfortunate that Bush is president as a general proposition, not as a claim from a speaker POV comparable to deictics and indexicals. It is not immediately clear why such elements should function as intervenors. While indexicals like *I*, *you*, *we* seem to all have the speaker POV, we suggest here that they each reflect slightly different POVs, as do expressives. That is, an expression like *we* actually assumes a shift in perspective from *I*. In addition, *I seem to me* suggests two guises, and an expressive like *Wow, I did it* assumes a slightly different perspective for *wow* than for *I*. The strongest hypothesis then is that (27) holds.

(27) Every perspectival shift can interfere with movement.

Before we hazard some possible proposals, we need to see if other phenomena fall into line.

4.4 Expletives

Consider other forms of coreference:

(28) a. ? *Only John thinks John can do it* (*no one else thinks they can do it*).
b. * *Only John thinks that the bastard can do it* (*no one else thinks they can do it*).

Here we cannot obtain a reading where there are a set of bastards each of whom thinks of himself that he can do it. Note again that the outside perspective that identifies John seems a little less particularized than the one that labels him a bastard.

4.5 Sequence of tense

It has been noted that Sequence of tense can affect binding as well, and therefore could affect extraction. Consider (29):

(29) a. *Every boy thought he could do it.*
b. *Every boy thought he can do it.*

The (b) example (following Nunes and Thompson 1995) favors the external reading over the bound reading. We find the same to be true of the cases in (30), where if the tense contrast is clear, the focal binding fails:

(30) a. *Only Mary thought Mary IS big.*
b. *Only I thought that I AM big.*
c. *Only I got a question that I CAN answer.*

While the pragmatics may force the bound variable reading through, speakers uniformly find the sentence "worse".

4.6 Indexicals against indexicals: experiencer datives

While the sentence in (31) seems fine, the judgment degrades sharply when another indexical intervenes, cf. (32).

(31) *Only you seem to think you can do it.*

(32) *Only you seem to me to think you can do it,*
a. *no one else* [*thinks you can do it*]
b. ?* *no one else* [*thinks they can do it*]

Note that the (a) example has a default interpretation of "seems to everyone in general", which is then shifted to the speaker point of view with the addition *to me*. This observation is modeled on Brandt's (2006) work on cipient datives in German where he argues that a separate truth interval is called for with the productively available cipient datives in German. Predictably, under our refined view of perspectives, even seemingly identical indexicals cause a problem:

(33) ?? *Only I seem to me to get my homework done.*

Here again we suggest that two guises are present: I see myself as if from outside. Therefore, it may still be that two different POVs are present.

Our goal is to formulate the syntactic dimension that would reflect this semantic interface with the least complex point of interaction between syntax and semantics, that is, the impact of the semantic representation would ideally be encapsulated in an efficiently viable form. We suggest that (34) and (35) hold:

(34) There exists a POV feature in a Spec position with an agreement index linked to its head constituent.
DP in Spec of T: $[_{Spec}$ $[+POV_{speaker}]$ $[_{D'}$ I $[+POV_{speaker}]]$

(35) There exists a POV for a general default (seat of knowledge) assumption, thereby allowing transition of movement through it if the moved element has a compatible index.

Therefore, if there is a contrary POV feature present, it will block movement.

4.7 Non-restrictive relatives and late merge

Finally, let us consider a rather subtle contrast which has been the source of arguments for late merge. Pairs like the following call for a contrastive judgment:

(36) a. *Only I got a question that I could answer* (*no one else got a question they could answer*).
b. *Only I got a question* <u>*which*</u> *I could answer* (*no one else got a question*).

The (b) example seems to break the connection again, though judgments can be confused by the overwhelming pragmatics favoring binding.

4.8 Summary

We have found five contexts where there seems to be intervenors that block association with focus:

I Late merged relative clauses
II Sequence of tense contexts
III Expletive expressives
IV Adverbial outside perspectives
V Dative indexicals

It is not necessarily the case that these cases deserve a unified explanation. We will take a few steps in that direction nonetheless.

5. Explanatory levels

The kinds of phenomena we have been examining have been largely the concern of those projecting semantic models (e.g. Potts 2005, Schlenker 2003, Nevins and Anand 2002). Our approach will be uncommitted about what kind of representation captures the meaning differences best. Instead, we argue that whatever semantic approach is taken, it must have a syntactic projection such that there are computable syntactic consequences. This is not the only possible position. One could argue that all of the sentences we have considered are syntactically well-formed and their deviation lies wholly in a semantic realm.

If, however, they pattern together with known syntactic variation, then a common representational level is called for. In this brief paper we will sketch the lines of an account with many details omitted. A vision of the overarching architecture of explanation should be useful in itself in giving a full picture of the set-induction phenomena.

5.1 The relative clause case

The relative clause case is at first puzzling: why should a restrictive *that*-clause tolerate binding, while a non-restrictive *which*-clause does not? We can utilize the focus theory if, first, we assume that non-restrictive relative *which*-clauses are merged late, as argued by Bhatt and Pancheva (2004), following Lebeaux's theory of late merge. Then we can argue that late merge is high in the tree, attached at the root CP and that therefore, the cyclic movement path is not formulable. This restriction is then reducible to the failure of c-command.

5.2 The Relativized Minimality option

One approach to intervention data would be to argue that the traditional mode of intervention is at work:

sure = a positive item like negation
unfortunately = an adverb

Each of these are blocked by Relativized Minimality since adverbs are blockers for adjunct phrases. We can, at a subtle level, see an effect for negation as well. The following sentences provide a contrast:

(37) a. *Only I got a question that I couldn't answer.*
b. ? *Only I didn't get a question that I could answer.*

The bound reading comes through for the (b) example, but it seems less acceptable than in the (a) example. Under a movement analysis, the focal element passes through the NegP node in the (b) example. However, Relativized Minimality turns on the argument/adjunct contrast. It seems like different features are primarily involved in blocking in the full range of cases. Instead, we think a related approach would offer a better route toward building a successful interface with the semantic generalizations that are evident here, while preserving Rizzi's suggestion that unlike elements of the same type can interfere with each other. Our goal will be to treat as syntactic features as few of the semantic features as possible. Our examples all show the impact of perspective:

indexicals	⇒	speaker perspective
expressives	⇒	speaker perspective
datives	⇒	seat of knowledge perspective
adverbs	⇒	seat of knowledge perspective
sequence of tense	⇒	seat of knowledge perspective

From these we can argue that time, location, and person features are derivative, possibly within a system of implicatures, from what we can call a formal POV perspective which implies not only a person, but also a time and a location. However, formally we aim to have a simple POV index which conflicts with any non-identical POV index. We are deliberately both vague and restrictive. These features can obviously form a part of world variables and other semantic terms, but our goal is to find an encapsulated formulation that is exclusively relevant to an efficient syntax. We call it "POV-External" as distinct from "Person-POV" and "Subject-POV". It includes small perspectival shifts, new guises that arise as implicatures in sentences like *I seem to me to be more prepared* (one of 705 such examples from Google) as distinct from *I seem to myself to be more prepared.*

5.3 Feature checking

In that vein, let us assume, following Hollebrandse (2000), Speas (2004) Tenny and Speas (1998) and (in part) Nevins and Anand (2002), that there is a POV operator that can occupy the Spec position of a CP, DP, ADVP. and VP. This feature is projected in the Spec position of the DP, ADVP or CP, and it can percolate to the relevant Spec of each maximal projection that focus movement passes through. If POV operators occupy a Spec position, then they will constitute positions through which the Focus element must move to associate with *only* at LF. Let us further assume that the POV feature functions as a [+uninterpretable] feature which can be valued by the head feature it is in agreement with and can be repeatedly checked by features that pass through that node. Theoretical work in the minimalist tradition has not been completely explicit about what happens to the Feature-checking mechanism when information passes through repeated CPs or other phases in long-distance rules. In a quite detailed account, Hornstein, Nunes, and Grohmann (2005: 295) argue that [+interpretable] phi-features are repeatedly checked in the Spec of participles and in Spec of T in passive constructions.[2] The POV feature, including Person, is a natural extension of the phi-feature type which, like traditional phi-features, can be spelled out on other categories besides nouns (determiners, adjectives, etc).

---

2 "The interesting point for our discussion is that the phi-features of *as alunas* must enter into a checking relation with both the participial T and the matrix T. This multiple checking thus confirms our previous conclusion that the phi-features of arguments are [+interpretable]." (p. 295)

We will assume that a focal phrase must move to have its features ultimately checked in the highest position where *only* also moves. In a sense, the POV feature is a "free rider", which must meet a compatibility requirement – agreement – with every intermediate node that is passed through.

We now make the prediction that only elements introducing an external perspective will interfere with set induction. We find our judgements to be weak, but they nonetheless proceed in the right direction: adverbs that do not introduce an external perspective (like *swiftly*) allow set-induction, while other POV-marked adverbs (like *unfortunately*) do interfere, cf. (38).

(38) a. *Only I swiftly got a question that I could answer.*
b. ?? *Only I unfortunately got a question that I could answer.*

We can represent the workings of the analysis in the following sketch of a tree. We will assume a minimal blocking representation under Relativized Minimality with the traditional configuration and not explore how Focus Phrases should be represented but assume blocking is not affected.

(39)

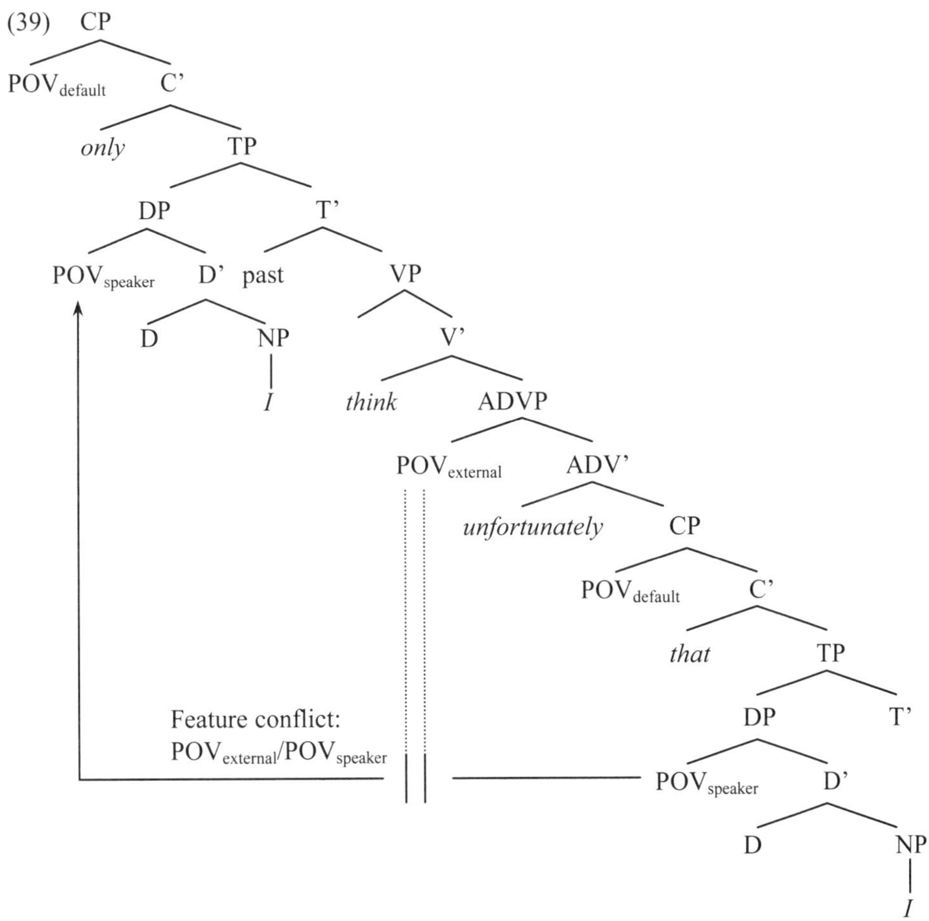

The POV-Ext of *unfortunately* will fail to be matched when the DP with a different Spec moves through the ADV-P Spec, either via adjunction of DP or by percolation of the POV-index. We leave out a full specification which engages other issues.

5.4 Other cases

Now we will indicate the direction of explanation for the other cases. The expressives likewise have a special ultra-internal POV, coming from an emotional sphere that is different from the intellectual one. The dative cases, likewise, suggest a second level, locative in Brandt's account, which is distinct from the primary level. And the sequence of tense cases suggest that the tense marker has a POV feature that is independently anchored in a second time dimension. All of these can be represented on the relevant Spec nodes and will produce an intervention if marked [+Distinct]. The consequence of the intervention is to block association with focus, which leaves a grammatical sentence that has not been linked. In a sense, then, the POV concept becomes, in its syntactic version, much like gender, which has been grammatized in a way that may not fully match its intuitive origins.

5.5 Conclusion

We have presented a new range of empirical data that expands what the focal binding linked to *only-I* involves. We have deliberately avoided a commitment to many aspects of the interfaces that link syntax, semantics, and the intentional system in order to seek a common representation for diverse information at the syntactic level. Needless to say, there are many respects in which the account needs refinement – it remains thus a sketch about how a notion of point of view can play a role in an efficient syntax and begin to build an appropriate interface.

References

Bhatt, R. and R. Pancheva. 2004. Late merger of degree clauses. *Linguistic Inquiry* 35, 1–45.

Brandt, P. 2006. Receiving and perceiving datives (cipients): A view from German. In: Hole, D., Meinunger, A. and W. Abraham (eds.), *Datives and Other Cases: Between argument structure and event structure*, 103–139. Amsterdam: Benjamins.

Hardt, D. (to appear). VP ellipsis and constraints on interpretation. In: K. Johnson (ed.), *Topics in Ellipsis*. Cambridge: Cambridge University Press.

Heim, I. 1991. The first person. Class handout, MIT.

Heim, I. 2005. Features of pronouns: the semantics of number, gender and person. Talk given at the École normale supérieure, Paris.

Hollebrandse, B. 2000. *The Acquisition of Sequence of Tense*. PhD dissertation, University of Massachusetts, Amherst.

Hornstein, N., J. Nunes, and K. Grohmann. 2005. *Understanding Minimalism*. Cambridge: Cambridge University Press.

Kratzer, A. 2006. Minimal pronouns. Talk given at the workshop *QP structure, Nominalizations, and the Role of DP.* Saarbrücken: Universität des Saarlandes.

Kratzer, A. 1998. More structural analogies between pronouns and tenses. In: D. Strolovitch & A. Lawson (eds.), *Proceedings of SALT VIII*, 92–110. Ithaca: Cornell University.

Lebeaux, D. 2000. *Language Acquisition and the Form of the Grammar.* Amsterdam: Benjamins.

Nevins, A. and P. Anand. 2002. Shifty operators in changing contexts: indexicals in Zazaki and Slave. In: R. Young (ed.), *Proceedings of SALT XIV*, 20–37. Ithaca: Cornell University.

Nunes, J. and E. Thompson. 1995. The discourse representation of temporal dependencies. In: P.M. Bertinetto, M. Bianchi, J. Higginbotham, and M. Squartini (eds.), *Temporal Reference, Aspect, and Actionality, vol. I: Semantic and Syntactic Perspectives*, 365–79. Torino: Rosenberg and Sellier.

Partee, B. 1989. Binding implicit variables in quantified contexts. In: C. Wiltshire, B. Music and R. Graczyk (eds.), *Papers from CLS 25*, 342–65. Chicago: Chicago Linguistics Society.

Potts, C. 2005. *The Logic of Conventional Implicatures*. Oxford: Oxford University Press.

Roeper, T. 1999. Notes on the *Only I* phenomenon. Manuscript, University of Massachusetts, Amherst.

Rooth, M. 1992. A theory of focus interpretation. *Natural Language Semantics* 1, 75–116.

Schlenker, P. 2003. A plea for monsters. *Linguistics and Philosophy* 26, 29–120.

Speas, M. 1999. Person and point of view in Navajo direct discourse complements. In: E. Jelinek (ed.), *WCCFL Papers in Honor of Ken Hale*, Cambridge, Mass.: MIT Press.

Speas, M. 2004. Evidentiality, logophoricity and the syntactic representation of pragmatic features. *Lingua* 114, 255–276.

Speas, M., and C. Tenny. 2003. Configurational properties of point of view roles. In: A. Di Sciullo (ed.), *Asymmetry in Grammar*, 315–344. Amsterdam: John Benjamins.

Terada, A. 2003. *A Study of Text Mining Techniques Using Natural Language Processing*. PhD dissertation, Tokyo Institute of Technology.

Conceptualizing Sentence Mood – Two Decades Later

Dietmar Zaefferer

0. Introduction

The *International Handbook of Contemporary Research on Semantics* (von Stechow and Wunderlich 1991) includes an article entitled *Theorien der Satzmodi* (Theories of Sentence Mood), which was written by Günther Grewendorf and Dietmar Zaefferer in 1984 as an outline of the state of the art on its topic at that time (Grewendorf and Zaefferer 1991). Two decades later, one of the coauthors invites the reader to take a look back not in anger, but with some curiosity about how much the authors were held captive by pictures that were taken for granted at the time of writing. Could we have gotten outside them or did they lie in our language, which kept repeating them inexorably?[1]

The answer advocated but only roughly outlined[2] on the following pages is that we could indeed have gotten outside these pictures or conceptualizations by scrutinizing more closely our specialized language, and so there was no deep reason for not revising them. I submit that three factors kept us from seeing things to be as simple as they are. The first was the tendency to use overspecified notions, especially with regard to illocution types. The second was an overly respectful attitude towards the venerable notion of truth, which was taken for granted in one of its uses instead of being analyzed into its readings. And the third and probably most serious factor was that we somewhat misconceptualized the relation between truth and speech acts: We rightly agreed with Austin and others in rejecting the descriptive fallacy, the conception of human language that reduces it to a tool for describing truly or falsely what there is, a conception whose correction resulted in the very birth of speech act theory, but in doing so we failed to grasp precisely enough how speech acts relate to truth value judgments.

To put it constructively, what I am arguing for here is a revised conceptualization of sentence mood and illocution types that is characterized by the following three features: (a) basic speech act concepts that are located at an adequate level of specification, which is claimed to be a very general one, (b) a less over-respectful and more pragmatic way of dealing with truth, replacing in part the solemn tradition-laden monosyllable *truth* and its kin with the less pretentious term *congruence* and its relatives, and (c) replacement of the unfortunate assumption of a very close connection between truth and linguistic action by a more differentiated picture.

---

1 The reader will recognize the allusion to the famous quote: "Ein *Bild* hielt uns gefangen. Und heraus konnten wir nicht, denn es lag in unserer Sprache, und sie schien es uns nur unerbittlich zu wiederholen." ('A *picture* held us captive. And we could not get outside it, for it lay in our language and language seemed to repeat it to us inexorably.') (Wittgenstein 1967, § 115).

2 For a fuller account cf. Zaefferer in preparation.

1. Preliminaries

In order to avoid possible confusion I will start with some terminological clarifications or rather ontological remarks.[3] One of the central notions both of philosophy of language and of linguistics is that of a linguistic sign. Linguistic signs are abstract entities since they have no spatiotemporal location, but they are abstracted from concrete entities or sign tokens, which do possess spatiotemporal location, but are ontologically quite heterogeneous: Written sign tokens are objects, spoken sign tokens are events. To keep things simple I will concentrate on the latter here and disregard the former. Apart from very exceptional cases like talking in one's sleep, the production of an oral linguistic sign token, an utterance, is not only an event, but also a special kind of event, an action, in other words, a case of goal-directed behavior. What could be the most general goal of performing such an action?

I submit that this is simply externalization. For instance when I inadvertently cut myself, no matter if someone else is present or not, by uttering a swear word I can externalize something that was only inside me before, and the very fact that I have externalized it may give me some relief. So my first thesis is that the least specific goal of overt linguistic action (the functions of internal talking will be addressed below) is to externalize, i.e., to make accessible in principle to others whatever content is associated with the linguistic sign a token of which is produced:

(A) The most general goal of overt linguistic action, as opposed to other kinds of action, is to externalize the content of the corresponding linguistic sign token.

This liberates linguistic action from definitional dependency on communication. Externalization is certainly necessary for communication (except for real mind reading), but certainly not sufficient. Next I assume a (hopefully) rather uncontroversial notion of linguistic sign:

(B) A linguistic sign is a type whose tokens are hybrids consisting of a perceivable aspect, their shape, and an inferable aspect, their content. Correspondingly, the abstract sign itself has a shape aspect, the type of the perceivable aspect of its tokens, and a content aspect, the type of the inferable aspect of its tokens.

(C) The content aspect of every linguistic sign can be partitioned into (a) its morpho-syntactic structure, (b) its lexical-conceptual elements and (c) the overall use potential that results from the interaction of (a) and (b).

This may seem a little implausible with very elementary signs like the English word *up*, which codes the concept of that direction which points away from the ground, a word which does not seem to have morpho-syntactic structure. But this term is intended here to include also part-of-speech properties, so in fact there are several related English words *up*, among them an adverb, but also a transitive and an intransitive verb meaning *rise* and *raise*, respectively, such that these different morpho-syntactic structures (cf. (a)

3 Ontological efforts are more than terminological clarifications insofar as they not only spell out how technical terms should be used, but also justify such normative proposals by an analysis of what the phenomena in the object domain really and essentially are (cf. Schalley and Zaefferer (eds.) forthcoming).

above) interact with the common lexical-conceptual content (cf. (b) above) to yield different use potentials (cf. (c) above).

There is one general principle of linguistic action that will prove useful in the following. It is a specialization of the general maxim of rational behavior 'Avoid unhelpful redundancy.' My formulation of this principle is inspired by a famous quote:[4]

(D) Whereof one needn't speak, thereof one may be silent.

This principle entails that in communication situations we use linguistic means only to fill the gap of what is not obvious from other aspects of the situation. Where there is no such gap, no linguistic action is needed and where the gap is small very little may be required. If for instance somebody enters an elevator, it is obvious that she or he wants to move in a vertical direction, so the yes-no interrogative reading of (1) (one of the two possibilities?) is redundant and only an alternative reading (which of the two possibilities?) makes sense. This latter reading has also a shorter coding, namely (2), and sometimes even (3) will suffice:

(1) *Do you want to go up or down?*

(2) *Up or down?*

(3) *Up?*

Given the elevator context, all three utterances constitute a complete speech act of the erotetic, i.e. knowledge elicitation kind. How do we know? The morpho-syntactic structure of (1) in its alternative reading, characterized by *do*-support, inverted subject and an intonation contour that rises on *up* and falls on *down*, tells us that this sentence is of the alternative interrogative type and the default illocution performed in uttering a sentence of this type is an alternative question. In the cases (2) and (3) only the intonation contour or its first half remain, but given the context and principle (D) that suffices to perform the same illocution. This takes us to the definition of sentence mood:

(E) A *formal sentence type* of language *L* is a morpho-syntactic sentence structure of *L*. An *L*-sentence is of this type exactly if it has this structure.

(F) A *sentence mood* of language *L* is a set of formal sentence types of *L* such that all sentences that are of one of these types share a non-empty part of their illocutionary use potential. An *L*-sentence is in this mood exactly if it is of one of the defining formal sentence types, i.e., if it has one of the corresponding structures.

(G) A *sentence mood meaning* of language *L* is what is coded by all sentences that are in the given mood of *L*, i.e., the shared part of their illocutionary use potentials.

The fact that all human languages have sentences that are in the declarative, the interrogative and the imperative sentence mood is a non-trivial, but not surprising finding of linguistic typology.[5] But what precisely are the meanings of these sentence moods and how do they relate to the prototypical illocution types of mainstream speech act theory?

---

4 "Wovon man nicht sprechen kann, darüber muß man schweigen." ('Whereof one cannot speak, thereof one must be silent.') (Wittgenstein 1960, sentence 7).

5 Cf. König and Siemund forthcoming.

2. A lean specification theory of linguistic action

Eleven theories of sentence mood had been selected for discussion in Grewendorf and Zaefferer 1991. They were divided into three groups according to the number of levels they assume for the analysis of the meanings of sentence mood on the one hand and the rest of the sentence on the other: Six one-level theories, five two-level theories, and two three-level theories. Not very surprisingly, the model-theoretic two-level approach developed by one of the co-authors ('illocutionary semantics') emerged as belonging to the most convincing competitors, with two reservations: the problems inherited from the possible worlds-based notion of proposition it used and the lack of an account for information structure.

In the meantime, the same author has developed a new concept of proposition called CAP (see below), a cognitivized version of the Austinian propositions devised by Barwise and Etchemendy (1987), and his work on modal categories (Zaefferer 2001, 2005) has revealed that some sentence mood meanings are decomposable into even more basic propositional operators. Since sentences in different moods can share the same propositional content, it comes as no surprise that sentence mood meanings in the sense of definition (G) above operate on propositional contents and are therefore in a sense modal operators. Whereas the old approach could only note the fact that declarative, interrogative, imperative and, somewhat less, exclamative sentence mood are universally most central, the new version proposes an explanation for this preference based on sentence mood meanings and some anthropological constants. The anthropological constants are derived from the assumption that humans are essentially mind sharing animals (Zaefferer forthcoming a) and consist in the four main kinds of mind sharing: emotion sharing, attention sharing, goal sharing, and knowledge sharing.

The aim of emotion sharing is most directly expressed by the use of interjections, autonomous linguistic signs without propositional content, and by optative sentences, forms that present their propositional content as the object of a wish of the agent. The aim of attention sharing is most directly expressed by the use of the index finger (pointing) with or without the utterance of imperatives like *Look!* or *Watch this!*.[6] The aim of goal sharing is most directly expressed by the use of imperative or more generally jussive sentences where the propositional content characterizes the goal to be shared (*Engine off!*) or the way it is to be reached (*Turn the engine off!*). The aim of knowledge sharing can be expressed directly by the use of declarative sentences, which by definition have a complete propositional content (*The engine is turned off.*), or more indirectly by the use of interrogative sentences, which have a somewhat incomplete propositional content (*Is the engine turned off?*). A more detailed presentation follows below.

But isn't this idea of mind sharing with its ramifications at variance with the assumption (A) stated above, which gets rid of the widespread definitional dependency of language use on communication by saying that the most general goal of language use is simply externalization and which therefore is also able to account for monological linguistic action? It would if it assumed that mind sharing is the most general goal of lan-

6 Research in developmental psychology has shown that infants point to achieve both: attention sharing and emotion sharing (Liszkowski et al. 2004).

guage use, but this assumption is not made. Mind sharing is conceptually derived, although it is of course central for both the phylogeny and ontogeny of language. The connection between monological and dialogical use of language is made by another function, one that is also involved in covert, internal talking, and this underlying function is concept activation.

A given concept is primarily activated by every occurrence of a phenomenon that falls under it and secondarily by association with another, generally or individually related concept. But language provides us humans with a third way of activating a concept, namely the occurrence of a perceivable pattern that is associated with this concept by the given language's coding conventions. So, internal talking serves the purpose of activating a concept by activating the neural routines that govern the production of one of the perceivable patterns that code this concept in the given language. The point of language-mediated mind sharing is that one person's externalization of a perceivable shape that is coupled with an inferable conceptual content permits the coactivation of (more or less) the same concept across the minds of agent and perceiver.

This seems trivial, but it leads to a very simple and far-reaching theory of sentence mood meaning: Sentence mood meanings are very general illocution types. In general, natural languages lack lexical means for expressing them. Statement, command, promise are much more specific illocution types, but this does not come as a surprise, since sentence moods are linguistic structures and structural meanings tend to be more general and harder to pin down than lexical meanings. The situation is similar with nominal cases and their meanings. What is the meaning of dative, for instance? Recipient is not completely off the mark, but it is far too strong because not every dative marks a recipient. Similarly with sentence structures: Of course, declarative sentences are often used to make statements, but there are many more uses of this kind of sentences, and there are other ways of making statements, too.

At this point it may be appropriate to mention Wittgenstein, the great pioneer of the meaning-as-use idea in philosophy of language. In his *Philosophical Investigations* he writes:

> But how many kinds of sentence are there? Merely assertive, interrogative, imperative? - There are *innumerable* such types: countless different ways of using what we call "sign," "word," "sentence." And this multiplicity is not stable, given for all time; new types of speech, new language games, as we might call them, arise and others become old and are forgotten. (Wittgenstein 1967: § 23)

Is this at variance with the theory of sentence mood meaning outlined above? Before answering this question it will be helpful to clarify a little what Wittgenstein meant and to look at the German original: "Wieviele Arten der Sätze gibt es aber? Etwa Behauptung, Frage und Befehl?" If *Arten der Sätze* is read as *Satzarten*, 'sentence types', then the two sentences are about form types and the second one can be read as *Etwa Behauptungssatz, Fragesatz und Befehlssatz?* ('Declarative sentences, interrogative sentences and imperative sentences?'). But then there is no contradiction with the subsequent sentence since the possibility that there are only three form types is easily compatible with the existence of innumerably many different types of using these form types. So it seems more plausible to assume that Wittgenstein meant 'kinds of sentence use' when he wrote *Arten der Sätze*, and there are of course not just the three kinds he mentions. Under neither interpretation is there a real contradiction with our picture.

And still, there is a problem. Our approach says not only that there is a very small number of sentence moods in human languages, it assumes also that these moods tend to have simple basic meanings, so there is a very small number of basic sentence mood meanings as well. And this seems to be at variance with Wittgenstein's claim. But again, at closer inspection the contradiction disappears: There is no contradiction between assuming a very small number of basic sentence mood meanings (and hence of most general use types) and an infinite number of special use types, just as there is no contradiction between saying that there are two kinds of natural numbers, namely odd and even, and claiming that there are infinitely many, like the primes between 900 and 1000 or the divisors of 84, etc. What Wittgenstein did not consider was the simple fact that every sentence has to go through the bottleneck of sentence mood before it can be uttered, and therefore any sentence meaning or use potential has to go through the bottleneck of sentence mood meaning before it can be realized. So human languages invariably impose structure on the multiplicity of uses: There are both innumerably many kinds of uses and a very small number of basic use types.

The arguably most central use-type has to do with knowledge, knowledge of some given propositional content that, for instance, an agent wants to share with her addressee. Suppose Maggie speaks English, her partner does as well, her stomach tells her that she is hungry, and she wants to share this knowledge. Then not all of the following sentences of English will serve her purpose equally well:

(4) *Am I hungry?*

(5) *I am hungry.*

(6) *Be hungry, Maggie!*

(7) *Boy, am I hungry!*

One cannot exclude that under appropriate circumstances an utterance of any one of the four will get the message across, but the most natural means is of course the declarative sentence (5). Since language use is not restricted to dialogical situations it seems advisable to characterize the sentence mood meaning of declarative sentences as follows:

(Dec) In uttering a declarative sentence with propositional content *p* the respective agent makes it inferable that her aim is activated knowledge of *p*.

The way (Dec) is formulated leaves open the identity of the agent whose activated knowledge state is aimed at, let's call him the intended knower, and so it is compatible with monological uses as well, as when upon receiving some signal from the stomach one silently 'says': *Oh, I am hungry*. (Dec) is also compatible with three different preceding states of the intended knower, which entail that there are three different ways of achieving the mood-specific aim of declaratives. If the corresponding knowledge is already activated, then the aim is reached by activation maintenance (keeping the knowledge active), if it is there in long-term memory but not activated, then it is reached by mere activation (reminding), and if it is not there at all, then it is reached by activation and creation of knowledge at the same time.

This illustrates nicely the first of the three abovementioned aspects of the approach advocated here: The adequate level of specification of the concepts that make up the present reconceptualization is claimed to be a relatively general one.[7] The second aspect, a more pragmatic way of dealing with the noble concept of truth, comes to bear with interrogatives. Although both (4) and (5) are about the actual food deprivation of the speaker, only the latter provides a good antecedent for reactions like (8) or continuations like (9):

(8) a. *Is that true?*
 b. *You are not telling the truth!*

(9) a. *Believe me, that's true!*
 b. *No, that's not true, just kidding!*

Although the reason seems to lie in the sentence mood meaning alone, since sentence mood is the only aspect that distinguishes (4) from (5), I have developed a theory of propositional content that challenges this traditional view by claiming (a) that the primary difference is in the propositional content (for a similar view cf. Ginzburg & Sag 2000), (b) that the modal force is basically the same, and (c) that differences in the modal force are due to different default ways of dealing with the different kinds of propositional content (Zaefferer 2004, in preparation). Here is the idea in a nutshell.

The basic picture goes back to J.L. Austin (1950), who proposed to think (a) of a (declarative) sentence as something that is correlated by descriptive conventions with a type, (b) of a statement made with this sentence as something that is correlated by demonstrative conventions with a token, and (c) of truth as a property a statement has just in case the given token is of the given type. Correspondingly, Barwise and Etchemendy (1987) call a pair of a situation and a situation type an Austinian proposition, and the theory outlined here defines a Cognitivized Austinian Proposition (CAP, for short) as a mental structure whose tokens consists of a representation of a situation token and a representation of a situation type.

Exemplifying the situation type by a rain situation and the situation token by the concrete situation of the Stanford Quadrangle on July 7, 2006, at 10 a.m. PDT, it seems plausible to assume that both the author and the reader of these lines can activate in their respective mind a token of the CAP that it is raining in the given situation (which may be true or false).

A crucial ingredient of the idea of cognitivizing Austinian propositions comes from the fact that human cognition treats situations as containers and that there are two ways of conceptualizing a container, inclusively, with all the inventory and events that are inside, and exclusively, as a mere frame or range of coordinates for possible inventories and events.

Assuming that the situation type and the relevant descriptive conventions associated with a given sentence are the same for both its declarative and its polar interrogative variants, it seems natural to locate the difference in a sortal restriction on the situation token to be provided by the demonstrative conventions for any felicitous utterance of

---

7 I am deliberately avoiding the term underspecification, which like overspecification seems to implicate that there is another, more adequate level of specification, and this is of course not what I want.

the respective sentence: The propositional content of a declarative consists of a type and an inclusive situation and is therefore either true or false, whereas the content of the corresponding polar interrogative pairs the same type with an exclusive situation, a mere range of coordinates, and therefore the question of truth cannot be answered before the corresponding inclusive situation is provided. Generalizing the term CAP to mean Cognitivized Austinian Propositional content we can now distinguish between full CAPs or simply propositions and near-CAPs or near-propositions.

Consequently, propositions possess truth values, since they are true for a given mind state if the given situation token is of the given situation type (according to each one's mental representation in the given mind state). By contrast, near-propositions lack truth values, since for the given mind state it is not possible to decide if the relevant situation token is of the given situation type (according to each one's mental representation in the given mind state), because the former is represented only by its frame. Accordingly, the propositional content of (5) is the full proposition that is true (for a given interpreting mind) just in case the current situation of the speaker (as represented by that mind) is of the type where the speaker is hungry in that situation (for that mind). The propositional content of (4), on the other hand, is that near-proposition that consists of the situation frame (in the given mind) that singles out the speaker's current situation and the situation type of the speaker being hungry (as represented by that mind).

The distinction between full and near-propositions is required for another distinction which will prove essential for our approach to sentence mood meanings, that between transparent and opaque knowledge representations. Let us suppose the speaker of (4) and (5) is indeed hungry in the respective utterance situation and the respective interpreting mind knows this. Then we can represent this knowledge by saying either (10) or (11):

(10) *He knows that Maggie is hungry.*

(11) *He knows whether Maggie is hungry.*

I call (10) a transparent knowledge representation because it makes its full content visible, i.e. that Maggie is hungry indeed, whereas (11) is opaque insofar as it does not reveal the precise content of the knowledge, only its dimension of variation. Now we are in a position to state the interrogative counterpart of (Dec):

(Int) In uttering an interrogative sentence with propositional content *p,* the respective agent makes it inferable that her aim is activated knowledge of *p*.

Note that the way (Int) is formulated only seemingly identifies the two sentence mood meanings, since propositional contents of interrogatives are sortally different from those of declaratives (namely they are near-propositions as opposed to full propositions). The claim of the theory advocated here is that all remaining differences result from this opposition: If a speaker provides a transparent representation of the knowledge that is to be activated, this means in general that she has this knowledge and wants to share it. If, by contrast, she provides an opaque representation of the knowledge that is to be activated, this means usually that she lacks this knowledge and wants it to be shared by someone who has it. This explains why the interpretations of the two sentence moods as attempts at knowledge sharing and requests for knowledge sharing, respectively, are generally seen as the default cases.[8] The advantage of (Dec) and (Int) is that they also cover non-default cases such as autosuggestive uses of declaratives or rhetorical, i.e., non-requestive uses of interrogatives.

Since the kind of illocutionary force defined here is very general and given the central role of knowledge in it I will speak of epistemic speech acts or epistemics for short. Consequently I will call transparent epistemics the uses of declaratives and opaque epistemics those of interrogatives, but I will also use the shorter and more familiar term assertives for the former and subsertives for the latter.[9]

Let's now look at imperatives like (6), repeated for convenience, (12) or (13):

(6) *Be hungry, Maggie!*

(12) *Fists up!*

(13) *Take your fists up, Maggie!*

Unlike interrogatives, imperatives code propositional contents which are not near- but full propositions. But unlike most declaratives, imperatives code propositions which cannot be naturally judged as true or false. When a bystander comments on an utterance of (13) *That's true*, what he means is that Maggie should take her fists up, und not that she does it. Still, the question of whether the represented situation token is of the given situation type is essential for understanding imperatives. The reason is that their propositional content directly defines their action goal:

(Int) In uttering an imperative sentence with propositional content *p* the respective agent makes it inferable that her aim is *p*.

Before tackling the question why truth value judgments are odd with imperatives in spite of the full propositions they contain, let us have a look at sentences like (7), repeated here for convenience, or (14), i.e., at sentences in the exclamative mood.

(7) *Boy, am I hungry!*

(14) *What a powerful punch she's got!*

With exclamatives truth value judgments are rather natural, which points at a mood meaning that includes that of declaratives. Two questions remain open: What is the precise propositional content of exclamatives, and what exactly distinguishes them from their closest declarative counterparts? To begin with the latter, there is no doubt that (7) conveys more information than (5), repeated below for convenience, in that it not only characterizes the speaker as being in a state of hunger, but also indicates that the degree of this hunger is unusually high, so (15) could be considered the closest declarative counterpart:

---

8 One of the central shortcomings of Searle's pentachotomy of basic illocutionary acts (Assertives, Directives, Commissives, Expressives and Declarations) is that it takes default cases to be basic categories (cf. Zaefferer forthcoming b).

9 The reason for introducing the neologism subsertive (contracted from subassertive) is that interrogatives are not only used for asking questions, but also for posing questions or approaching assertions.

(5) *I am hungry.*

(15) *I am unusually hungry.*

The difference between (5) and (15) now helps to answer the first question. If we compare (16) and (17) as possible reactions to an utterance of (7), the oddity of the latter indicates that the propositional content of (7) is identical with that of (5) and not with that of (15).

(16) *That's not true, I just saw you eat a whole pizza!*

(17) *That's not true, you are just as hungry as you are every day at this time!*

This diagnosis leads to the assumption that unusually high degree is not part of the propositional content, but of what is emotionally expressed in the use of exclamatives and what distinguishes them from declaratives that express the same information neutrally as in (15). So the meaning of the exclamative sentence mood should be explicated as follows:

(Exc) In uttering an exclamative sentence with propositional content *p* the respective agent makes it inferable that her aim is not only activated knowledge of *p*, but also emotional expression of the unusually high degree to which *p* holds.

To conclude this section (and to prepare the following one) I now return to the question of why truth value judgments are odd with imperatives (let's call this the imperative gap) and, more generally, to a closer examination of the relation between truth and linguistic action. Two decades ago, I had a rather technical solution for the problem: The primary bearers of truth values are propositions. If a proposition is true (at a given index), then the embedding speech act is also called true if it is an assertive speech act, and it is called fulfilled if it is a directive speech act etc. This is not completely wrong, but it does not explain anything and it is not completely right either.

My proposal for an explanation of the imperative gap contains a somewhat speculative element, an instance of a phenomenon, however, whose existence is well attested in the history of lexical meanings. The phenomenon is the existence of synchronous restrictions on the use of a lexical item that seem ill-motivated by its current meaning, but are explainable as remnants of the etymological source of this item. The etymological source of *true* and *truth* is the social quality of steadfast loyalty, i.e., of staying with a person (immovably like a tree) and not going away.[10] My guess is that one aspect of this source is still active in the contemporary uses, where for instance a predication is called true if it preserves essential traits of what it is about and does not deviate from it, in other words, truth presupposes a temporal order where first there is something to be conserved and then there is something that conserves it.[11] If we hear that a person's claim is true to the facts, then we assume that the facts were first, and the claim came later. What about true predictions? Some people say that if there will be rain tomorrow

10 Interestingly, the antonym *false* has also a social origin: It goes back to Latin *falsus* from *fallere* 'deceive'.

11 On her web page the Secretary of the State of Michigan answers the question *What is a true copy of a document?* as follows: A true copy is an exact copy of a document with no alterations or changes.

then it is a fact already today that there will be rain tomorrow, so the corresponding prediction is true already today, although it will turn out to be so only tomorrow. Most people, however, prefer to speak of correct predictions instead of true predictions, much in line with my claim.

But if my proposal that the concept *true* presupposes the previous existence of some original which can be loyally stayed with or else disloyally deviated from, then it is obvious that this concept cannot be applied to imperatives, even if their goal is reached and the corresponding proposition is realized by the addressee, because the crucial difference between an imperative utterance and a prediction is that the former makes sense only if the corresponding prediction would presumably not be correct without the directive intervention. But if there are contexts that prevent the embedded propositions from being properly called 'true', then we need a more general concept for the evaluation of propositions. Given the notion of dyadic Cognitivized Austinian Propositions introduced above it seems natural to apply the concept of congruence to this case:[12]

(H) A proposition consisting of a situation token *s* and a situation type *T* is *congruent* if *s* is of type *T* and *disgruent* else.

Next we will have a closer look at the second part of the old assumption that propositions have truth values and some embedding speech acts inherit them. The first part turned out to be only partially correct, so what about the second part? In order to find that out let's look at a little syllogism.

A good candidate for a speech act that can be properly called true or false seems to be an assertion.[13] Assuming that every assertion is a speech act and that every speech act is an act we can be sure that every assertion is an act. Nobody will deny either that there are true assertions and false assertions. But then we are forced – unless we want to get into troubles with first order logic – to accept that there are true acts and false acts. And this (after exclusion of the readings 'real' for *true* and 'fake' for *false*) sounds like a clear case of category mistake, no less clear than the assumption that there are true commands. To perform an act is to do something, so if there were true acts, *What he did is true* should be a flawless sentence of English.[14]

So we have arrived at the somewhat surprising result that neither speech acts nor their propositional contents, at least not all of them, are good true value bearers. This finding seems important enough to be highlighted and so we perform an act of christening and give it a decent name:

(TBF) The *truth bearers fallacy* consists in the erroneous assumption that the entities which can most properly and invariably be judged as true or false are propositions and assertive speech acts.

But why have we been deceived by this fallacy in the first place, why has this picture kept us captive for more than two decades? With respect to propositions the explanation

---

12 Another way of viewing this move is interpreting it as a terminological proposal that eliminates the problematic tension between the ordinary language concept of true and the homonymous technical term by giving the latter a new label.

13 Recall that all assertions are assertives, but not vice versa.

14 Cf. Siebel (2002: 137): "Telling about an action that it is true or false is a category mistake."

is easy: Many propositions can indeed properly be called true or false, and for those that cannot an extension of the meaning into a technical term is what has been assumed by formal semanticists anyway.

But why did we, in the little syllogism above, accept the premise that there are true assertions and false assertions? The reason lies, as often with language, in a little ambiguity: The noun *assertion* (and similar nouns such as *statement*, *affirmation* etc.) has at least two readings. As *nomen actionis* it codes an event type, a speech act that can be successfully performed, or not performed at all, etc.; as *nomen acti* it codes an object type, a social object that can be accepted, or doubted, or rejected, among other things, namely that kind of object that results from successful performances of assertions in the first sense. The trick in the syllogism is to make the subject accept the first premise with *assertion* in the first sense (let's call this the production reading) and the second premises with *assertion* in the second sense (let's call this the product reading). If the subject does not notice the trick, he is trapped in the conclusion.

Let us now turn to an application of our findings to an old and much debated issue of speech act theory, the proper treatment of explicit performative speech acts.

3. How explicit performatives are really performed

The use of the term *performatives* as a shorthand for explicit performative speech acts has always struck me as somewhat peculiar. Acts are always performative, and trivially so: How else could I realize an act than by performing it? But only linguistic acts can be explicit in the sense of self-referentially indicating their own type. Maybe a technical term like autotyping speech acts would be more adequate. But in order to avoid upsetting the community more than I have done already in the preceding sections, I will stick to the traditional terminology here.

Günther Grewendorf has contributed already in the seventies and on several occasions to the debate on the proper treatment of explicit performatives, the last time in Grewendorf 2002. There he argues that the two most visible theories on the market, Searle's performatives-as-declarations approach (Searle 1989) and Bach and Harnish's performatives-as-statements view (Bach and Harnish 1979, Harnish 1997, 2002), are both mistaken. To show this he attempts to refute three claims that are shared by both approaches, their discrepancies notwithstanding. The three claims are (a) that explicit performatives are statements, (b) that explicit performatives can be true or false, and (c) that the propositional content of an explicit performative includes both the meaning of the matrix clause with the performative verb and the meaning of the embedded clause this verb takes as argument.

Without going into the details I will make a few remarks on these three claims from the point of view outlined above. Concerning the first claim I agree with Grewendorf if the term *statement* is taken in the narrow sense. Statements in the narrow sense are prototypical constative utterances which fit perfectly with the concept of true in the narrow sense in that they imply a temporal order: First there is an independently existing state of affairs and then there is somebody who states its existence. Since in explicit performatives the relevant state of affairs comes only into existence in virtue of the speech act in question, such a speech act cannot be a statement by definition. From the point of view of the new picture the first claim should be replaced by the following statement: (a') explicit performatives are transparent epistemics, i.e., assertives. This follows from the fact that only declarative sentences can be used in explicitly performed speech acts and from our definitions above.

Concerning the second claim, that performatives can be true or false, our findings explain the judgments of both sides: In the strict etymologically based sense this has to be rejected because there is no preceding state of affairs of which the speech act in question can give a true picture. But in a more technical sense there is a propositional content that is true if the speech act succeeds and false otherwise. In the new framework the issue disappears because here the propositional content of an explicit performative is congruent if it succeeds and disgruent otherwise.

This takes us to the third claim, the one about the propositional content of explicit performatives. Here Grewendorf's position is that the speaker only phatically says that he does what he does and not rhetically and that therefore the performatively used verb form does not contribute to the propositional content. Here I strongly disagree, since on this view the propositional contents of utterances of (18) and (19) would be identical.

(18) *I ask you to buckle up as we make our descent to Heathrow airport.*

(19) *Please buckle up as we make our descent to Heathrow airport.*

My claim is that the *please* in (19) does not contribute to the proposition in question, but the *I ask you to* in (18) does. And the different anaphoric accessibilities created by utterances of (18) and (19) show clearly that (18) encodes two propositions, an embedding and an embedded one, whereas (19) encodes only one. Consider the following reply by an unruly passenger:

(20) *You can do that as often as you want, I won't do it.*

If (20) is used as a reply to (18), then *do that* clearly takes the predication of the embedding proposition, *ask you to buckle up*, as antecedent, whereas *do it* refers back to *buckle up*, the predicate of the embedded proposition. If however (20) is used as a reply to (19), then *do it* still refers back to *buckle up*, but the antecedent for *do that* is much harder to find and must be coerced as something like *say ,Please buckle up as we make our descent to Heathrow airport'.*

Ellipsis data confirm this diagnosis. (23) is fine as a follow-up to (21), but impossible as a withdrawal after an utterance of (22):

(21) *I command you to jump out the window.*

(22) *Jump out the window!*

(23) *No, I don't, just kidding.*

This proves that the proposition that includes the performative verb is anaphorically accessible, so it exists and it is the (highest) propositional content of the corresponding utterances.

What is my answer now to the topic question of this section? How are explicit performatives really performed? They are performed in uttering a declarative sentence if and only if the content of this utterance is a proposition whose token is the current speaker situation, whose type is that of an illocution performance or its result and which is congruent. For the sake of illustration let's look at a final example, one that deviates a little from the canonical case insofar as it mentions neither agent nor addressee.

(24) *This is a reminder that the use of cell phones is not permitted on our flights.*

Suppose the purser has already informed the passengers about the content of the *that*-clause in (24). Then they know that the use of cell phones is not permitted on the flights of the airline in question. By definition a reminder is a reactivation of already accessible knowledge. Now the purser utters a token of (24). Then by (Dec) the passengers are entitled to infer that he aims at knowledge that what he is referring to is a reminder with the given content. Since there is nothing else he could be referring to than what he is doing right now, passengers can ask themselves if what the purser does reactivates their knowledge that the use of cell phones is not permitted on the flights of the given airline. The answer is positive and so the passengers cannot help but contribute their share to the purser's aim being achieved: They now possess activated knowledge that what the purser did was perform a reminder of the prohibition in question.

So the new picture both integrates and rejects elements from all three approaches mentioned above: It agrees with Grewendorf in rejecting the claim that explicit performatives are statements in the narrow sense and therefore good truth value bearers, but it rejects his idea that the performative verb (or better, as we have just seen, performative predicate) does not contribute to the corresponding proposition. It agrees with Bach and Harnish to the extent that their notion of statement is weakened and assimilated to the notion of a transparent epistemic alias assertive, but it rejects their assumption that explicit performatives are derived and indirect speech acts. It agrees with Searle in subsuming the explicit performatives under the declarations, illocutions whose performance makes their propositional content congruent, but it rejects his claim that derivatively they are indirect statements as well.

4. Conclusion

In this contribution to a special birthday of my former coauthor I have outlined some of the changes my conceptualization of sentence moods and illocutions has undergone in the two decades since we jointly wrote *Theorien der Satzmodi*. How much have we been held captive by pictures that were taken for granted at that time? Not more and not less than others by other pictures. How difficult was it to get outside? Research is hard work and progress is slow, but as long as we are not satisfied with the pictures we have, we better keep revising and redrawing them.

I would like to conclude with some comments on the view implicated in the paragraph from Wittgenstein's *Philosophical Investigations* I quoted in footnote 1 above and which I repeat here in its English version: "A *picture* held us captive. And we could not get outside it, for it lay in our language and language seemed to repeat it to us inexorably."

I claim that no human language has the power ascribed to it by Wittgenstein in this paragraph, for two reasons: First, for every 'picture' provided by a language for a given phenomenon there are innumerably many alternative ways of paraphrasing or rephrasing it. Second, every human language has the means for introducing new expressions that allow a compact and easy-to-handle encoding of the revised picture. If this is correct, and I think it is, then the opposite of Wittgenstein's position comes out true: Whenever a picture threatens to keep us captive, language gives us the means to redraw and reshape it until we reach a new and hopefully more satisfying view of what it is a picture of. One purpose of this paper was to provide evidence that supports this claim.

I am confident that the Cognitivized Austinian Propositional contents, the analysis of sentence mood meanings proposed here and the discovery of the truth bearers fallacy will contribute to a more adequate picture of what happens when we use this specifically human mind sharing, mind stirring, and mind structuring tool we call language. And this concludes my presentation of the way I am conceptualizing sentence mood – two decades later.

References

Austin, J. L. 1950. Truth. *Proceedings of the Aristotelian Society*. Supplementary Volume xxiv. Reprinted in: *Philosophical Papers*, 3rd ed., Oxford: Oxford University Press 1979, 117–33.

Bach, K. and R. M. Harnish, 1979. *Linguistic Communication and Speech Acts*, Cambridge MA: MIT Press.

Barwise, J. and J. Etchemendy. 1987. *The Liar. An Essay in Truth and Circularity*. Oxford: Oxford University Press.

Ginzburg, J. and I. Sag. 2000. *Interrogative Investigations: The Form, Meaning and Use of English Interrogatives*. Stanford: CSLI.

Grewendorf, G. 2002. How performatives don't work. In: G. Grewendorf and G. Meggle (eds.) (2002), 25–39.

Grewendorf, G. and D. Zaefferer. 1991. Theorien der Satzmodi. In: Stechow and Wunderlich (eds.) (1991), 270–286.

Grewendorf, G. and G. Meggle (eds.). 2002. *Speech Acts, Mind, and Social Reality. Discussions with John R. Searle*. Dordrecht: Kluwer.

Harnish, R. M. 1997. Performatives and standardization: a progress report. In: E. Rolf (ed.), *Pragmatik*, Sonderheft 8 of *Linguistische Berichte*, 161–175.

Harnish, R. M. 2002. Are performative utterances declaratives? In: Grewendorf and Meggle (eds.) (2002), 41–64.

König, E. and P. Siemund. (to appear). Speech act distinctions in Grammar. In: T. Shopen (ed.), *Language Typology and Syntactic Description*. Cambridge: Cambridge University Press.

Liszkowski, U., M. Carpenter, A. Henning, T. Striano, and M. Tomasello. 2004. Twelve-month-olds point to share attention and interest. *Developmental Science* 7, 297–307.

Schalley, A. C. and D. Zaefferer (eds.). (forthcoming). *Ontolinguistics. How Ontological Status Shapes the Linguistic Coding of Concepts.* Berlin: de Gruyter.

Siebel, M. 2002. What is an illocutionary point? In: G. Grewendorf and G. Meggle (eds.), *Speech Acts, Mind and Social Reality. Discussions with John R. Searle*, 124–140. Dordrecht: Kluwer.

Stechow, A. von and D. Wunderlich (eds.). 1991. *Semantik. Ein internationales Handbuch der zeitgenössischen Forschung / Semantics. An International Handbook of Contemporary Research.* Berlin: de Gruyter.

Wittgenstein, L. 1960. *Tractatus logico-philosophicus. Logisch-philosophische Abhandlung.* Frankfurt/Main: Suhrkamp.

Wittgenstein, L. 1967. *Philosophische Untersuchungen.* Frankfurt/Main: Suhrkamp.

Zaefferer, D. 2001. Modale Kategorien. In: M. Haspelmath, E. König, W. Oesterreicher, W. Raible (eds.), *Language Typology and Language Universals / Sprachtypologie und sprachliche Universalien / La typologie des langues et les universaux linguistiques. An International Handbook / Ein internationales Handbuch / Manuel international* (HSK 20.1), 784–816. Berlin: de Gruyter.

Zaefferer, D. 2004. Interrogative mood as content-type-driven epistemo-volitional modality. ESSLLI-Workshop 'Syntax, semantics and pragmatics of questions', Nancy.
(http://amor.rz.hu-berlin.de/~h2816i3x/ESSLLI_Questions_Zaefferer.pdf).

Zaefferer, D. 2005. A general typology of modal categories. *Tidsskrift for Sprogforskning* 3/2, temanummer *Modalitet.*

Zaefferer, D. (forthcoming a). Language as mind sharing device. Mental and linguistic concepts in a general ontology of everyday life. In: Schalley and Zaefferer (eds.).

Zaefferer, D. (forthcoming b) Deskewing the Searlean picture. A new speech act ontology for linguistics. *Proceedings of the Thirty-Second Annual Meeting of the Berkeley Linguistics Society.*

Zaefferer, D. (in preparation). Types, moods, and force potentials. Towards a comprehensive account of German sentence mood meanings. *Theoretical Linguistics*, special issue on Sentence Type Meanings and V-to-C.

The Values of Semantics

Thomas Ede Zimmermann

0. Introduction

This paper is about (what I call) *realistic* semantic theories – theories that are essentially mappings: every expression α (of a given language) is assigned its semantic value $[\![\alpha]\!]$ – its denotation, intension, character, sense, content, update potential, or what have you. According to the view that I am concerned with, the (conventional) *meaning* of an expression is its semantic value. In other words, linguistic meanings are conceived of as the objects a semantic theory assigns to expressions, and that is all there is to (conventional) meaning. On this view, a semantic theory is correct just in case it gets the meaning map right, i.e., if it associates with each (analyzed) expression the (literal) meaning that expression actually has. Two semantic theories agree on their subject matter – literal meaning in a given language – just in case they assign precisely the same meaning to each expression of the language.

Two theories may thus disagree in the way they specify the semantic values; for instance, one may be formulated in terms of compositional meaning combinations while the other may be formulated in a more 'representational' way. Yet, as long as they assign the same meaning to each expression of the object language, they are in agreement on their subject matter. On the other hand, if two theories agree on which expressions are synonymous with which, they may still disagree about what these meanings are – and hence disagree on their subject matter.

Some qualifications may be necessary in order to turn this caricature into a truly standard view of semantic theory, but I think they are largely negligible. For instance, it is usually recognized that there may well be more to conventional meaning than what is accounted for by realistic meaning assignments – expressive meaning being a case in point.[1] So what I am concerned with might not cover the whole realm of linguistic meaning, but only part of it – but certainly a crucial part. More importantly, there may be some tolerance as to the precise nature of semantic values; for instance, it may be felt immaterial whether a truth value is taken to be a natural number or a set of variable assignments or sequences; consequently, two semantic theories that only disagree on what truth values are do not really disagree on their subject matter. Hence, the view of semantics that I am interested in is slightly more sophisticated, but in a way that is largely, though maybe not entirely, irrelevant to my concern. Finally, some semantic theories assign more than one value to each expression; and some of them may occasionally assign less than one. However, these exceptions are readily dealt with by combining the various semantic values into one 'super-value' and replacing gaps by dummy values, respectively.

---

1 See, however, Potts (2003) for the opposite view.

Even given these amendments, I think there is something wrong with the general picture outlined in the first two paragraphs, and in this paper I will try to pinpoint where exactly it does not match actual semantic practice, and the semanticists' (implicit) understanding of it.

1. Puzzling values

Let me start with a puzzle due to Friederike Moltmann (2004a; 2000b). Suppose we have a (realistic) semantic theory that tells us among other things:

(1) a. $[\![$*ordinariness*$]\!] = P_o$

(2) a. $[\![$*honesty*$]\!] = P_h$

– where P_o and P_h are the properties of ordinariness and being honest, respectively (whatever they may be). In other words, according to this theory, *bare adjective nominalizations* (to use Moltmann's term) refer to properties. Given these properties, it seems we may also refer to them by definite descriptions, or (again following Moltmann) by *explicit property terms*. We would thus also have:

(1) b. $[\![$*the property of ordinariness*$]\!] = P_o$

(2) b. $[\![$*the property of honesty*$]\!] = P_h$

To be sure, the (b) equations require the object language to contain some mild form of philosophical jargon, but then neither ontological scruples nor stylistic purism have kept us from using such jargon in formulating semantic theory itself, and there is no good reason to put our own language out of the reach of semantic analysis. By the transitivity (and symmetry) of identity the (a) and (b) equations thus lead to:

(1) c. $[\![$*ordinariness*$]\!] = [\![$*the property of ordinariness*$]\!]$

(2) c. $[\![$*honesty*$]\!] = [\![$*the property of being honest*$]\!]$

But now it seems we're in trouble if we try to apply (1c) and (2c) to:[2]

(3) *Ordinariness is boring.*

(3') *The property of ordinariness is boring.*

(4) *John is looking for honesty.*

(4') *John is looking for the property of being honest.*

Assuming compositionality (and thus substitutivity), it would seem that our analysis (1a) forces us to treat (3) and (3') as synonymous; and similarly for analysis (2a) and the pair (4) and (4'):

---

2 Moltmann (2004*a*: 7, 19; 2004*b*: 744). Moltmann uses the examples to argue against (1a) and (2a). I do not wish to debate her argument or its conclusion.

(1) d. $[\![$*Ordinariness is boring*$]\!]$ = $[\![$*The property of ordinariness is boring*$]\!]$

(2) d. $[\![$*J. is looking for honesty*$]\!]$ = $[\![$*J. is looking for the property of being honest*$]\!]$

However, there is an obvious difference in content (and maybe even in truth value) between (3) and (3') – as there is an obvious difference in content (and maybe even in truth value) between (4) and (4'). So it appears that we should either give up compositionality or else our theory.

Appearances are deceiving, though: *tertium datur*. In fact, the refutation of (1a) and (2a) by (3) and (4) is reminiscent of a classical puzzle about basic (extensional) quantifier semantics, according to which we may safely assume equations like:[3]

(5) a. $[\![\textit{nothing}]\!]^{\ldots} = \{\varnothing\}$

(6) a. $[\![\textit{Jones}]\!]^{\ldots} = \{M \subseteq U \mid j \in M\}$

As before, we ought to be able to extend the semantic analysis to explicit set terms:[4]

(5) b. $[\![\textit{the set of memberless sets}]\!]^{\ldots} = \{\varnothing\}$

(6) b. $[\![\textit{the set of sets that contain Jones as a member}]\!]^{\ldots} = \{M \subseteq U \mid j \in M\}$

And as before, the (a) and (b) equations allow us to envoke the transitivity of identity, thus arriving at:

(5) c. $[\![\textit{nothing}]\!]^{\ldots} = [\![\textit{the set of memberless sets}]\!]^{\ldots}$

(6) c. $[\![\textit{Jones}]\!]^{\ldots} = [\![\textit{the set of sets that contain Jones as a member}]\!]^{\ldots}$

It is now easy to find examples that are analogous to (3) and (4) in that the replacement of a quantifier by a corresponding explicit set term does not preserve truth conditions, or even truth-values, let alone synonymy:

(7) a. *Nothing is easy.*

(7) b. *The set of memberless sets is easy.*

(8) a. *Jones is an infinite set.*

(8) b. *The set of sets that contain Jones as a member is an infinite set.*

However, it is clear that there is no conflict between the differences observed in (7) and (8) on the one hand, and the assumed equations (5) and (6) on the other hand: contrary to appearances, the (b) variants of (7) and (8) do not derive from the (a) variants by

---

3 This type of puzzle was brought to my attention by Friedrich Kambartel (personal communication) in the late 1970s. Note that extensions may not count as semantic values of their own, but must be derived from semantic values – contents, intensions, characters, or whatever – by some operation on certain parameters indicated by '…'; still, in the case at hand, the pertinent semantic values stand in a one-one correspondence to extensions and may thus be safely identified with them.

4 There is a technical obstacle here: according to standard type-theoretic implementations of quantifier theory, the (a) and (b) equations in (5) or (6) cannot simultaneously be true. What we need is a more liberal framework that allows for higher-order objects to occur in lower-order domains. Such frameworks have been developed (cf. Cresswell 1973) and are usually assumed to be (relatively) consistent (cf. Glubrecht *et al.* 1983: 441). Hence logic should not keep us from accepting equations like (5) and (6).

substituting co-designative expressions. As sentences of English, (7a) and (7b) – and likewise (8a) and (8b) – are individuated by their morphosyntactic structure; the surface strings above are only used to communicate them (which in turn only works because they do not happen to be ambiguous in any obvious way). However, while the surface strings of the (b) variants may be derived from (7a) and (8a) by substitution, the corresponding structures do not stand in such a relation. But then compositionality does not relate to surface forms but to structures, or, as I will from now on say, *expressions*.

Let me go into some detail. In the case of (7), one should first note that, according to (5b), the value of the set-description is the set described by it. Hence (5b) is a 'naive' Fregean analysis of the description at hand, i.e. one in accordance with Frege (1892). But then the difference between (7a) and (7b) is one in logical form, the former being a case of quantification and the latter an instance of predication. In other words, it is the subject that is used to classify whatever the predicate denotes in (7a), whereas in (7b), it is the other way round:

(9) a. Quantification
$[\![NP_{quant}\ VP]\!]^{\cdots} = 1 \Leftrightarrow [\![VP]\!]^{\cdots} \in [\![NP_{quant}]\!]^{\cdots}$

b. Predication
$[\![NP_{ref}\ VP]\!]^{\cdots} = 1 \Leftrightarrow [\![NP_{ref}]\!]^{\cdots} \in [\![VP]\!]^{\cdots}$

Hence even though, by hypothesis, the two subjects are co-designative, there is a difference in structure, which incidentally correlates with a difference in categorization. However this is specified, it is part of the structure underlying (7a) that the subject *nothing* is a quantified NP. Moreover, it is part of the lexical specification of the subject of (7b) that it is a referential NP. But then replacing the subject of (7a) by the subject of (7b) would result in a structural description that violates the lexical specification of the latter; in particular, it would not describe an expression of English. And so the puzzle about (7a) dissolves.

It ought to be noted that the Fregean interpretation (5b) was as crucial for the emergence of the puzzle as it was for its solution. On a Russellian account, i.e. one based on Russell (1905), (7a) and (7b) would not have differed in (overall) logical form, but their subjects would not have been co-designative anyway, i.e., (5c) would not have held. For, according to a Russellian treatment of the definite description in (7b), we would have had (5b') instead of (5b):

(5) b'. $[\![\textit{the set of memberless sets}]\!]^{\cdots} = \{P \subseteq U \mid \{\varnothing\} \in P\}$

The equations in (6) are different in that they mix a quantificational account of proper names (along the lines of Montague 1970) with a Fregean interpretation of definite descriptions. And in fact, the solution to the puzzle (6) requires names to be treated as ordinary quantifiers and definite descriptions as singular terms. Hence, and that is the lesson for the time being, the puzzle about (8) just would not arise in this form because standard semantic theories rarely satisfy both equations under (6) at the same time; and even where they do, it could be solved by (9) and the outlandish classification of the subjects indicated.

Let us now return to the initial puzzle about (1) and (3). Since it is of the same format as the ones about referential vs. quantificational noun phrases, it ought to be susceptible to the same kind of solution in terms of logical form. So in analogy to the semantic split between quantification vs. predication, we might assume that the morpho-syntactic distinction between bare adjective nominalizations and explicit property terms reflects a constructional difference between a generic and a predicational reading. Simplifying somewhat, let us assume that (3) expresses that anyone and anything ordinary is boring; whereas in (3') being boring is attributed to a certain abstract object. We thus arrive at the analyses:

(10) a. Subsumption
$[\![NP_{ban}\ VP]\!] = [\![NP_{ban}]\!] \sqsubseteq [\![VP]\!]$

b. Predication
$[\![NP_{ept}\ VP]\!] = [\![VP]\!]([\![NP_{ept}]\!])$

In (10a) '$\sqsubseteq$ ' stands for property inclusion as a proposition-forming operator; in (10b) attribution of a property is taken to be a form of functional application. As before, the solution of the puzzle requires a categorial distinction – NP_{ban} and NP_{ept} in this case – separating the two co-designative expressions. It should be noted that, according to (10b), explicit property terms in subject position are interpreted like referential noun phrases in (9b); in fact, (10b) is the interpretation expected on a Fregean account of definite descriptions – which is what explicit property terms are, after all. (Obviously, they could also have been given a Russellian interpretation.) On the other hand, even if (10a) were – somewhat inadequately – construed as a case of quantification, the quantifier lies in the relation $\sqsubseteq$ of sub-propertyhood, not in the subject. Hence (10a) is not the same as (9a); but it probably generalizes to other forms of noun phrases like certain generic uses of bare plurals.[5]

Now for (4). The formulation of the puzzle suggests a property analysis of opaque verbs like *seek*, according to which the verb expresses a relation between an individual and a property that may be denoted by an indefinite object, or even a definite description.[6] In the former case, an indefinite noun phrase like *a black flower* is taken to denote the property of being a black flower; in the latter case, a definite noun phrase like *the boss* denotes the property of being the one and only boss. Since the object in (4') is a definite description, it will denote the (unique) property of being identical to P_h – which is certainly distinct from P_h itself, the property of being honest: the former, but not the latter (I hope) is necessarily instantiated by exactly one object. But then, still following the property analysis of opacity and its underlying dynamic interpretation of descriptions, (2b) does not hold – and there is no substitution puzzle. (2a) does hold though, and in fact we may add:

5 I am oversimplifying. To begin with, (10a) does not fully capture the intended meaning; cf. Moltmann (2004*b*: 754). Moreover, (10a) and (10b) do not cover all possible readings. For instance, the most prominent readings of (3') and (4') are also attested for the unprimed variants.

6 Cf. Zimmermann (1993). Hence this treatment of definite descriptions is neither Fregean nor Russellian but *dynamic* in the loose sense usually associated with the kind of semantic frameworks developed by Heim (1982), Kamp (1981), and others.

(11) $[\![\textit{honesty}]\!] = [\![\textit{an honest individual}]\!] = P_h$

If we now assume – possibly somewhat simplifying again – that (4) expresses that John is looking for someone honest, i.e. an honest individual – (4) can be paraphrased by the unspecific reading of:

(12) *John is looking for an honest individual.*

Given the property analysis of opacity, the (assumed) equivalence between (4) and (12) may be accounted for by:[7]

(13) Unspecificity
$[\![V_{op}\ NP_{ban}]\!] = [\![V_{op}]\!]([\![NP_{ban}]\!])$

Note that (13) is just the ordinary unspecific construal; so by allowing *honesty* and *an honest person* to share their category – if only in certain environments – we may actually conceive of (12) as the result of substituting the latter for the former in (4) and then obtain the desired (assumed) synonymy by (11).

Given the property analysis of opacity, then, there is no puzzle about (4) and (4'). In fact, whereas (4) comes out as equivalent to (12), (4') expresses something rather strange, viz. that Jones is after some abstract object – be it via representation (in a dictionary, say), be it literally and fictionally, searching Plato's Heaven. I think this is as it should be.

Following the quantifier analysis of opacity (first proposed in Montague (1969)) instead, the equations in (2) ought to come out as correct, at least if we follow the 'naive' Fregean approach to definite descriptions. However, in this case the logical form strategy may be adapted. Without going into the details, let me just propose a rather straightforward rule of construal that takes care of bare adjective nominalizations in the object positions of opaque verbs:

(14) Opaque Properties
$[\![V_{op}\ NP_{ban}]\!] = [\![V_{op}]\!](\hat{P}\,(\exists y)[\,[\![NP_{ban}]\!](y) \wedge P\{y\}])$

2. Rôles of values

The substitution puzzles of the kind discussed here turned out to be illusive. Why bother? Because, I would like to argue, they shed light on an important feature of realistic semantic theories, and so do their solutions. What the puzzles have in common is that they all rest on a shift of subject matter, from ordinary individuals to semantic values. As I said, I do not think there is anything illicit or paradoxical with this move as such: granting that the technical problems can somehow be overcome, ontological worries should not count as long as the theory itself relies on the existence of the objects in question. It is not the fact that semantic values serve as the subject matter of the object language that creates the puzzles; rather, it is the fact that they enter semantic theory twice over, thereby playing different, mutually incompatible rôles.

---

7 (13) makes use of the usual Schönfinkel-Curry reduction of binary relations to nested unary functions.

To make the point I am after more tangible, let me lay out the common structure of the puzzles in form of a simple piece of logical deduction. In each case, the starting point was a semantic theory assigning values v to expressions A of a given language, or fragment thereof:

(15a) $[\![A]\!] = v$

Next, the object language is extended so as to include part of the meta-language that was used to describe the (relevant) value of A – whereupon the extended language is interpreted accordingly, i.e. by extending the old value assignment to the new expressions "*v*", to which I will refer as *explicit value terms*:

(15b) $[\![\text{"}v\text{"}]\!] = v$

Hence, given the transitviy (and symmetry) of identity, there is no way to escape:

(15c) $[\![A]\!] = [\![\text{"}v\text{"}]\!]$

By (15c), A and its value description "*v*" would seem to be synonymous and thus intersubstitutable *salva significatione* in all contexts X:

(15d) $[\![X]\!] = [\![X[^{A}/_{\text{"}v\text{"}}]]\!]$

However, suitable choice of X may show that the synonymy predicted in (15d) does not bear scrutiny. It is important to realize that it is not the introduction of explicit value terms as such that creates the problem, but the concomitant presence of such X. For it may well happen that the newly introduced term behaves sufficiently similar to the old onc, thereby confirming equation (15c). For instance, it does seem possible to replace *John* by the more explicit term *the extension of "John"* (once we have agreed that the latter does describe its extension[8]), without thereby causing a meaning shift as dramatic as those observed in the cases considered so far; I suspect that any meaning differences may be explained in pragmatic terms.

Looking at the structure of the argument, it might seem that the only way to save a semantic equation like (15a) is to deny (15b). Indeed, this is how the puzzle about (5a) was solved from the point of view of a Russellian analysis of descriptions: (5b) gets replaced by (5b'), and thus (5c) no longer follows; neither is there a problem with contexts X of the form '___ *is easy*'. However, denying (15b) is not the only way to go about. For, as we have seen in connection with the Fregean analysis of descriptions, (15b) and thus (15c), may be upheld, while (15d) is denied. More accurately, what was denied was that (7a) and (7b) instantiate the same schema, that they result from each other by swapping their subjects. Hence, in the present general setting, what may be denied is a presupposition of (15d), viz. that the result $[\![X[^{A}/_{\text{"}v\text{"}}]]\!]$ of replacing A by "*v*" in X is in the domain of the value assignment $[\![\cdot]\!]$ – i.e. that $[\![X[^{A}/_{\text{"}v\text{"}}]]\!]$ is an expression given that X is.[9] So there are basically two ways of protecting (15a) from the

8 Sceptics may add a reference to a pertinent theory *T*: ... *the extension of "John" according to T.*

9 In an algebraic meta-theoretic setup in the tradition of Montague (1970) and Janssen (1997), the domain of the interpretation function $[\![\cdot]\!]$ is most naturally understood to be restricted to grammatical material, i.e. expressions, which, unlike the class of all structural descriptions, are not necessarily closed under substitution, as we have seen in connection with (7).

invasion (15c) of the meta-language: the *R* strategy of denying (15b); and the *F* strategy of rejecting the presupposition (15d) is based on.

In order to see the difference between the two strategies more clearly, it is instructive to consider the case of truth values, identifying them with the numbers 0 and 1. Given this identification,[10] we find that both the numeral *zero* and the description *the number of self-distinct objects* have the same extension as the sentence:

(16) *Nothing is identical with itself.*

Moreover, given the analytic character of these extensional equivalences, the expressions presumably coincide in whatever semantic values we might be interested in:

(17) $[\![$*zero*$]\!]$ = $[\![$*the number of self-distinct objects*$]\!]$ = $[\![(16)]\!]$

We can certainly avoid the ensuing embarrassment concerning substitution of, say, the subject of (18) with (16) by giving up our identification of truth values with numbers.

(18) *The number of self-distinct objects is zero.*

Insisting on the numerical reconstruction of truth values, we could have the numeral *zero* refer to some *ersatz* number 0*, which must then not be confused with the true number 0, the hallmark of falsity. We could, and if we did, we would thereby strictly adhere to the *R* strategy. However, such a maneuvre appears to be unnecessary because presumably the result of replacing the subject of (18) by (16) ought to be ruled out on purely syntactic grounds. In particular, it would be uninterpretable, just as the *F* strategy would have it.

Neither strategy is without problems. While the *R* strategy needs to specify the semantic value of the newly introduced explicit value term – just saying what it is *not* obviously is not enough –, the *F* strategy needs to make a category distinction between the original expression *A* and the corresponding value term "*v*". Of course, inasmuch the descriptive apparatus for individuating morpho-syntactic structure is arbitrary, it is always possible to come up with *some* such distinction. But as long as the latter is only motivated by the attempt to escape the unwanted consequence (15d), it may rightly be felt to be *ad hoc*.

Whichever of the two strategies is applied, setting the system (15) of equations right cannot be the end of the story. Following the *F* strategy, the very acceptance of (15c) presents a problem: since the original expression and the corresponding explicit value term receive the same semantic values, we must take precautions so as to not predict that they are synonymous – which, as we saw, they need not be. Following the *R* strategy, the very rejection of (15b) presents a problem: the obvious semantic link between the explicit value term and the semantic value of the original term are still in need of explanation.

---

10 … which, incidentally, is not completely arbitrary but may be grounded in a satisfaction-based account of truth values as sets of 0-tuples on the one hand, and von Neumann's set-theoretic reconstruction of ordinal numbers as the sets of their predecessors on the other; cf. Zimmermann (2002: 32ff.) for an exposition of this (I believe) folklore view. Still, the considerations to come only depend on there being some fixed identification of the truth values, not this particular one.

Neither of these tasks needs to be particularly delicate or difficult. Take (5), as a case in point. According to the *F* strategy, the quantifier *nothing* and the description *the set of memberless sets* coincide in their extensions. Yet, whereas the former *expresses* its extension, the latter *denotes* it. In other words, the semantic value the two expressions have in common plays (at least) two distinct rôles. In particular, as the difference between (9a) and (9b) illustrates, the way it interacts with other semantic values depends on whether it acts as the value of the quantifier, or as the value of the description. Hence a semantic distinction is made, but it is not made in terms of semantic values. A similar observation can be made about the semantic value(s) of sentences and their explicit descriptions.[11] Again the value of, say, (16) and the subject of (18) may be the same, but it plays different rôles, depending on whose value it is taken to be: as the semantic value of the latter (description), it may act as the 'logical subject' of a predication; as the semantic value of the former (sentence), it may be negated. One conclusion to be drawn at this point is that, even within a realistic framework (in the present sense), sameness of semantic values should not be taken to imply synonymy; for the latter relation to hold, the values would also have to play the same rôle.

According to the *R* strategy, the semantic distinction between the original expression and the explicit value term is already reflected in their semantic values. Hence, it would seem, the rôles they play in semantic theory need not be distinguished otherwise. And indeed, taking up the Russellian stance on (5), the original expression *nothing* and the explicit value term *the set of memberless sets* are both quantifiers, thus behaving alike as far as their interaction with other semantic values is concerned. However, the distinct values and the common theory-internal rôle they play cannot be all there is to their meanings. For although the values are distinct, there is a clear connection between them, which semantic theory should, and does, make out. After all, *vis-à-vis* the quantifier *nothing*, the explicit value term *the set of memberless sets* is not just any old quantifier, but a very special one, viz. (as the term suggests) one that describes its semantic value. And this close relationship between the two expressions does not go unnoticed by semantic theory: the value of the quantifier can be derived from that of the description in a canonical way – i.e. the way semantic referents of (non-empty) Russellian descriptions are defined in general.[12] But this relationship, though definable in semantic terms, does not itself feature in semantic theory, where semantic values are the only game in town. The relation between the two values – and the relation between the value of a description and its semantic referent in general – is called in when it comes to *applying* semantic theory to other areas in the analysis of cognition and communication. Hence, the definability of semantic reference is crucial, though not part of (realistic) semantic theory itself.

---

11 Frege – possibly blinded by his ontological distinction between functions and saturated objects – seemed to have thought of the difference between sentences and descriptions of (or names for) truth values as an artefact of natural language. The analytic tradition has not followed its founding father in this respect; nor will I.

12 By taking the sole member of the intersection, that is. The term *semantic referent* in this connection is due to Kripke (1977), where it is used in building the bridge to (broadly Gricean) pragmatics.

In a sense, then, both strategies are missing something. In the case of the *F* strategy, what is missing is a specification of the internal rôles played by – at least some – semantic values; without it, no adequate account of synonymy appears possible. In the case of the *R* strategy, what is missing is a specification of the external rôles played by – at least some – semantic values; without it, no adequate account of the application of semantic theory appears possible. So whether we take the *R* strategy or the *F* strategy, there appears to be more to meaning – even literal meaning – than semantic values. Since realistic semantic theories are usually uncommitted (or mixed) as to the two strategies, they are missing something on either count. In other words – and as announced in the introduction – they should not be construed as merely pairing expressions with meanings; rather, they ought to be conceived of as pairings of expressions with their values plus specifications of the internal and external rôles these values play.

What are external and internal rôles? Despite the similarity in terms, they must be individuated in quite different ways. A semantic value *v* a given semantic theory assigns to a particular expression *A* can play an external rôle if there is an interfacing theory (of some neighbouring discipline) that relies on *A* having certain properties which can be systematically derived from *v*. For example, pragmatic reasoning frequently relies on assumptions about the proposition *p* (literally) expressed by a given sentence *A*: a scalar implicature strengthens *p* by negating its expressive alternatives; irony somehow reverses *p* on the (minimal) assumption that the speaker obviously believes *p* to be false; whether a speaker can be construed as making a promise, and if so which promise, is partly determined by the propositional content of the sentence uttered, i. e., *p*; etc.[13] In that case *p* ought to be somehow identifiable with the help of *v* and possibly some other relevant (e.g., contextual) factors. If so, *v* plays the external rô le of (partly) determining the literal content of *A* in the sense of pragmatic theory. On the other hand, *v*'s internal rôle consists in the way *v*, as a semantic value of *A*, interacts with the semantic values of the other expressions. Obviously their internal and external rôles impose some restrictions on the particular nature of semantic values – e.g. as regards fine-grainedness. But they do not determine them completely.

3. More puzzles

Rather than making the above ideas precise and turn them into a general semantic meta-theory, let me discuss a few more puzzles that hopefully give a rough idea of the direction such an enterprise would take. A full treatment must be left to another occasion.

For the first puzzle to arise, we imagine two alternative semantic theories, T_e and T_{ett} of English that only differ in their treatment of proper names. While T_e assigns to each name its bearer as its extension – as illustrated in (19), T_{ett} treats names in the above-mentioned quantificational way (6a), repeated in (20):

---

13 Cf. Horn (1972: 112); Grice (1989 [1978]: 54f.); and Searle (1969: 57ff.), respectively.

(19) $[\![Jones]\!]^{\cdots} = j$

(20) $[\![Jones]\!]^{\cdots} = \{M \subseteq U \mid j \in M\}$

Note that, in order for T_e and T_{ett} to not differ in other respects than the equations (19) vs. (20), their descriptive coverage has to be somewhat limited. Granting this, it is clear that whatever T_e says is mimicked by something T_{ett} says, and vice versa. For instance, T_e analyzes sentences with *Jones* in their subject position as predications that are true just in case the extension of *Jones* according to (19) is a member of the extension of the predicate; T_{ett} analyzes the same sentences as quantifications that are true just in case the extension of *Jones* according to (20) contains the extension of the predicate as a member. Hence both theories assign the same truth conditions to such sentences – and similarly for other expressions in which proper names occur. Since T_e and T_{ett} agree on all other expressions anyway, the two theories appear to be very similar – so similar indeed that it is tempting to think of them as notational variants of each other. In particular, it is not clear how one theory could be right without the other one; or in what sense T_e and T_{ett} may describe different languages.[14] Yet, although they agree on most predictions, they seem to differ in some. In fact, if the semantic values assigned to expressions are meant to be their (literal) meanings, then T_e and T_{ett} disagree on what proper names like *Jones* mean.

The disagreement does not go very deep. In fact, not only agree T_e and T_{ett} on their predictions concerning expressions that are not proper names; they also agree on how the semantic values of proper names interact with those of other expressions. In other words, they also agree on the *internal* rôle these values play in semantic theory. Moreover, it is clear that they would also agree on the *external* rôles played by these values: T_e directly connects a name with its bearer, which may also be defined in terms of T_{ett} – as we have seen in the preceding section: Montague-lifted names are Russellian descriptions, after all. It is this agreement on the external and internal rôles played by the semantic values they respectively assign, that makes the two theories equivalent. The agreement does not concern the semantic values as such, only the rôles they play according to T_e and T_{ett}. For there is nothing in the nature of a (Russellian) quantifier that makes it equivalent, or boil down to, an individual. It thus turns out that, to account for the equivalence of T_e and T_{ett} – and likewise to account for the equivalence of any theory and one of its functor/argument flip-flop variants – one must abstract from the specific values they assign and look at the internal and external rôles these values play.

The second puzzle compares two (imaginary) theories $T^{\hat{}}$ and $T^{\check{}}$ that completely agree as to which values they assign to which expressions but disagree as to the external status of these values.[15] Its bite lies in its meta-theoretic setting, the theory of extension and intension. For dramatic effect, we take $T^{\hat{}}$ to be an adequate account of English

---

14 This is what I take to be a wide-spread attitude among semanticists towards the theoretical variation under discussion (which I believe has been around as a *gedankenexperiment* in the semantic community since the early 1980s). However, McGinn (2000: 63ff.) argues in favour of the naive treatment (19), because it is more adequate in its ontological attributions (and not just because it is simpler). I think this is wrong: the ontological difference pertains to the semanticists' theoretical entities and does not, or not necessarily, extend to the empirical realm of the theory.

15 Similar puzzles have been discussed by Parsons (1972) and Lewis (1974).

within such a framework. We can then derive $T^{\vee}$ by assigning to each expression A a rigid (= constant) intension that maps any point i in Logical Space onto A's intension according to $T^{\wedge}$:

(21) $[\![A]\!]^{T^{\vee}}(i) = [\![A]\!]^{T^{\wedge}}$

What is the difference between the two theories? Well, according to $T^{\wedge}$, (4) expresses the proposition p that is true of all points at which John is looking for an honest individual. According to $T^{\vee}$, (4) does not express a proposition but a propositional concept,[16] viz. the concept of being p. Since the latter is rigid, one may think of the sentences according to $T^{\vee}$ as denoting the propositions they express according to $T^{\wedge}$. So what's the difference? In particular, what would be the difference between two languages conforming to $T^{\wedge}$ and $T^{\vee}$, respectively? None, it seems. However, $T^{\wedge}$ and $T^{\vee}$ assign neither the same intensions not the same extensions to the expressions of English; all that can be said is that the intensions assigned by $T^{\wedge}$ are the extensions (rigidly) assigned by $T^{\vee}$. And what is worse: since the extensions in $T^{\vee}$ behave just like the intensions in $T^{\wedge}$, the internal rôles played by the $T^{\wedge}$-intensions do not seem to match the internal rôles of the $T^{\vee}$-intensions; and similarly for the corresponding extensions (which are derived values anyway). What the puzzle about $T^{\wedge}$ and $T^{\vee}$ shows is that semantic values cannot be identified across theories. Both $T^{\wedge}$ and $T^{\vee}$ assign pairs of values to expressions, and the overall behaviour of these values – their internal and external rôles – are the same. Hence, the theories can be regarded as equivalent.

Finally, the distinction between values and their (external and internal) rôles may also help in clearing up a foundational problem in dynamic semantics where truth-conditionally identical sentences (and discourses) are not necessarily assigned the same semantic values.[17] The following pair is a case in point:

(22) *Someone is walking through the park.*

(23) *It is not the case that no one is walking through the park.*

The fact that (22) but not (23) can, easily and coherently, be followed by *He is whistling* is reflected in a difference in their dynamic semantic values. The puzzle arises by adding the (not uncommon) assumption that this difference correlates with a communicative effect: informing someone by way of (22) would put him/her in a different *information state* than using (23) would. But then it is not clear what to make of the difference between the two states. In particular, it does not appear to correspond to a difference in information content, or to some cognitive difference in subjects. The present view allows for a differentiation between the values assigned to sentence pairs like (22) and (23) and the rôles these values play in semantic theory. For even if the values are distinct and play distinct internal rôles, they may well end up playing the same external rôle when it comes to making the connection with a theory of information exchange.

---

16 in the sense of Stalnaker (1978), where propositional concepts (= functions from Logical Space to propositions) are employed for pragmatic purposes.

17 The puzzle has been discussed in Zimmermann (1999). See Spohn (1998), Stalnaker (1998), and Dekker (2004) for relevant background and alternative views.

4. Conclusion

There is more to semantics than assigning semantic values. This is the moral to be drawn from the above discussion. As a consequence, semantic theories should be formulated in a way that explicitly takes into account the internal and external rôles played by the semantic values they assign. Though I will not carry out this programme here and now, I would like to end with a short speculation on how this task should be attacked.

One lesson to be learnt from the puzzle (21) about the relativity of extension and intension is that semantic values ought to be primarily individuated externally, which could be done by a map from expression-value pairs to non-semantic entities. In the case at hand, these may be propositions; in other cases they may be the referents of noun phrases; the potential information states of communicators; the time of utterance; the world(s) at large; etc. Which non-semantic entities serve as external points of reference depends on which interfaces the semantic theory is supposed to share with its neighbours, and what these neighbours expect semantic theory to contribute. The latter expectation also determines which expressions are going to play an external rôle in the first place. It is, e.g., by no means obvious that, say, logical words like *every* and *or* play any rôle of their own, apart from contributing to the values of other expressions. This rôle, however, is the prime (and some would say: the only) internal rôle played. The very concept of contributing semantic values has been studied extensively over the last few years.[18] I suspect some of the methods and results can directly go into the framework envisaged above.[19]

References

Cresswell, M. J. 1973. *Logics and Languages*. London: Methuen.

Dekker, P. 2004. Grounding dynamic semantics. In: A. Bezuidenhout and M. Reimer (eds.), *Descriptions and Beyond*, Oxford: Oxford University Press.

Fernando, T. 2005. Compositionality inductively, co-inductively and contextually. In: M. Werning, E., Machedry, and G. Schurz (eds.), *The Compositionality of Meaning and Content, Volume 1: Foundational Issues*, 87–96. Frankfurt: Ontos.

Frege, G. 1892. Über Sinn und Bedeutung. *Zeitschrift für Philosophie und philosophische Kritik (NF)* 100, 25–50. [Translated as: On Sense and

[18] See Hodges (2001), Westerståhl (2004), and Fernando (2005) for the kind of study I have in mind.

[19] Different parts of the above material were presented to audiences in Batumi (Sixth International Tblisi Symposium on Language, Logic and Computation, September 2005), Stuttgart (Institutskolloquium, Institut für Maschinelle Sprachverarbeitung, November 2005), Frankfurt (Logisch-semantisches Kolloquium, November 2005), and Leipzig (Graduiertenkolleg 'Universalitä t und Diversitä t', February 2006). I am grateful for the critical and constructive feedback I received from them. In particular, I would like to thank Agnes Bende-Farkas, Adrian Brasoveanu, Hannes Dölling, Hans Kamp, Cécile Meier, Barbara Partee, Christian Plunze, and Hong Zhou. Thanks are also due to Patrick Brandt for suggesting improvements of the pre-final version. I take full responsibility for remaining inaccuracies and errors.

Reference (1892). In: P. T. Geach & M. Black (eds.), 1960. *Translations from the Philosophical Writings of Gottlob Frege*, 56–78. Oxford: Blackwell.]

Glubrecht, J.-M., A. Oberschelp, and G. Todt. 1983. *Klassenlogik*. Mannheim/Wien/Zürich: Bibliographisches Institut.

Grice, P. 1989 [1978]. Further notes on logic and conversation. In: P. Grice, *Studies in the Way of Words*, 41–57. Cambridge, Mass.: Harvard University Press. [Originally published in P. Cole (ed.), 1978. *Syntax and Semantics 9: Pragmatics*. New York: Academic Press.]

Heim, I. 1982. *The Semantics of Definite and Indefinite Noun Phrases*. Ph.D. Dissertation, University of Massachusetts at Amherst.

Hodges, W. 2001. Formal features of compositionality. *Journal of Logic, Language and Information* 10, 7–28.

Horn, L. 1972. *On the Semantic Properties of Logical Operators in English*. Ph.D. Dissertation UCLA, Los Angeles.

Janssen, T.M.V. 1997. Compositionality (with an appendix by Barbara H. Partee). In: J. v. Benthem and A. G. B. t. Meulen (eds.), *Handbook of Logic and Language*, 417–473. Amsterdam: Elsevier.

Kamp, H. 1981. A theory of truth and semantic representation. In: J. Groenendijk, T.M.V. Janssen, and M. Stokhof (eds.), *Formal Methods in the Study of Language. Part 1*, 277–322. Amsterdam: Mathematical Centre Tracts.

Kripke, S. A. 1977. Speaker's reference and semantic reference. In: P. French, T. Uehling, and H. Wettstein (eds.), *Midwest Studies in Philosophy, Vol II: Studies in the Philosophy of Language*, 255–276. Minneapolis: University of Minnesota Press.

Lewis, D. 1974. Tensions. In: M. K. Munitz and P. K. Unger (eds.), *Semantics and Philosophy*, New York: New York University Press.

McGinn, C. 2000. *Logical Properties: Identity, Existence, Predication, Truth*. Oxford: Oxford University Press.

Moltmann, F. 2004a. Properties and kinds of tropes: new linguistic facts and old philosophical insights. *Mind* 113, 1–41.

Moltmann, F. 2004b. Two kinds of universals and two kinds of collections. *Linguistics and Philosophy* 27, 739–776.

Montague, R. 1970. Universal Grammar. *Theoria* 36, 373–398.

Parsons, T. 1972. Some problems concerning the logic of grammatical modifiers. In: D. Davidson and G. Harman (eds.), *Semantics of Natural Language*, 127–141. Dordrecht: Reidel.

Potts, C. 2003. *The Logic of Conventional Implicatures*. Ph.D. Dissertation University of California at Santa Cruz.

Russell, B. 1905. On denoting. *Mind* 14, 479–493.

Searle, J. R. 1969: *Speech Acts: An Essay in the Philosophy of Language*. Cambridge, Mass.: Cambridge University Press.

Spohn, W. 1998. The intentional versus the propositional conception of the objects of belief. In: C. Martínez, U. Rivas, and L. Villegas-Forero (eds.), *Truth in Perspective*, 271–291. Aldershot: Ashgate.

Stalnaker, R. 1978. Assertion. In: P. Cole (ed.), *Syntax and Semantics 9: Pragmatics*, 315–332. New York: Academic Press.

Stalnaker, R. 1998. On the representation of context. *Journal of Logic, Language and Information* **7**, 3–19.
Westerståhl, D. 2004. On the compositional extension problem. *Journal of Philosophical Logic* 33, 549–582.
Zimmermann, T. E. 1993. On the proper treatment of opacity in certain verbs. *Natural Language Semantics* 1, 149–179.
Zimmermann, T. E. 1999. Remarks on the epistemic rôle of discourse referents. In: L. Moss, J. Ginzburg, and M. de Rijke (eds.), *Logic, Language, and Computation. Vol. 2*, 346–368. Stanford: CSLI. [Reprinted in: H. Kamp & B. Partee (eds.), 2004. *Context-Dependence in the Analysis of Linguistic Meaning*, 521–538. Amsterdam: Elsevier.]
Zimmermann, T. E. 2002. Semantik/Pragmatik: Begleitskript zur Einführung in die Sprachwissenschaft. Johann Wolfgang Goethe-Universität Frankfurt am Main. http://web.uni-frankfurt.de/fb10/zimmermann/Semantik.pdf

List of Publications

Monographs

Grewendorf, G. 1975. *Argumentation und Interpretation. Wissenschaftstheoretische Untersuchungen am Beispiel germanistischer Lyrikinterpretationen*. Kronberg: Scriptor.

Grewendorf, G., F. Hamm, and W. Sternefeld. 1987. *Sprachliches Wissen. Eine Einführung in moderne Theorien der grammatischen Beschreibung*. Frankfurt/Main: Suhrkamp (Korean translation: 1991).

Grewendorf, G. 1988. *Aspekte der deutschen Syntax. Eine Rektions-Bindungs-Analyse*. Tübingen: Narr.

Grewendorf, G. 1989. *Ergativity in German*. Dordrecht: Foris.

Grewendorf, G. 1995. *Sprache als Organ – Sprache als Lebensform*. Frankfurt/Main: Suhrkamp (Greek translation: 1999).

Grewendorf, G. 2002. *Minimalistische Syntax*. Tübingen: Francke-Verlag (UTB).

Grewendorf, G. (forthcoming) 2006. *Noam Chomsky*. (Beck'sche Reihe *Große Denker*). Munich: Beck.

Edited Books

Grewendorf, G. and G. Meggle (ed. and introduced). 1974. *Linguistik und Philosophie*. Frankfurt/Main: Athenäum (Reprinted as: *Linguistik und Philosophie*. Weinheim: Beltz Athenäum, 1995).

Grewendorf, G. and G. Meggle (eds.). 1974. *Sprache und Ethik*. Frankfurt/Main: Suhrkamp.

Grewendorf, G. (ed.). 1979. *Sprechakttheorie und Semantik*. Frankfurt/Main: Suhrkamp.

Grewendorf, G. and W. Sternefeld (eds.). 1990. *Scrambling and Barriers*. Amsterdam: Benjamins.

Grewendorf, G. (ed.). 1992. *Rechtskultur als Sprachkultur. Zur forensischen Funktion der Sprachanalyse*. Frankfurt/Main: Suhrkamp.

Grewendorf, G. and G. Meggle (eds.). 2002. *Speech Acts, Mind, and Social Reality*. Dordrecht: Kluwer.

Grewendorf, G. (ed.). 2004. *Linguistik in Deutschland zwischen den Zeiten*. Linguistische Berichte 200. Hamburg: Buske.

Articles

Grewendorf, G. 1972. Sprache ohne Kontext. Zur Kritik der performativen Analyse. In: D. Wunderlich (ed.), *Linguistische Pragmatik*, 144–182. Frankfurt/Main: Athenäum.

Grewendorf, G. and G. Meggle. 1974. Zur Struktur des metaethischen Diskurses. In: G. Grewendorf and G. Meggle (eds.), *Sprache und Ethik*, 7–31. Frankfurt/Main: Suhrkamp.

Grewendorf, G. 1976a. Ansätze einer linguistischen Pragmatik. In: E. v. Savigny (ed.), *Probleme der sprachlichen Bedeutung*, 189–199. Kronberg: Scriptor.

Grewendorf, G. 1976b. Fortschritte der Sprechakttheorie. In: E. v. Savigny (ed.), *Probleme der sprachlichen Bedeutung*, 101–123. Kronberg: Scriptor.

Grewendorf, G. 1977a. Something new about J. L. Austin? *Studies in Language* 3, 423–436.

Grewendorf, G. 1977b. Präsuppositionen bei disjunktiven Fragen. *Linguistische Berichte* 52, 13–31.

Grewendorf, G. 1978a. Universalien. In: H. Rademacher and E. Braun (eds.), *Wissenschaftstheoretisches Lexikon*, 619–622. Graz/Vienna/Cologne: Styria.

Grewendorf, G. 1978b. Zur Semantik von Wertäußerungen. *Germanistische Linguistik* 2–5, 155–192.

Grewendorf, G. 1978c. Probleme der logischen Analyse von Fragen. *Papiere zur Linguistik* 19, 5–57.

Grewendorf, G. 1978d. Nicht-empirische Argumente. Zur Problematik ihrer wissenschaftstheoretischen Untersuchung. *Zeitschrift für allgemeine Wissenschaftstheorie* 7, 21–40.

Grewendorf, G. 1978e. Ist Wittgensteins Privatsprachenargument trivial? In: *Wittgenstein und sein Einfluß auf die gegenwärtige Philosophie*, Akten des 2. Internationalen Wittgenstein-Symposiums, 363–365. Vienna: Hölder-Pichler-Tempsky.

Grewendorf, G. 1979a. Linguistik und Deutschlehrerausbildung. Vorschläge zum linguistischen Grundstudium. In: *Sprachwissenschaft im Unterricht, Linguistische Arbeiten und Berichte (LAB)*, Berlin, Heft 12, 133–167.

Grewendorf, G. 1979b. Explicit perfomatives and statements. *Journal of Pragmatics* 3, 431–445.

Grewendorf, G. 1979c. Haben explizit performative Äußerungen einen Wahrheitswert? In: G. Grewendorf (ed.), *Sprechakttheorie und Semantik*, 175–196. Frankfurt/Main: Suhrkamp.

Grewendorf, G. 1979d. Explizit performative Äußerungen und Feststellungen. In: G. Grewendorf (ed.), *Sprechakttheorie und Semantik*, 197–216. Frankfurt/Main: Suhrkamp.

Grewendorf, G. 1979e. Die Funktion der Theorie in den Natur- und Textwissenschaften, Paper presented at the evang. Akademie Pullach. Ms.

Grewendorf, G. 1979f. Is Wittgenstein's Private Language Argument trivial? *Ratio* 2, 149–161.

Grewendorf, G. 1980a. Sprechakttheorie. In: *Lexikon der germanistischen Linguistik* (LGL), 2. edition, 287–293. Tübingen: Niemeyer.

Grewendorf, G. 1980b. Oberflächensyntax und das Problem der Bedeutung. In: H.-H. Lieb (ed.), *Oberflächensyntax und Semantik*, 108–121. Tübingen: Niemeyer.
Grewendorf, G. 1980c. Analytische Philosophie. In: J. Speck (ed.), *Handbuch wissenschaftstheoretischer Begriffe*, Band 1, 20–24. Göttingen: Vandenhoeck & Ruprecht.
Grewendorf, G. 1980d. Funktionale Satzperspektive und deutsche Wortstellung. *Linguistische Berichte* 66, 28–40.
Grewendorf, G. 1980e. Argumentation in der Sprachwissenschaft. *Zeitschrift für Literaturwissenschaft und Linguistik* (LiLi) 10, 129–151.
Grewendorf, G. 1981a. Answering as decision making: A new way of doing pragmatics. In: H. Parret, M. Sbisà, and J. Verschueren (eds.), *Possibilities and Limitations of Pragmatics*, 263–284. Amsterdam: Benjamins.
Grewendorf, G. 1981b. Grammatische Kategorie und pragmatische Funktion. *Lunder germanistische Forschungen*, 233–247.
Grewendorf, G. 1981c. Der implizite Charakter von Sprachregeln und seine Konsequenzen für die Fremdsprachendidaktik. *Linguistik und Didaktik* 45/46, 20–34.
Grewendorf, G. 1981d. Pragmatisch sinnvolle Antworten. Ein entscheidungstheoretischer Explikationsvorschlag. In: D. Krallmann and G. Stickel (eds.), *Zur Theorie der Frage. Forschungsberichte des Instituts für deutsche Sprache*, 95–118. Tübingen: Narr.
Grewendorf, G. 1982a. Deixis and Anaphorik im deutschen Tempus. *Papiere zur Linguistik* 1, 47–83.
Grewendorf, G. 1982b. Zur Pragmatik der Tempora im Deutschen. *Deutsche Sprache* 3, 213–236.
Grewendorf, G. 1982c. Logische Kommunikation und kommunikative Logik. Zur Kritik einer formalen Pragmatik von Begründungsdiskursen. In: C. F. Gethmann (ed.), *Logik und Pragmatik*, 95–124. Frankfurt/Main: Suhrkamp.
Grewendorf, G. 1982d. Behaupten und Zustimmen. In: C. F. Gethmann (ed.), *Logik und Pragmatik*, 125–142. Frankfurt/Main: Suhrkamp.
Grewendorf, G. 1983a. What answers can be given? In: F. Kiefer (ed.), *Questions and answers*, 45–84. Dordrecht: Reidel.
Grewendorf, G. 1983b. Reflexivierung in deutschen A.c.I.-Konstruktionen. Kein transformationsgrammatisches Dilemma mehr. *Groninger Arbeiten zur Germanistischen Linguistik* (GAGL) 23, 120–196.
Grewendorf, G. 1983c. Bairisch als pro-drop Sprache. Paper presented at the annual meeting of the Deutsche Gesellschaft für Sprachwissenschaft. Ms.
Grewendorf, G. 1984a. On the delimitation of semantics and pragmatics: The case of assertions. *Journal of Pragmatics* 8, 517–538.
Grewendorf, G. 1984b. Reflexivierungsregeln im Deutschen. *Deutsche Sprache* 1, 14–30.
Grewendorf, G. 1984c. Besitzt die deutsche Sprache ein Präsens? In: G. Stickel (ed.), *Pragmatik in der Grammatik* (Jahrbuch 1983 des Instituts für deutsche Sprache), 224–242. Düsseldorf.
Grewendorf, G. 1984d. In support of small clauses. Zur Satzwertigkeit des A.c.I. im Deutschen. Paper presented at the University of Vienna. Ms.
Grewendorf, G. 1985a. Vertrauen ins Mißtrauen. Linguistische Analysen zur Argumentation des Bundesverfassungsgerichts anläßlich des Urteils zur vorzeitigen Auf-

lösung des 9. Deutschen Bundestages. In: J. Dyck et al. (eds.), *Rhetorik. Ein internationales Jahrbuch*, 101–115. Stuttgart: Frommann-Holzboog.
Grewendorf, G. 1985b. Anaphern bei Objekt-Koreferenz im Deutschen. Ein Problem für die Rektions-Bindungs-Theorie. In: W. Abraham (ed.), *Erklärende Syntax des Deutschen*, 137–171. Tübingen: Narr.
Grewendorf, G. 1985c. Sprache als Organ und Sprache als Lebensform. Zu Chomskys Wittgenstein-Kritik. In: D. Birnbacher and A. Burckhardt (eds.), *Sprachspiel und Methode. Zum Stand der Wittgenstein-Diskussion*, 89–129. Berlin and New York: de Gruyter.
Grewendorf, G. 1986a. Computereinsatz in der Linguistik-Ausbildung. In: *Informationsverarbeitung in den Geistes- und Sozialwissenschaften*, IBM-Hochschul-Kongreß Baden-Baden 1986, Munich: IBM.
Grewendorf, G. 1986b. Relativsätze im Deutschen: Die Rattenfängerkonstruktion. *Linguistische Berichte* 105, 409–434.
Grewendorf, G. 1987. Kohärenz und Restrukturierung. Zu verbalen Komplexen im Deutschen. In: B. Asbach-Schnitker and J. Roggenhofer (eds.), *Neuere Forschungen zur Wortbildung und Histographie. Festgabe für Herbert Brekle zum 50. Geburtstag*, 123–144. Tübingen: Narr.
Grewendorf, G. and C. Poletto. 1989. La costruzione cleft: analisi contrastiva. *Rivista di Grammatica Generativa* 14, 105–142.
Grewendorf, G. 1990a. Verbbewegung und Negation im Deutschen. In: *Groninger Arbeiten zur Germanistischen Linguistik* (GAGL) 30, 57–125.
Grewendorf, G. 1990b. Small pro in German. In: G. Grewendorf and W. Sternefeld (eds.), *Scrambling and Barriers*, 295–315. Amsterdam: Benjamins.
Grewendorf, G. 1990c. XY ... Ungelöst. Zu Theorie und Praxis forensischer Linguistik. In: H. Kniffka (ed.), *Texte zu Theorie und Praxis forensischer Linguistik*, 247–288. Tübingen: Niemeyer.
Grewendorf, G. and W. Sternefeld. 1990. Scrambling theories. In: G. Grewendorf and W. Sternefeld (eds.), *Scrambling and Barriers*, 3–37. Amsterdam: Benjamins.
Grewendorf, G. 1991. Unpraktische Sprachethik. *Ethik und Sozialwissenschaften* 2, Heft 2.
Grewendorf, G. and C. Poletto. 1991. Die Cleft-Konstruktion im Deutschen, Englischen und Italienischen. In: G. Fanselow and S. Felix (eds.), *Strukturen und Merkmale syntaktischer Kategorien*, 174–216. Tübingen: Narr.
Grewendorf, G. and J. Sabel. 1991. Scrambling and incorporation. *Sprachwissenschaft in Frankfurt*, Arbeitspapier Nr. 2, 1–48.
Grewendorf, G. and D. Zaefferer. 1991. Theorien der Satzmodi. In: D. Wunderlich and A. v. Stechow (eds.), *Semantik. Ein internationales Handbuch der zeitgenössischen Forschung*, 270–286. Berlin and New York: Walter de Gruyter.
Grewendorf, G. 1992a. Rechtskultur als Sprachkultur. Der sprachanlytische Sachverstand im Recht. In: G. Grewendorf (ed.), *Rechtskultur als Sprachkultur*, 11–41. Frankfurt/Main: Suhrkamp.
Grewendorf, G. 1992b. Parametrisierung der Syntax. In: L. Hoffmann (ed.). *Deutsche Syntax*, 11–73. Berlin and New York: Walter de Gruyter.
Grewendorf, G. 1993a. Die "kognitive Revolution" in der Sprachwissenschaft und ihr Einfluß auf die germanistsche Linguistik. In: *Kultureller Wandel und die Germanistik in der Bundesrepublik. Vorträge des Augsburger Germanistentags*

1991, Band 2: Germanistik und Deutschunterricht im historischen Wandel, 77–91. Tübingen: Niemeyer.
Grewendorf, G. 1993b. Der Sprache auf der Spur: Anmerkungen zu einer Linguistik nach Jäger Art. *Zeitschrift für Sprachwissenschaft* 12.1, 113–132.
Grewendorf, G. 1994a. Kohärente Infinitive und Inkorporation. In: A. Steube and G. Zybatow (eds.), *Zur Satzwertigkeit von Infinitiven und Small Clauses*, 31–50. Tübingen: Niemeyer.
Grewendorf, G. 1994b. Sprachfähigkeit in der Informationsgesellschaft. In: L. Jäger and B. Switalla (eds.), *Germanistik in der Mediengesellschaft*, 215–244. Munich: Fink.
Grewendorf, G. 1994c. Variable binding and reconstruction. *Sprachwissenschaft in Frankfurt*, Arbeitspapier Nr. 12, 1–51.
Grewendorf, G. 1994d. Interview with Noam Chomsky: On linguistics and politics. *Linguistische Berichte* 153, 10–15. Also published in: *Protosoziologie* 6, 1994, 293–303.
Grewendorf, G. 1994e. Infinitive coerenti e incorporazione. In: R. Dolci and G. Giusti (eds.), *Studi di grammatica tedesca e comparativa*, 67–97. University of Venice.
Grewendorf, G. and J. Sabel. 1994. Long scrambling and incorporation. *Linguistic Inquiry* 25, 263–308.
Grewendorf, G. and C. Wilder. 1994. Universal grammar in German and Scandinavian linguistics. In: C. P. Otero (ed.): *Noam Chomsky. Critical Assessments*. vol. I, 784–817. London and New York: Routledge.
Grewendorf, G. 1995a. Präsens und Perfekt im Deutschen. In: *Zeitschrift für Sprachwissenschaft*, 14.1, 72–90.
Grewendorf, G. 1995b. German. A grammatical sketch. In: J. Jacobs, A. v. Stechow, W. Sternefeld, and T. Vennemann (eds.), *Syntax: An International Handbook of Contemporary Research*, 1288–1319. Berlin: Walter de Gruyter.
Grewendorf, G. 1996. Does second language grow? *Behavioral and Brain Sciences* 19, 727–728.
Grewendorf, G. and J. Sabel. 1996. Multiple specifiers and the theory of adjunction: On scrambling in German and Japanese, *Sprachwissenschaft in Frankfurt*, Arbeitspapier Nr. 16.
Grewendorf, G. and J. Sabel. 1997. *Wh*-scrambling in the minimalist framework. In: C. Wilder, H.-M. Gärtner, and M. Bierwisch (eds.), *The Role of Economy Principles in Linguistic Theory*, 58–80. Berlin: Akademie Verlag.
Grewendorf, G. 1998. Causatives in Germanic and Romance languages. Ms., University of Frankfurt.
Grewendorf, G. 1999a. The additional-wh effect and multiple wh-fronting. In: D. Adger et al. (eds.), *Specifiers. Minimalist approaches*, 146–162. Oxford: Oxford University Press.
Grewendorf, G. 1999b. Das funktionalistische Paradox. Zum Problem funktionaler Erklärungen in der Linguistik. In: H. E. Wiegand (ed.), *Sprache und Sprachen in den Wissenschaften. Geschichte und Gegenwart*, 313–336. Berlin: de Gruyter.
Grewendorf, G. and J. Sabel. 1999. Scrambling in German and Japanese: Adjunction versus multiple specifiers. *Natural Language and Linguistic Theory* 17, 1–65.

Grewendorf, G. 2000a. Die sprachliche Pflege des Rechts. Linguistische Aspekte der Rechtsprüfung. *Zeitschrift für Literaturwissenschaft und Linguistik* (LiLi) 118: *Sprache des Rechts*, 96–114.

Grewendorf, G. 2000b. Gibt es syntaktische Spuren? In: Ch. J. Baldauf. (ed.), *2. Symposion Autorenerkennung*. Tagungsband des Bundeskriminalamts, 83–96. Wiesbaden.

Grewendorf, G. 2000c. Eine minimalistische Analyse von Kausativkonstruktionen. Ms., University of Frankfurt.

Grewendorf, G. 2001a. Multiple *wh*-fronting. *Linguistic Inquiry* 32, 87–122.

Grewendorf, G. 2001b. Η θέση της Άρνησης στη Γερμανική Γλώσσα. ('On the postion of the negation in German') In: E. KOTZIA and E. ΕΥΘΥΜΙΟΥ (eds.), *Γενετική γραμματική και συγκριτική ανάλυση*, 69–112. Athen: Kastaniotis.

Grewendorf, G. 2002a. How performatives don't work. In: G. Grewendorf and G. Meggle (eds.), *Speech Acts, Mind, and Social Reality. Discussions with John R. Searle*, 25–39. Dordrecht: Kluwer.

Grewendorf, G. 2002b. "Kein Patent auf Leben!" Zur Pragmatik des Patentrechts. *Zeitschrift für Literaturwissenschaft und Linguistik* (LiLi) 128: *Sprache des Rechts* (2), 8–33.

Grewendorf, G. 2002c. Left dislocation as movement. In: S. Mauck and J. Mittelstaedt (eds.), *Georgetown University Working Papers in Theoretical Linguistics*, vol. 2, 31–81.

Grewendorf, G. 2003a. Dynamic binding and the problem of object-related anaphors. In: L. Gunkel, G. Müller, and G. Zifonoun (eds.), *Arbeiten zur Reflexivierung*, 91–113. Tübingen: Niemeyer.

Grewendorf, G. 2003b. Improper remnant movement. *Gengo Kenkyo: The Journal of the Linguistic Society of Japan* 123, 47–94.

Grewendorf, G. 2004a. Performativität und Deklarativität. In: J. Fohrmann (ed.), *Rhetorik*, 490–501. Stuttgart: Metzler.

Grewendorf, G. 2004b. Interview with Christoph Schwarze, "E la nave va: Linguistische Berichte vom Wandel der Zeit." In: G. Grewendorf (ed.), *Linguistik in Deutschland zwischen den Zeiten. Linguistische Berichte* 200, 470–480. Hamburg: Buske.

Grewendorf, G. 2005a. The asymmetry of short and long wh-movement in German. In: H.-G. Obenauer (ed.), *L'Architecture Propositionelle. La Syntaxe de la Périphérie Gauche, Recherches Linguistiques de Vincennes* 33, 38–54.

Grewendorf, G. 2005b. The discourse configurationality of scrambling. In: J. Sabel and M. Saito (eds.), *The Free Word Order Phenomenon: Its Syntactic Sources and Diversity* (Mouton de Gruyter. Studies in Generative Grammar 69), 75–135. Berlin: de Gruyter.

Grewendorf, G. 2005c. Recent developments in the theory of phases. Ms., University of Frankfurt.

Grewendorf, G. and P. Öhl. 2005. Komplementierertypen und Grammatikalisierung. Ms., Universiät Frankfurt.

Grewendorf, G. and C. Poletto. 2005. Von OV zu VO: ein Vergleich zwischen Zimbrisch und Plodarisch. In: E. Bidese, J. Dow, and T. Stolz (eds.), *Das Zimbrische zwischen Germanisch und Romanisch* (Diversitas Linguarum, vol. 9), 114–128. Brockmeyer-Verlag.

Grewendorf, G. (to appear). Review of: A. Belletti (ed.), *Structures and Beyond. The Cartography of Syntactic Structures*, vol. 3. *Language*.
Grewendorf, G. (forthcoming). The left clausal periphery: clitic left dislocation in Italian and left dislocation in German. In: W. Frey and B. Shaer (eds.), *The Left Clausal Periphery*.

Translations

Grewendorf, G. and G. Meggle. 1974. G. H. v. Wright, *Erklären und Verstehen*. Frankfurt/Main: Athenäum.

Manuscripts

Grewendorf, G. 1971. *Untersuchungen zur Unterscheidung der deskriptiven und wertenden Komponente in der Bedeutung von Werläußerungen*. MA-thesis, University of Munich.
Grewendorf, G. 1986. *Lokale Bereiche*. Ms., University of Frankfurt.
Grewendorf, G. and F. Hamm 1987. *Computerlinguistik*. Ms., University of Frankfurt.

List of Supervised Dissertations

Fritz Hamm (February 1989). Natürlich-sprachliche Quantoren. Modelltheoretische Untersuchungen zu universellen semantischen Beschränkungen.

Joachim Sabel (April 1995). Restrukturierung und Lokalität. Universelle Beschränkungen für Wortstellungsvariationen.

Young-Me Park (July 1997). Zur Theorie der A'-Bewegung. Eine universalgramatische Analyse von Topikalisierungsphänomenen.

Hans-Martin Gärtner (November 1997). Generalized transformations and beyond.

Katharina Hartmann (December 1998). Right node raising and gapping. Interface conditions on prosodic deletion.

Jochen Zeller (December 1999). Particle verbs, local domains, and the theory of lexical licensing.

Melanie Wratil (November 2004). Die Syntax des Imperativs.

Eric Fuß (February 2005). The rise of agreement. A formal approach to the syntax and grammaticalization of verbal inflection.

Sonja Ermisch (November 2005). Issues in the left periphery. A typological approach to topic and focus constructions.